Dieter Radaj

Heat Effects of Welding

Temperature Field, Residual Stress, Distortion

With 265 Figures

Springer-Verlag
Berlin Heidelberg New York
London Paris Tokyo
Hong Kong Barcelona Budapest

Dr.-Ing. habil. DIETER RADAJ

Professor of Mechanical Engineering
Braunschweig Technical University

Senior Research Manager
Daimler-Benz Corp., Stuttgart

ISBN-13: 978-3-642-48642-5 e-ISBN-13: 978-3-642-48640-1
DOI: 10.1007/978-3-642-48640-1

Library of Congress Cataloging-in-Publication Data
Radaj, Dieter. - Heat effects of welding ; temperature field, residual stress, distortion / Dieter Radaj. p. cm. Includes biographical references and index. ISBN-13: 978-3-642-48642-5 1. Welded joints. 2. Heat transmission. 3. Residual stresses. I. Title TA492.W4R26 1992 671.5'2--dc20 92-6376 CIP

Softcover reprint of the hardcover 1st edition 1992

Typesetting: Macmillan, Bangalore, India.
61/3020- 5 4 3 2 1 0 - Printed on acid-free paper

Foreword

Almost all welding technology depends upon the use of concentrated energy sources to fuse or soften the material locally at the joint, before such energy can be diffused or dispersed elsewhere. Although comprehensive treatments of transient heat flow as a controlling influence have been developed progressively and published over the past forty years, the task of uniting the results compactly within a textbook has become increasingly formidable. With the comparative scarcity of such works, welding engineers have been denied the full use of powerful design analysis tools.

During the past decade Dr Radaj has prepared to fulfil this need, working from a rich experience as pioneer researcher and teacher, co-operator with Professor Argyris at Stuttgart University in developing the finite element method for stress analysis of aircraft and power plant structures, and more recently as expert consultant on these and automotive structures at Daimler-Benz. His book appeared in 1988 in the German language, and this updated English language edition will significantly increase the availability of the work.

Transient heat flow studies are facilitated by existence of an array of closed solutions, the mastery of which confers a compact and thorough understanding of such as fusion welding power input requirements, welding speeds and fused widths, and post-fusion cooling rates. When this core of knowledge is assimilated, the welding engineer can rapidly choose or optimize a suitable process for any new application, geometrical shape or material, without wastefully protracted experimentation. Section 2 of the textbook introduces these aspects exhaustively, but with enviable lucidity.

Despite a profoundly beneficial influence in stabilizing the performance of welding processes, the transient distributions of temperature, together with thermal expansion, are also the causes of dual problems, endemic with all welded joints, namely residual stresses and distortion. It is in this domain that the finite element and comparable numerical methods of analysis have come into their own, since the calculations acquire three sequential phases, namely distributions of temperature with time, elastic thermal stresses, and plastic deformations. The finally resulting residual stresses have important influences on buckling of thin sections, and propensity towards cracking in thick sections. Dr Radaj has presented a timely and comprehensive treatment of these effects in

Section 3, to provide not only methods of prediction, with numerous results, but of experimental measurement for purposes of validation. Section 4 of the book brings together the more important aspects of stabilizing and reducing residual stresses by heat or mechanical treatments, and of distortions by appropriate joint design. In these respects the author has selected for summary purposes from the worldwide contributions made over many years to Commission X of the International Institute of Welding, with terms of reference Residual Stresses; Stress Relief; Brittle Fracture.

It can be said without hesitation that the author has contributed a masterly work of timeless value, with the virtue and integrity of a seamless garment, superior to the reference book that a committee of authors would alternatively produce, Those welding engineers who master its contents will find that they have acquired an invaluable means to increase professional competence, and will join in offering thanks to Dr Radaj for completing the self-imposed task of creating the means.

A A Wells
Chairman 1978–90, Commission X
International Institute of Welding

Preface

Welding is the most important method of joining components made of metallic materials. The technological core of most welding processes is the melting of the surfaces to be joined, either with or without filler metal (surfacing welds included), followed by subsequent cooling. Melting is achieved by local heat input, keeping heat diffusion into the component and heat dissipation into the surroundings as low as possible.

The effects of local heating comprise the formation of weld pool and bead or nugget shape, the generation of weld imperfections and defects, the initiation of hot and cold cracks mainly in the partly molten zone (intensified by hydrogen diffusion), the microstructural changes in the heat-affected zone of the base metal connected with hardening or softening, and the generation of residual stresses and distortion in the whole structure connected with mainly negative effects on strength.

The above phenomena constitute what is termed "weldability", a property of the structure to be welded influenced by design, material and manufacturing measures. Weldability has been mainly treated on an empirical basis supported by welding tests. However, theoretical developments in thermodynamics, continuum mechanics and microstructural kinetics have made a more systematic approach possible in recent years. This includes the weldability-related optimization of design, material and the welding process. The systematic approach based on theory and mathematics is termed "weldability analysis".

Based on the systematic approach, the contents of this book proceed from the temperature field, treat residual stresses and distortion on that basis and end with the practitioners' measures to reduce the former. A preliminary chapter introduces the field and a final chapter reviews the strength effects of welding.

The book is based on my regular lectures at the Braunschweig Technical University. Additional research work was performed as part of my responsibilities at the Daimler-Benz Corporation. The English manuscript was carefully typed by Hedwig Schiebel. The diagrams were meticulously ink-drawn by Helga Schmidt. Shicheng Zhang has assisted me in reading the page proofs. All this support is gratefully acknowledged. I am particularly indebted to Alan Wells for his appreciative foreword.

Stuttgart, April 1992 Dieter Radaj

Contents

List of symbols

The list of symbols used in equations, figures and text is arranged first according to the Latin and then according to the Greek alphabet. The upper case letter is always quoted first, the lower case second. Within each letter group, parameters with the same dimensions are arranged consecutively.

A uniform and unique notation, which is additionally comprehensible to the English reader, could not always be introduced because of different habits in the various fields of engineering science compiled here in one book Some symbols are multiply assigned if mistaken usage can be excluded. The symbols with multiple assignments are listed separately if the meaning is different and in the same line if the meaning is similar.

The former notations σ_Y, $\sigma_{0.1}$, σ_U are retained instead of using the newer standard notations R_e, $R_{p0.1}$, R_m because they are consistent with the stress symbols and are therefore more descriptive.

A	[mm²]	Cross-section area of girder, rod or plate
A_w, A_{wm}	[mm²]	Cross-section area of weld or pass, the same in model
A_f	[mm²]	Cross-section area of flange
A_e	[mm²]	Cross-section area of electrode
A_d, A_m	[mm²]	Cross-section area of deposited and fused (i.e. melted) metal
A	[mm²/N]	Parameter in residual stress solution (Fig. 110)
A	[mm²/N]	Parameter for strain evaluation (eq. (198))
A^*	[mm²/N]	Above parameter modified for strain gauges
A	[1/s]	Material characteristic value for creep (eq. 238)
a, a^*	[mm²/s]	Thermal diffusivity, the same of reference material
a, a_0	[mm]	Weld or throat thickness, the same for root pass
a_l	[mm]	Leg length of fillet weld
a, a^*	[mm]	Semi-axis of ovaloid in positive and negative x-direction
a_f	[mm]	Spacing between specimen and film
a	[mm²/N]	Material characteristic value for creep (eq. (238))
B	[mm²/N]	Parameter for strain evaluation (eq. (199))
B^*	[mm²/N]	Above parameter modified for strain gauges
B	[(mm²/N)ⁿ]	Material characteristic value for creep (eq. (232))
b, b^*	[1/s]	Heat transfer coefficient for plate and rod
b	[mm]	Semi-axis of ovaloid in y-direction
C_0	[J/mm²sK⁴]	Radiation coefficient
$[C_T]$	[J/K]	Heat capacity matrix
dC^*	[–]	Differential coefficient (eq. (133))
C	[N/mm²]	Material characteristic value for hardening (eq. (125))
C	[–]	Material characteristic value for creep (eq. (235))
C	[cm³/100g]	Hydrogen concentration
c	[J/gK]	Specific heat capacity
c	[mm]	Semi-axis of ovaloid in z-direction
c	[–]	Carbon content

D	[mmN]	Plate flexural rigidity
D	[1/s]	Material characteristic value for creep (eq. (234))
D	[cm²/s]	Hydrogen diffusion coefficient
d	[mm]	Diameter
d_0	[mm]	Diameter of constant density heat source or drilled hole
d_n	[mm]	Diameter of normal distribution heat source
d_e	[mm]	Diameter of electrode rod
d_c	[mm]	Diameter of electrode covering
d_{150}	[mm]	Diameter of area heated to 150°C
d_r	[mm]	Diameter of roller
d_A	[Å]	Interplanar spacing of atoms in crystal lattice
E, E^*	[N/mm²]	Elastic modulus, the same for reference material
E_0, E_{20}	[N/mm²]	Elastic modulus at 0 or 20°C
E_T	[N/mm²]	Elastic modulus at temperature T
e	[mm]	Eccentricity, distance from centre of gravity
e^*	[mm]	Interspace between weld spots or tack welds, the pitch
e	[–]	Euler constant (e = 2.7183), base of natural logarithms
F	[N]	Tensile force
$F_s, \bar{F}_s$	[N]	Longitudinal shrinkage force, active shrinkage force
F_{st}	[N]	Transverse shrinkage force
F_{st}^*	[N/mm]	Transverse shrinkage force per unit length of weld
F_r	[N]	Roller compressive force
ΔF	[N]	Range of force or twice the amplitude of force
F	[1/s^m]	Material characteristic value for creep (eq. (233))
$f(\sigma)$	[–]	Creep function dependent on stress
G	[N/mm²]	Shear modulus
G	[–]	Material characteristic value for creep (eq. (234))
$g(t)$	[–]	Creep function dependent on time
g	[mm²/s]	Gravitational acceleration
H, H_0	[N/mm²]	Hardening modulus, the same at 0°C
ΔH	[J/kmol]	Creep activation energy
H_p	[–]	Holloman-Jaffe index (eq. (228))
H_v	[cm³/100g]	Hydrogen source (or sink) density
h	[mm]	Thickness of plate, shell, weld, weld pass, height of bar or flange
h_m	[mm]	Above thickness in model
h_{rb}	[mm]	Height of rib
h_{tot}	[mm]	Total plate thickness
h_{wb}	[mm]	Web thickness
h_{fl}	[mm]	Flange thickness
h_0	[mm]	Initial height of bar, depth of blind hole
Δh	[mm]	Height of relieved layer
h_d	[mm]	Height of deposited bead
h_m	[mm]	Depth of fused metal or weld pool
h_m^*	[mm]	Above depth for deep penetration weld (Figs. 54 and 55)
h_f	[mm]	Distance of flange from bending centre line
h_{te}	[mm]	Height of tension zone (Fig. 165)
h_r	[mm]	Height of weld reinforcement
h_t	[mm]	Transition thickness of plate (eqs. (93), (97) and 173))
h_1, h_2	[mm]	Total thickness of weld passes (Fig. 196)
h'_1, h'_2	[mm]	Total thickness of weld passes (Fig. 196)
$h(T)$	[–]	Creep function dependent on temperature
I	[mm⁴]	Moment of inertia
I_y, I_z	[mm⁴]	Moment of inertia relative to y- and x-axis
I, I_{eff}	[A]	Current, effective current
i_m	[J/g]	Specific heat content of melted mass, i.e. enthalpy
Δi	[J/g]	Change of specific heat content, i.e. of enthalpy
i	[–]	Sequence of numbers, i = 1, 2, 3, for tensor indexing

i	[–]	Number of weld passes (Fig. 196)
j	[–]	Sequence of numbers, j = 1, 2, 3, for tensor indexing
j	[–]	Number of weld passes (Fig. 196)
j	[A/mm^2]	Current density in cross section of electrode
K	[N/mm^2]	Compression modulus
K_t	[mm^2/N]	Factor of transformation plasticity (eq. (128))
K	[–]	Material characteristic value for creep (eq. (236))
$[K_T]$	[J/sK]	Heat conduction matrix
K_{Ic}, K_{Ic}^*	[N/mm$^{3/2}$]	Fracture toughness, the same of reference material
k, k_1, k_2, k_3	[1/mm^2]	Concentration factor of normal distribution heat source
k	[J/mm^3]	Proportionality factor (eq. (143))
k^*	[–]	Geometry correction factor on $\Delta t_{8/5}$
k_t	[–]	Factor for arc operating time (eq. (104))
k_m	[–]	Correction factor from measurement (eq. (104))
k	[–]	Geometry factor for buckling (eq. (187))
k_b	[–]	Factor for backtilting (eq. 181)
k_n, k_i	[–]	Correction factor for multipass and intermittent welding
l	[mm]	Length of weld, plate, shell, strip, straining length
Δl	[mm]	Elongation of plate, length of rod element, longitudinal shrinkage
l_{pl}	[mm]	Length of plastic zone
l_{st}, l_i	[mm]	Length or interspace length of string weld
l_w, l_w^*	[mm]	Length of weld, length of weld pool
l_c	[mm]	Length of flame core
l_e	[mm]	Exposed length of electrode
l^*	[mm]	Length of heat strip
l_t	[mm]	Distance of transition zone from heat source
l_e, l_c	[mm]	Length of end and centre section of slot weld (Chapter 3.3.2)
l_a	[mm]	Length of electric arc
l_m	[mm]	Length of molten pool
l_r	[mm]	Length of ripple lag
M_1, M_2	[N]	Bending moments per unit length of perimeter (Fig. 141)
M^*	[N/mm^2]	Coefficient in eq. (133)
ΔM_z	[mmN]	Increment of resultant moment relative to z-axis
M_{st}	[mmN]	Transverse shrinkage moment
M_{st}^*	[N]	Above moment per unit length of weld
M_{sl}	[mmN]	Longitudinal shrinkage moment
m	[–]	Hardening exponent (eq. (125))
m	[–]	Material characteristic value for creep (eq. (233))
m	[–]	Number of buckling waves
m_i, m_j	[–]	Correction factor on angular shrinkage (eq. (180))
$\dot{m}_e$	[g/s]	Melting rate at electrode
$\dot{m}_d$	[g/s]	Deposition rate at workpiece
n	[–]	Material characteristic value for creep (eq. (232))
n	[–]	Number of plates or weld passes, order of interference line
n^*	[–]	Geometry factor in $\Delta t_{8/5}$
P	[mm]	Perimeter of rod
p	[–]	Correction factor on efficiency (eq. (70))
p_0	[N/mm^2]	Radial pressure related to σ_0 (eq. (158))
p_a	[N/mm^2]	Pressure of electric arc
p, p_0	[N/mm^2]	Partial pressure of hydrogen, ambient air pressure
p^*	[N/mm]	Pressure intensity in circumferential line of cylindrical shell
p^*	[W/mm^2]	Power density
Q	[J]	Net heat amount, heat content
$\bar{Q}$	[J]	Gross heat amount
Q_{li}	[J]	Limit heat content
Q_v	[J/mm^3]	Volume-specific heat

$\dot{Q}_v$	[J/mm³s]	Volume-specific heat flow or source density
$\{Q\}$	[J/s]	Column-vector of nodal point heat sources
q	[J/s]	Net heat flow, net heat input (or output)
$\bar{q}$	[J/s]	Gross heat flow, gross heat input (or output)
q_0, q_{li}	[J/s]	Initial heat flow, limit heat flow
q_m	[J/s]	Heat flow or heat input in model
q_e	[J/s]	Net heat output at electrode
q_c	[J/s]	Lower calorific value of gas supply
q_l	[J/mms]	Heat input per unit length of weld and unit time
q^*	[J/mm²s]	Heat flow density, heat source density (per area)
q^*_{max}, q^*_{min}	[J/mm²s]	Maximum and minimum value of the above density
q^*_c	[J/mm²s]	Heat flow density of convection
q^*_r	[J/mm²s]	Heat flow density of radiation
$q^*_{i,i}$	[J/mm³s]	Tensor of heat flow derivatives
q, q^{**}_{max}	[J/mm³s]	Heat source density (per volume), maximum value of this
$q_w, \bar{q}_w$	[J/mm]	Net and gross heat input per unit length of weld
q_m	[J/g]	Heat quantity per unit of melted-off mass
q	[1/s]	Material characteristic value for creep (eq. (234))
R, R_m	[mm]	Distance in space from centre, the same in model space
R	[mm]	Radius of shell, ring, rod or weld spot
R_i, R_o	[mm]	Inner Radius, outer radius
ΔR_x	[N]	Increment of resultant force in x-direction
R	[N/mm]	Transverse stiffness, restraint intensity
R	[Ω]	Electrical resistance
R	[J/kmolK]	Gas constant
r, r_m	[mm]	Distance in plane from centre, the same in model plane
Δr	[mm]	Distance from centric weld in annular plate
r_0	[mm]	Radius of constant density heat source
r_m	[mm]	Radius of cylindrical melted pool
r	[mm]	Radius of interference ring
Δr_x	[mm]	Increment of displacement in x-direction
S	[mm³]	Area moment of cross section part
S	[cm³/100g]	Hydrogen solubility
T	[K]	Temperature, temperature rise
T_0	[K]	Initial, ambient or working temperature
T_{max}	[K]	Maximum temperature, peak temperature
T_{bmax}	[K]	Maximum temperature at bottom of plate
T_u	[K]	Upper temperature (Fig. 168)
T_a	[K]	Annealing temperature
T_a	[K]	Austenitizing temperature
T_{amax}	[K]	Austenitizing peak temperature
T_m, T^*_m	[K]	Melting temperature, the same for reference material
T^{**}_m	[K]	Temperature of melted-off droplet
T_1	[K]	Cooling-down temperature of first layer
T_p	[K]	Preheating temperature
T_r, T_{rmax}	[K]	Temperature of resistance heating, maximum value of the same
T_{li}	[K]	Limit temperature
T_{tl}, T_{tu}	[K]	Lower and upper transformation temperature
T^*_m	[K]	Temperature of weld at start of transverse shrinkage
T_{NDT}	[K]	Transition temperature of Drop Weight Test
$\{T\}$	[K]	Column vector of nodal point temperatures
$\{\dot{T}\}$	[K/s]	Column vector of nodal point temperature derivatives
ΔT	[K]	Temperature difference, temperature increment
ΔT_{el}	[K]	Elastically sustainable temperature difference
ΔT_0	[K]	Global temperature rise or heating-up
$\partial T/\partial n$	[K/mm]	Temperature gradient perpendicular to isothermal face
t	[s]	Time, time coordinate
t_e	[s]	Time after end of welding

t_w	[s]	Time after start of welding
t_s^*	[s]	Time after start of shrinkage
t_c	[s]	Current flow time
t_{cM}, t_{cF}	[s]	Cooling time: end of 100% martensite, start of ferrite formation
Δt	[s]	Time difference, dwell time
Δt_0	[s]	Time interval between line and normal source (Fig. 24)
$\Delta t_{8/5}$	[s]	Cooling time from 800 to 500°C
Δt_a	[s]	Austenitizing time
Δt_{a1}, Δt_{a2}, Δt_{a3}	[s]	Austenitizing time, one-, two- and three-dimensional heat diffusion
Δt_1	[s]	Dwell time above T_1 (Fig. 104)
t_a	[s]	Austenitizing time (Fig. 104)
t_1^*	[s]	Cooling time from A_{c3} to M_s (Fig. 104)
Δt_m	[s]	Dwell time in model
t_s	[s]	Saturation time at switch-off of heat source
t_a	[h]	Annealing time
t_a^*	[h]	Heating and annealing time
U	[V]	Voltage
u, u_m	[mm]	Displacement, the same in model
u_r	[mm]	Radial displacement
u	[–]	Variable in $E_i(-u)$ and $\phi(u)$ (eqs. (42) and (43))
V, ΔV	[mm³]	Volume, volume change
$\dot{V}_{Ac}$	[l/h]	Acetylene consumption
v, v_m	[mm/s]	Welding speed, the same in model
v	[mm/s]	Flow velocity
v_e	[mm/s]	Melting-off or feed velocity of electrode
v_{em}	[mm/s]	Above velocity in model
v_{mf}	[mm/s]	Melting-off velocity in friction welding
w	[mm]	Deflection
w_0	[mm]	Initial deflection, constriction in centre section of cylindrical shell
Δw	[mm]	Buckling height, buckling depth
w	[mm]	Width of plate or strip
w_m	[mm]	Width of molten pool or fused zone
w_m^*	[mm]	Above width with deep penetration weld (Figs. 54 and 55)
w^*	[mm]	Transverse deformation of weld
w_{pl}	[mm]	Width or half-width of plastic zone
w_t	[mm]	Width of tension zone ($0 \leq \sigma_{te} \leq \sigma_Y$)
w_{tY}	[mm]	Width of tension yield zone ($\sigma_{te} = \sigma_Y$)
w_{t0}	[mm]	Width of tension zone in infinitely wide plate
w_0	[mm]	Width of stress-free high-temperature zone
w_g, w_b	[mm]	Width of welding groove, width of bead
w_a	[mm]	Width of annealing zone or heated zone
w_r, w_g	[mm]	Width of roller, width of gap
w_{tr}	[mm]	Width of transformation zone
X	[–]	Volumetric content of microstructural phase
x	[mm]	Cartesian coordinate, weld longitudinal direction
x_m	[mm]	Above parameter in model
Δx	[mm]	Distance or displacement in x-direction, groove longitudinal off-set
y	[mm]	Cartesian coordinate, weld transverse direction, distance from weld
y_m	[mm]	Above parameter in model
y_1	[mm]	Meridional coordinate (Fig. 145)
y_0	[mm]	Distance of heat strip from axis z of cross-section
Δy	[mm]	Distance or displacement in y-direction; groove transverse off-set
z	[mm]	Cartesian coordinate, thickness direction of plate, axis of electrode
z_m	[mm]	Above parameter in model
z_0	[mm]	Distance of heat strip from y-axis of cross section
Δz	[mm]	Groove transverse off-set in z-direction perpendicular to plate

α, α^*	[1/K]	Thermal expansion coefficient, the same for reference material
α_m	[1/K]	Mean value of α in temperature range
α_1	[1/K]	Maximum value of α at $\gamma\alpha$-transformation
α	[°]	Measuring direction at drill hole
α	[–]	Groove included angle
α_c	[J/mm²sK]	Coefficient of convective heat transfer
α_r	[J/mm²sK]	Coefficient of radiation heat transfer
α_t	[J/mm²sK]	Coefficient of heat transfer from electrode to ambient air
α_e	[g/Ah]	Melting rate per unit of amperage
α_d	[g/Ah]	Deposition rate per unit of amperage
β	[–]	Principal stress direction
$\Delta\beta$	[–]	Angular shrinkage, shrinkage angle
$\Delta\beta^*$	[–]	Tilting shrinkage, tilting angle
$\Delta\beta_b^*$	[–]	Back-tilting shrinkage, back-tilting angle
γ	[–]	Scale factor of model
Δ_{tr}, Δ_{up}	[mm]	Transverse shrinkage, upset displacement
Δ_e, Δ_{emax}	[mm]	Edge transverse displacement, maximum value of this
Δ_{ti}	[mm]	Transverse shrinkage of intermittent welds
Δ_{t1}, Δ_{t2}	[mm]	Transverse shrinkage in phase 1 and 2
Δ_{tr}^*	[mm]	Transverse expansion (Fig. 189)
δ_c	[mm]	Critical crack opening displacement
δ	[–]	Volume change by transformation
$E_i(-u)$	[–]	Integral exponential function (eq. (42))
$\varepsilon, \varepsilon_{ij}$	[–]	Strain, strain tensor
ε	[–]	Degree of blackness
$\{d\varepsilon\}$	[–]	Column vector of differential components of strain tensor
$\varepsilon_{tot}, \varepsilon_{eq}$	[–]	Total strain, equivalent strain
$\varepsilon_T, \varepsilon_{T\,ij}$	[–]	Thermal strain, thermal strain tensor
$\{d\varepsilon_T\}$	[–]	Column vector of differential components of thermal strain tensor
ε_T^*	[–]	Shrinkage strain (eq. (149))
$\varepsilon_t, \varepsilon_1$	[–]	Tangential strain, longitudinal strain
ε_{max}	[–]	Maximum strain in blind hole drilling method
$\varepsilon_{T\,max}$	[–]	Maximum value of thermal strain
$\Delta\varepsilon_{nt}$	[–]	Non-thermal high-temperature transverse strain
$\varepsilon_{el}, \varepsilon_{e\,ij}$	[–]	Elastic strain, elastic strain tensor
ε_p^*	[–]	Effective plastic strain (eq. (124))
$\varepsilon_p, \varepsilon_{p\,ij}$	[–]	Plastic strain, plastic strain tensor
$\varepsilon_c, \varepsilon_{c\,ij}$	[–]	Creep strain, creep strain tensor
$\varepsilon_{d\,ij}, \varepsilon_{v\,ij}$	[–]	Deviatoric strain tensor, volumetric strain tensor
$\varepsilon_{tr}, \varepsilon_{tr\,ij}$	[–]	Transformation strain, transformation strain tensor
$\varepsilon_{tp}, \varepsilon_{tp\,ij}$	[–]	Transformation plastic strain and strain tensor
ε_Y	[–]	Yield limit strain, strain at initial yielding
$\varepsilon_0, \varepsilon_{0\,max}$	[–]	Weld-logitudinal initial strain, maximum value of the same
ε_{0t}	[–]	Weld-transverse initial strain, transverse dislocation
ε_0	[–]	Back-strain at blind hole subjected to depth-constant stress
$\varepsilon_r, \varepsilon_R$	[–]	Radial strain, residual strain
$\varepsilon_x, \varepsilon_y, \varepsilon_z$	[–]	Strain in x-, y- and z-direction
$\varepsilon_{xz}, \varepsilon_{yz}$	[–]	Strain directed 45° relative to x- and z- or y- and z-axis
$\varepsilon_{00}, \varepsilon_{45}$	[–]	Strain directed 0° or 45° at drill hole
$\varepsilon_{90}, \varepsilon_{135}$	[–]	Strain directed 90° or 135° at drill hole
$\dot{\varepsilon}_{e\,ij}$	[1/s]	Tensor of elastic volumetric strain rates
$\dot{\varepsilon}_{vp\,ij}$	[1/s]	Tensor of viscoplastic strain rates
η_h	[–]	Heat efficiency at weld pool (eq. (1))
η_e	[–]	Heat efficiency at electrode (eq. (58))
η_m	[–]	Fusion efficiency of welding process (eq. (70))
η_t	[–]	Thermal efficiency of base metal melting (eqs. (71) and (72))
η^*	[–]	Shape factor of transverse weld (eq. 153)

θ	[–]	Angle of inclination of force-displacement curve of plate (Fig. (172))
θ	[–]	Angle of inclination of dilatometer curve versus *T*-axis (Fig. (109))
θ	[–]	Meridional angle (Fig. 145)
θ	[–]	Slope angle of weld reinforcement
θ	[–]	Temperature parameter (eq. (100))
θ_a	[–]	Austenitizing temperature parameter (Fig. 99)
θ	[–]	Plate thickness parameter (eq. (92))
ϑ	[–]	Bragg angle (eq. (209))
κ	[–]	Equilibrium coefficient (eq. (107))
λ	[J/mmsK]	Thermal conductivity
λ	[1/mm]	Cylindrical shell parameter (eq. (160))
λ	[Å]	Wave length of X-rays
$d\lambda$	[mm^2/N]	Differential scalar factor (eq. (122))
λ_σ	[–]	Welding suitability index in respect of residual stress (eq. (223))
λ_ε	[–]	Welding suitability index in respect of distortion (eq. (224))
μ	[–]	Fullness ratio of fusion zone in cross section
μ_t	[–]	Stiffness factor of transverse shrinkage (eq. (166))
μ_l	[–]	Stiffness factor of longitudinal shrinkage (eq. (149))
ν	[–]	Poisson's ratio
ξ	[–]	Heat dissipation factor (eq. (115))
ξ_2, ξ_3	[–]	Heat flow parameter, two- and three-dimensional heat diffusion
Π	[–]	Harmonic (or potential) function
π	[–]	Archimedes constant ($\pi = 3.14159$)
ϱ	[mm]	Curvature radius
ϱ	[g/mm^3]	Density
$\varrho_1, \varrho_2, \varrho_3$	[–]	Distance parameter, one-, two- and three-dimensional heat diffusion
ϱ^*	[mmΩ]	Specific electrical resistance
σ	[N/mm^2]	Normal stress
$\{d\sigma\}$	[N/mm^2]	Column vector of differential stress components
σ_x, σ_y	[N/mm^2]	Normal stress in *x*- and *y*-direction
σ_x, σ_{x0}	[N/mm^2]	Residual stress in *x*-direction, the same on surface
σ_r, σ_t	[N/mm^2]	Radial and tangential residual stress
σ_{r1}, σ_{t1}	[N/mm^2]	Value of σ_r and σ_t in diagram point A_1 (Fig. 199)
$\sigma_{d\,ij}, \sigma_{v\,ij}$	[N/mm^2]	Deviatoric and volumetric stress tensor
σ_l, σ_{tr}	[N/mm^2]	Longitudinal and transverse residual stress
σ_{lb}, σ_{la}	[N/mm^2]	Longitudinal residual stress before and after annealing
$\sigma_{trb}, \sigma_{tra}$	[N/mm^2]	Transverse residual stress before and after annealing
$\sigma_{l\,max}$	[N/mm^2]	Maximum longitudinal residual stress
σ_{tr0}	[N/mm^2]	Transverse residual stress in surface of multi-pass weld (Fig. 171)
$\sigma_{tr}, \sigma_{tr}^*$	[N/mm^2]	Transverse residual stress in plate, the same in weld
σ_{te}, σ_c	[N/mm^2]	Tensile and compressive residual stress (eq. (137))
σ_t^*, σ_c^*	[N/mm^2]	Above stresses after stress relief
σ_1, σ_2	[N/mm^2]	First and second principal stress
$\sigma^I, \sigma^{II}, \sigma^{III}$	[N/mm^2]	Residual stress of first, second and third order (Fig. 5)
$\sigma_I, \sigma_{II}, \sigma_{III}$	[N/mm^2]	Superimposed load stress (Fig. 252)
$\Delta\sigma_I, \Delta\sigma_{II}, \Delta\sigma_{III}$	[N/mm^2]	Stress increase as a consequence of yield (Fig. 252)
$\sigma_{\parallel}, \sigma_\perp$	[N/mm^2]	Residual stress in direction of rolling and transverse to it
$\sigma_a, \sigma_{ci}, \sigma_{me}$	[N/mm^2]	Axial, circumferential and meridional residual stress
$\sigma_{ciC}, \sigma_{aC}$	[N/mm^2]	Circumferential and axial residual stress in cylindrical shell
$\sigma_{ciS}, \sigma_{meS}$	[N/mm^2]	Circumferential and meridional residual stress in spherical shell
$\sigma_{a0}, \sigma_{ci0}$	[N/mm^2]	Axial and circumferential residual stress in centre cross section
σ_R, σ_L	[N/mm^2]	Residual stress, load stress
σ_R^*	[N/mm^2]	Residual stress after unloading
σ_{RL}	[N/mm^2]	Residual and load stress superimposed with yielding
$\sigma_{R\,max}$	[N/mm^2]	Maximum residual stress

σ_C, σ_{Rn}	[N/mm²]	Constraint stress, reaction stress (Fig. 7)
σ_0	[N/mm²]	Circumferential initial stress, i.e. residual stress source (eq. (158))
σ_0, σ_{0T}	[N/mm²]	Initial stress (eq. (226), Fig. 235), the same at temperature T
σ_{01}	[N/mm²]	Weld-longitudinal initial stress, i.e. residual stress source
σ_m	[N/mm²]	Stress in model
σ_{cr}	[N/mm²]	Critical radial stress, buckling stress
σ_s	[N/mm²]	Shrinkage stress
σ_{eq}	[N/mm²]	Equivalent stress, mostly after von Mises
σ_Y, σ_Y^*	[N/mm²]	Yield limit or yield stress, the same of reference material
$\sigma_{Y\,max}$	[N/mm²]	Maximum yield stress
σ_U, σ_{Ru}	[N/mm²]	Ultimate strength, rupture stress
$\sigma_{0.1}$, $\sigma_{0.2}$	[N/mm²]	0.1% and 0.2% offset yield stress
$\sigma_{y0.1}$	[N/mm²]	Yield limit or 0.1% offset yield stress
$\sigma_{0.1\,0}$	[N/mm²]	0.1% offset yield stress at 0°C
σ_{Y450}	[N/mm²]	Yield limit at 450°C
σ_Y, σ_{Yw}	[N/mm²]	Yield limit of base metal and of weld or filler metal
σ_T	[N/mm²]	Thermal stress from elastic suppression of thermal strain
σ_{Ywp}	[N/mm²]	Yield stress of weaker phase
σ^*	[N/mm²]	Surface tension
τ_{xy}	[N/mm²]	Shear stress in x- and y-direction
τ_{max}	[N/mm²]	Maximum shear stress
τ^*	[N/mm²]	Octahedral shear stress
τ_R	[N/mm²]	Residual shear stress
τ_Y	[N/mm²]	Shear stress yield limit, yield shear stress
$\tau_{0.4}$	[N/mm²]	0.4% offset yield shear stress
τ_s	[N/mm²]	Surface shear stress
τ_1, τ_2, τ_3	[–]	Time parameter, one-, two- and three-dimensional heat diffusion
$\Delta\tau_{a1}$, $\Delta\tau_{a2}$, $\Delta\tau_{a3}$	[–]	Austenitizing time parameter, dimensionality of heat diffusion
$Y_0(u)$, $Y_1(u)$	[–]	Bessel function of second kind and zero or first order
$\Phi(u)$	[–]	Gaussian probability integral
$\varPhi$, $\varPhi^*$	[–]	Angle of inclination of force-displacement curve of weld (Fig. 172)
φ	[–]	Angle of bending shrinkage by longitudinal weld (Fig. 192)
$\Delta\varphi$	[–]	Angle of bending shrinkage by transverse weld (Fig. 193)
φ	[–]	Azimuth angle
$\Delta\varphi_z$	[–]	Increment of twisting angle relative to z-axis
ψ	[–]	Angle of inclination of X-ray beam versus surface normal line
ψ	[–]	Heat saturation function
ψ_1, ψ_2, ψ_3	[–]	Heat saturation for one-, two- and three-dimensional heat diffusion
ψ_d	[–]	Loss factor of deposited weld
ω	[–]	Cooling rate parameter

1 Introduction

1.1 Scope and structuring of contents

This monograph on heat effects of welding refers to temperature fields, residual stresses and distortion occurring during and after welding and thermal cutting. Welding is the non-detachable joining or coating of components or base materials under the (mostly local) application of heat or pressure, with or without the use of filler material (definition according to German Standard DIN 1910 [341]). Joining is preferably performed with the welding zone being in the plastically deformed or liquid state. Thermal cutting is the separation of components or base material through the local application of heat. This monograph refers primarily to (gas and arc) fusion welded joints (seam welds) and to (resistance) pressure welded joints (spot welds). Account is also taken of friction and stud welded joints, claddings and flame cuts. The applications under discussion comprise metals with emphasis on steels and light alloys.

Due to the highly localized transient heat input, considerable residual stresses (welding residual stresses) and deformations (welding distortion, welding shrinkage, welding warpage) occur during and after welding. In contrast to load stresses, i.e. internal forces being in equilibrium with external forces, residual stresses are internal forces occurring without external forces. The sections and chapters which follow deal mainly with "macroscopic" residual stresses, which are of relevance for the engineer and which can be described in terms of continuum mechanics, whereas the "microscopic" residual stresses between or in the crystallites are ignored. Warpage is a phenomenon of structural instability as a result of shrinkage or distortion.

Welding residual stresses and welding distortion may greatly impair manufacturing and strength. Measures are, therefore, taken to minimize welding residual stresses and welding distortion, or to eliminate them after welding. In manufacturing, welding deformations jeopardize the shape and dimensional tolerances required. Joint misalignment and increased groove gaps render manufacturing more difficult. Welds, especially tack welds, may rupture partially or completely as a result of residual stresses produced during welding. Residual stresses relieved during machining result in unacceptable distortions of the workpiece. Welding residual stresses may cause brittle fractures in the finished structure. Tensile residual stresses reduce fatigue strength and corrosion resistance. Compressive residual stresses diminish the stability limit. In contrast, the positive effects of welding residual stresses (compression is favourable in respect of fatigue and corrosion; tension is favourable in respect of stability) are of secondary practical significance.

This state of affairs results in structuring of the material into the following principal sections:

- welding temperature field analysis,
- welding residual stress and distortion analysis,
- reduction of welding residual stress and distortion,
- effects of welding on strength.

The analysis of welding residual stress and of welding distortion, viewed in a historical perspective, developed to a large extent independently of each other although, viewed from the physical aspect, they are closely related. This is based first of all on the fact that welding residual stresses are primarily of interest as a basis for assessing strength, whereas welding deformations, by contrast, are primarily considered as phenomena impairing manufacture. However, a methodical difference also exists in respect of physics in that it is generally necessary to develop a finite element model of the structure or continuum for assessing welding residual stresses, whereas simpler approaches within the framework of common engineering theories of structures very often suffice for assessing welding deformations. The assumptions which have to be introduced in the latter case relate, in particular, to the magnitude of the weld shrinkage force and the extension of the plastic zone. Welding residual stress analyses, though, can provide reference values of these parameters. If, on the other hand, an approach based on measurement is adopted, this then necessitates sophisticated measuring techniques for welding residual stresses which are practicable only under laboratory conditions, whereas welding deformations can be measured under workshop conditions using simple equipment.

1.2 Weldability analysis

The multitude of influencing parameters and their non-linear, transient and temperature-dependent effect prevent valid statements on welding residual stresses and welding deformation in individual cases to be made, and render generalizations controversial. The practician uses a standardized definition of terms relating to weldability as a classification system which is of some value considering the effect of welding residual stress and welding deformation (albeit in a more linguistic than scientific manner). The scientific analyst, on the other hand, has been able to reduce the complexity of the phenomena by decoupling into thermodynamic, mechanical and microstructural processes.

The term "weldability of the component" as defined by German Standard DIN 8528 [342] is subdivided into the elements of "welding suitability of the material", "welding reliability of the design" and "welding feasibility in manufac-

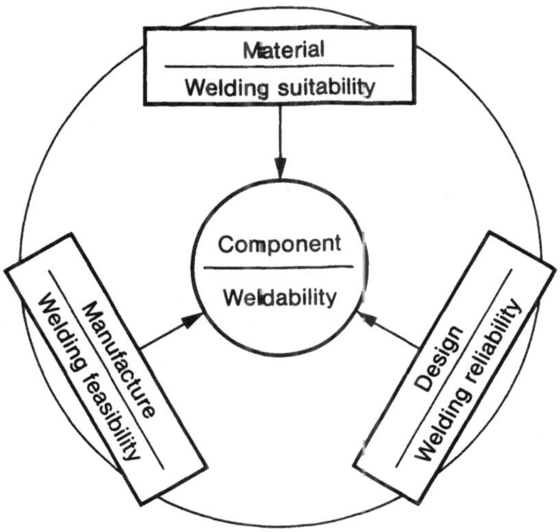

Fig. 1. Definition of the term "weldability" according to German standard DIN 8528

ture" (Fig. 1). Welding residual stresses and welding deformations constitute an important partial aspect of weldability. They may give rise to hot cracks, cold cracks, brittle fracture and early instability, affect serviceability as a result of distortion or warpage and impair manufacture; for example, as a consequence of transverse movement of the groove edges. Novel welding suitability indices, composed of common thermal and mechanical material characteristic values, are given in this book for the relative assessment of welding suitability of the material from the aspect of welding residual stresses and welding deformations. The welding reliability of the design is assessed primarily on the basis of welding residual stress analysis, whereas welding feasibility in manufacture is assessed primarily on the basis of welding deformation analysis. The results of such analyses have to be compared with corresponding limit values (strength values and dimensional tolerances). Suggestions for improvements and possibilities for optimization of design, materials and manufacturing are derived from comparative analyses.

In line with the classification presented above, the parameters which have an influence on weldability can be grouped in the following manner, closely reflecting current practice:

- material-related, including type (chemical), composition and microstructure of base and filler metal;
- design-related, including shape, dimensions, support conditions, loading of the structure and type, thickness and arrangement of the welds;
- manufacture-related, including welding method, welding velocity, welding performance, groove shape, welding sequence, multiple layer, tacking, clamping, preheating and post-weld heat-treatment.

Weldability in a narrower sense is understood in respect of the strength properties required [138] (Fig. 2). The strength of the welded joint is subjected to

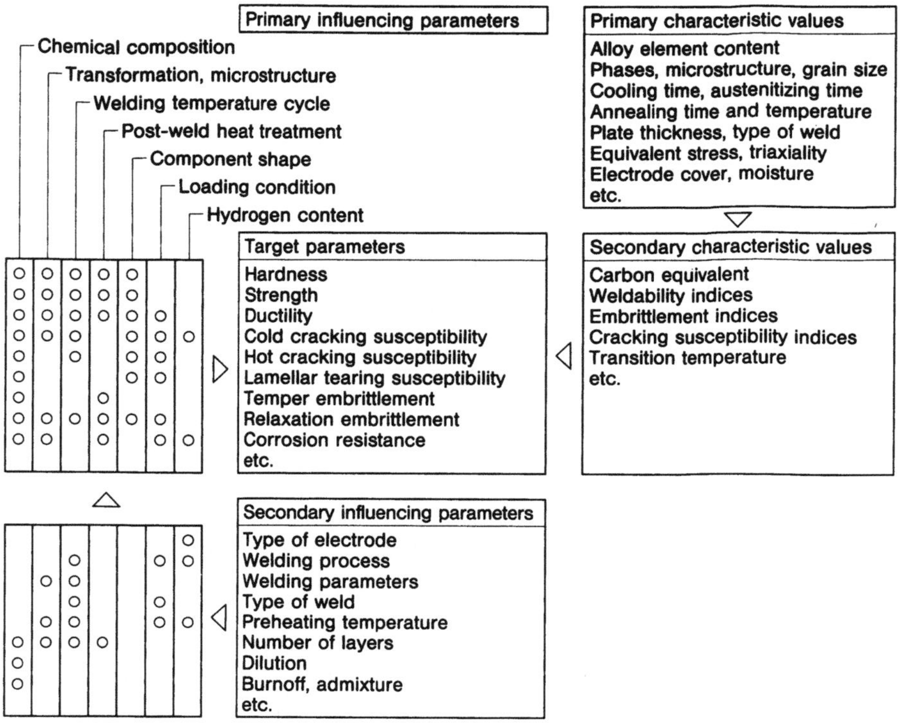

Fig. 2. Parameters with major influence on strength of welded joints (primarily heat affected zone) constituting part of their weldability; after Buchmayr [124]

primary influence parameters such as chemical composition or temperature cycle, which are determined to some extent by secondary influence parameters such as electrode type or preheating temperature. The strength behaviour is described by primary or physical characteristic values which may enter into secondary or technological characteristic values. The complexity of the term "weldability" even when restricted to strength properties is made clear by the graph which contains only an incomplete listing of variables.

On the other hand, the decoupling according to Fig. 3 of the processes underlying weldability into temperature field, stress and deformation field and microstructural state field [194] has proven its worth for numerical-analytical treatment, especially of welding residual stresses and welding deformations. Arrows indicate mutual influences: the arrows with a solid line signify a strong influence; arrows with a broken line represent a weak influence (often of negligible relevance in engineering terms). A point worth stressing is the need to incorporate microstructural transformations in the analyses, which depend not only on the chemical composition of the material, but also on its thermal history (particularly the welding-related history). The influence of microstructure makes itself noticeable especially in the heat-affected and fusion zone of the weld.

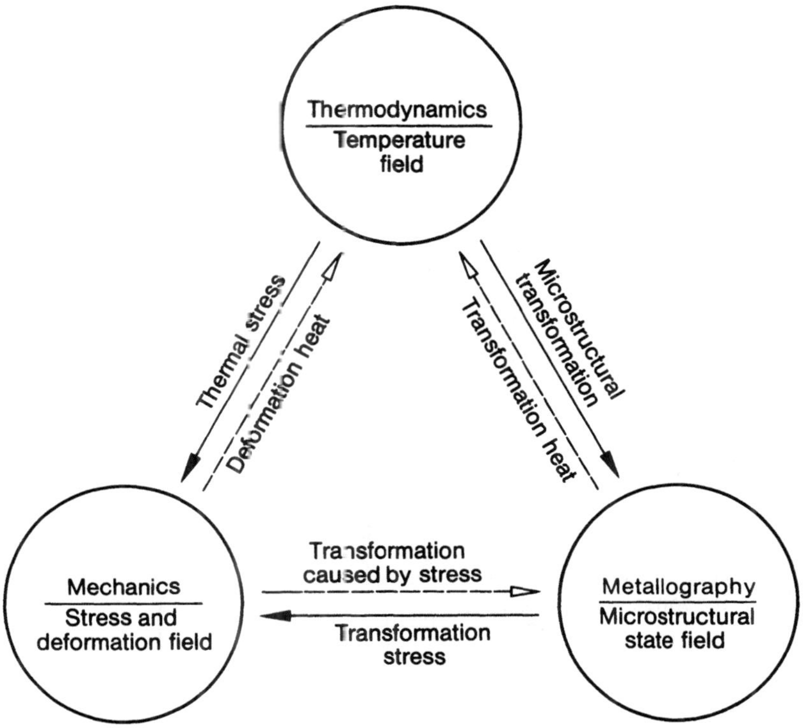

Fig. 3. Decoupling and mutual influencing of temperature field, stress and deformation field and microstructural state field; after Karlsson [194]

The graph of Fig. 3 has been extended [138] in Fig. 4 to emphasize the influence of transformation behaviour and to indicate the essential input and output parameters referring to finite element analysis. Of course, hardness is neither the sole nor the most important output parameter characterizing weldability, but it is the parameter which can directly be determined based on transformation behaviour.

1.3 Residual stresses

Residual stresses are internal forces without external forces acting. As constraint stresses, they are in equilibrium only with themselves. Reaction stresses originating from self-equilibrating support forces may superimpose on the constraint stresses. The total residual stresses superimpose on the stresses from external load, i.e. the load stresses. Residual stresses are acting temporarily or permanently.

A distinction is made between first, second and third order residual stresses (Fig. 5). First order residual stresses, σ^{I}, extend over macroscopic areas and are

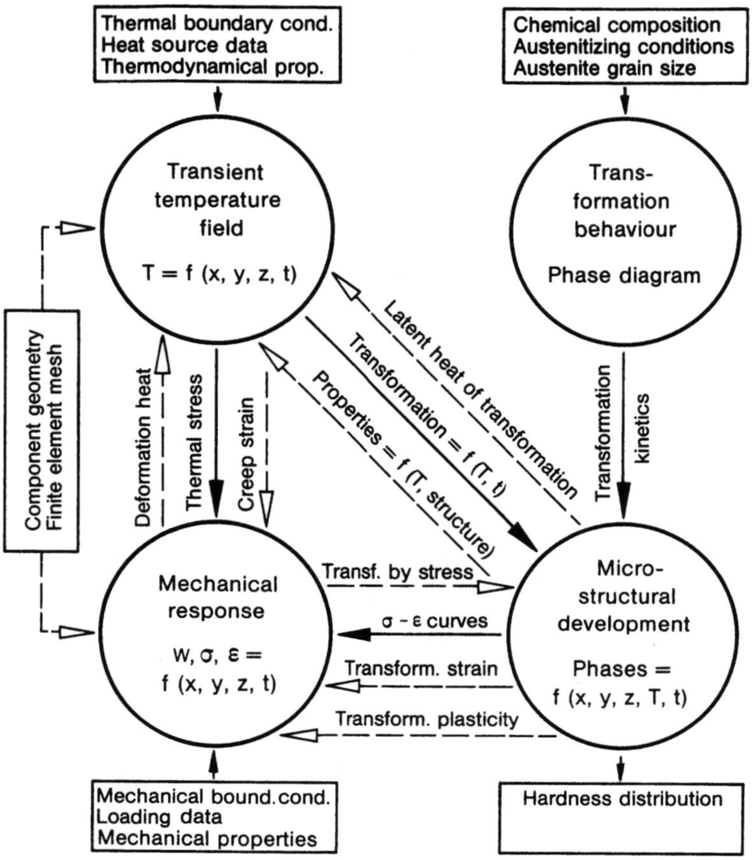

Fig. 4. Extended graph of Fig. 3 emphasizing the influence of microstructural transformation; after Buchmayr and Kirkaldy [138]

averaged stresses over several crystallites. Second order residual stresses, σ^{II}, act between crystallites or crystallite subregions (size approx. 1–0.01 mm) and are averaged within these areas (for instance residual stresses around piled-up dislocations or secondary phases). Third order residual stresses, σ^{III}, act between atomic areas (size approx. 10^{-2}–10^{-6} mm; for instance, the residual stresses around a single dislocation). The presentations which follow refer exclusively to first order residual stresses, i.e. to the macroscopic residual stresses, which are of particular relevance for engineering purposes.

Residual stresses are the result of inhomogeneous permanent (i.e. plastic) deformation which, at the material element, can be subdivided as follows:

- volumetric (or "dilatoric") strain as a result of thermal expansion, chemical conversion, microstructural transformation or change in state;
- distortional (or "deviatoric") strain (i.e. shear strain) as a result of (time-independent) plastic, and also (time-dependent) viscoplastic deformation.

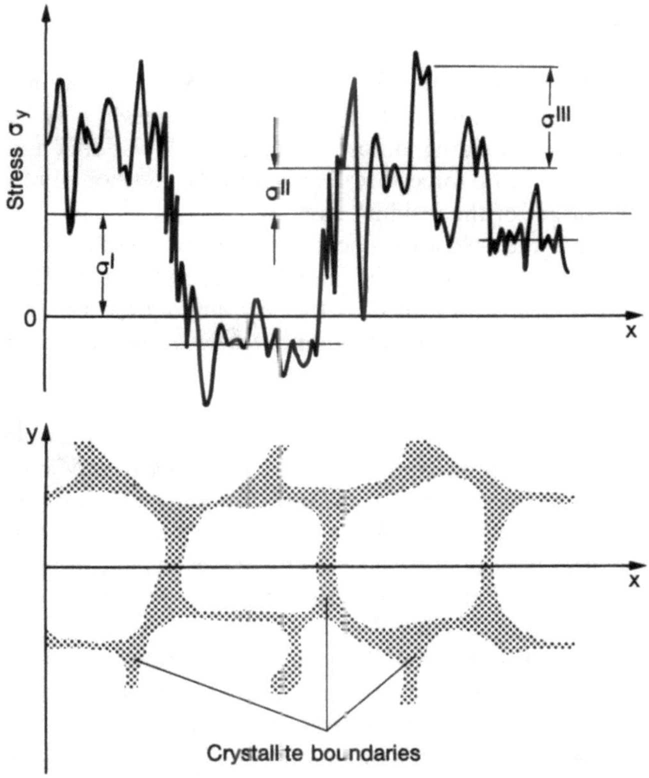

Fig. 5. First, second and third order residual stresses (σ^{I}, σ^{II}, σ^{III}) in crystallite structure, acting in direction y, plotted over coordinate x, σ^{II} recorded as crystallite mean value; after Macherauch and Hauk [15] (ibid., Vol. 1, p. 11)

Such permanent deformations at the material element relative to a compatible initial state are also termed "initial deformations", "residual deformations" or "extra deformations" and introduced as such into analysis. They can be handled alternatively as "initial stresses" or "residual stress sources".

Residual stresses may also arise as a consequence of a change in connectivity of the component by welding, i.e. as a result of (macro-)dislocations. Such stresses occur for example in a ring which is cut open, shortened and then closed again.

The residual stresses caused by inhomogeneous thermal expansion (or contraction) are termed "thermal stresses". Elastic thermal stresses disappear after removing the inhomogeneous temperatures by which they have been caused. For this reason, many authors do not classify them as residual stresses. Where major differences in temperature exist, the thermal stresses give rise to plastic deformations. After removal of the temperature differences, residual stresses remain. The residual stresses caused by microstructural transformation are termed "transformation stresses".

1.4 Welding residual stresses

Welding residual stresses and residual stresses at flame cut edges are thermal stresses (primarily cooling stresses), on which transformation stresses may be superimposed. In the case of cold welding, diffusion welding, roll-cladding and explosion cladding, the cold working forces are, exclusively or additionally to the above heat effects, the origin of the residual stresses.

During the welding process, the weld area is heated up sharply relative to the surrounding area (Fig. 6), and fused locally. The material expands as a result of being heated. The heat expansion is restrained by the surrounding colder area, which gives rise to (elastic) thermal stresses. The thermal stresses partly exceed the yield limit, which is lowered at elevated temperatures. Consequently, the weld area is plastically hot-compressed and, after cooling down, too short, too narrow or too small relative to the surrounding area. It thus displays tensile residual stresses; the surrounding area compressive residual stresses. Microstructural transformation during cooling (for instance the $\gamma\alpha$-transformation described below) involves an increase in volume. If this occurs at a (lower) temperature at which the yield limit is sufficiently high, this results in compressive residual stresses in the weld area and tensile residual stresses in the surrounding area.

The rule of thumb which applies is that, in the areas of the component which have cooled down last, tensile stresses occur where thermal stress dominates, and compressive stresses where transformation stress dominates.

Welding residual stresses are generated in components which, as a rule, are already subjected to residual stresses and they are changed when the part undergoes further processing or is operated in service. Manufacturing processes which cause residual stresses are casting, hot and cold working, machining, coating, surface treatment as well as heat treatment and age hardening. Residual

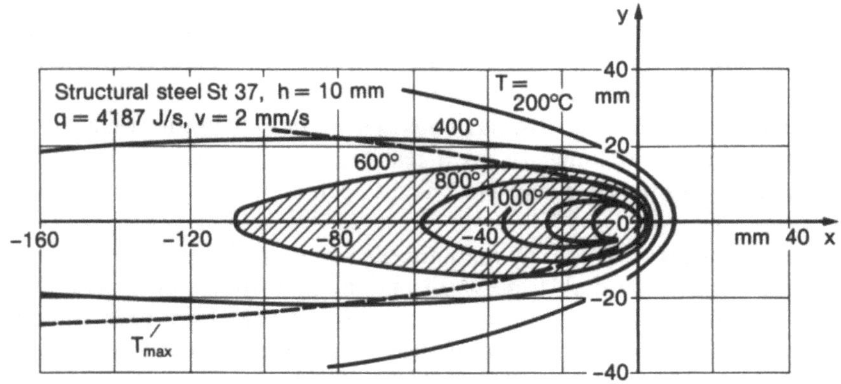

Fig. 6. Temperature field isotherms around welding heat source moving uniformly and linearly in an infinite plate, modelled as line source perpendicular to plane of plate, quasi-stationary field in moving coordinate system *xy*, the area of negligible yield stress shaded, the position of local maximum temperature indicated as broken line; after Rykalin [1]

stresses are also generated by improper assembly. Residual stresses undergo changes as a result of local or global yield under non-recurring (static or dynamic) or recurring (for instance cyclic) loading.

Thermal and transformation stresses do not occur in the homogeneous material provided all the areas of the component heat up and cool down simultaneously, i.e. provided no temperature differences occur at any time. Existing residual stresses can be strongly diminished under temperatures close to and above the recrystallization temperature (approximately half the melting temperature measured in degrees [K]) by the reduction of yield limit and elastic modulus as well as by stress relaxation and creep (stress relieving under high temperature, stress relief annealing). Following this, the component must be cooled down slowly and uniformly. Recrystallization involves a partially favourable, partially unfavourable change of the mechanical material characteristic values. Residual stresses occur in the inhomogeneous material (for instance when joining materials of different type) even then if the material is heated up and cooled down slowly and uniformly.

Existing residual stresses are also reduced if load stresses are superimposed on the residual stresses in such a way that the yield limit is exceeded locally, which results in a favourable stress redistribution (mechanical or thermal stress-relieving at low temperature: cold stretching, flame stress relieving, vibration stress relieving).

1.5 Welding residual stress fields

Residual stresses in the longitudinal direction of the weld, in the transverse direction and in the direction of plate thickness occur as a result of similar mechanisms.

Weld-longitudinal stresses are generated in accordance with the mechanism of a weld seam which contracts longitudinally. The tensile stresses are limited to a narrow area close to the weld, their maximum value being at or above the yield limit. Lower compressive stresses exist in the surrounding region, dropping off rapidly the further away they are from the weld.

Weld-transverse stresses in the plane of the plate are generated in accordance with the mechanism of a weld seam which contracts transversely, especially when the plate is restrained. They are not restricted to a narrow area close to the weld, but also comprise the surrounding region. They are supported in a similar way as external forces and, provided the support has sufficient elasticity, remain below the yield limit.

Weld-transverse stresses in the direction of plate thickness may at least then be generated if the plate thickness is sufficiently large. They result in the unsafe triaxial tensile stress state.

The residual stresses in the interior layers of multi-pass welds are changed (generally reduced) by thermomechanical processes.

The residual stress state around weld spots in sufficiently extended plates is generated axisymmetrically in accordance with the mechanism of a spot which contracts radially. Nearby edges may restrict the axisymmetry.

The welding residual stresses just described are changed in the case of ferritic steels by the microstructural $\gamma\alpha$-transformation (face centred γ-phase mixed crystal, austenite; body centred α-phase mixed crystal, ferrite), i.e. by the transformation of austenite into pearlite, bainite or martensite. In certain critical temperature ranges which depend on the reached maximum temperature and mean cooling rate, these steels undergo microstructural transformations which involve a volumetric expansion. This influence is of lesser significance in the case of mild steel because the transformation occurs for the greater part at high temperatures where the yield limit is low. This influence is significant in the case of higher alloy steels because, in the case of such steels, the temperature range of transformation, at the cooling rates which generally occur in practice, can be shifted to relatively low temperatures with correspondingly high yield limit. An especially marked form of the $\gamma\alpha$-transformation at low temperature is the formation of martensite, i.e. the hardening in the fusion and heat-affected zone of mild and low-alloy steels. The maximum values of the thermal stresses in the centre plane of the weld are reduced by the superimposition of the transformation stresses and possibly reversed into compressive stresses. Austenitic steels do not undergo any transformation in the considered temperature range.

The residual stresses in flame cut edges are determined by the transformation of ferrite into martensite. Provided carburization and cooling rate are sufficiently high, high compressive residual stresses occur.

Residual stresses in explosion or roll-bonded claddings are generated under local plastic deformation resulting from compressive cold work.

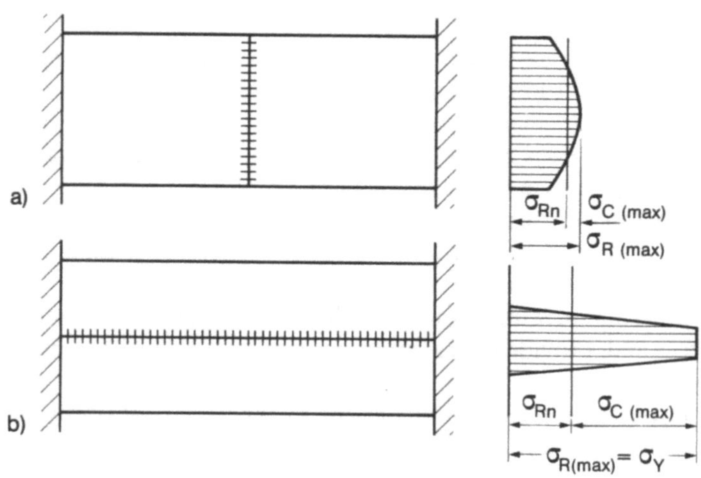

Fig. 7. Reaction stresses σ_{Rn} (as a consequence of the plate transverse edges being restrained) and constraint stresses σ_C at transverse weld (**a**) and at longitudinal weld (**b**)

Residual stresses without external support forces are termed "constraint stresses". Residual stresses as a result of external support forces are termed "reaction stresses" (in special cases also "misalignment stresses"). The constraint stresses are in equilibrium with themselves. The reaction stresses are in equilibrium with the reaction forces at the supports. Provided a component is supported in such a way that it can deform freely, only constraint stresses occur. Where component supports restrain the deformation, reaction stresses superimpose. This superimposition is presented schematically in Fig. 7 for the centric transverse and longitudinal weld.

A distinction is made between the temporary welding residual stresses during welding and the permanent welding residual stresses following complete temperature equalization. The former determine the weldability; the latter are of interest with respect to the strength of the structure.

1.6 Type examples

To illustrate what has been said so far, a number of type examples of (primarily) measured welding residual stress fields are presented below.

A frequently investigated specimen is the rectangular plate with centric jointing weld, which can be manufactured quickly and at low cost. The published experimental results (and comparative analytical results) are mostly based on a plane stress field assumed as uniform over the thickness of the plate. With single-side welding, this does not adequately correspond to reality in view of the superimposed bending effect. Such a result is presented in Fig. 8 (cutting method for residual stress measurement and finite difference method for numerical

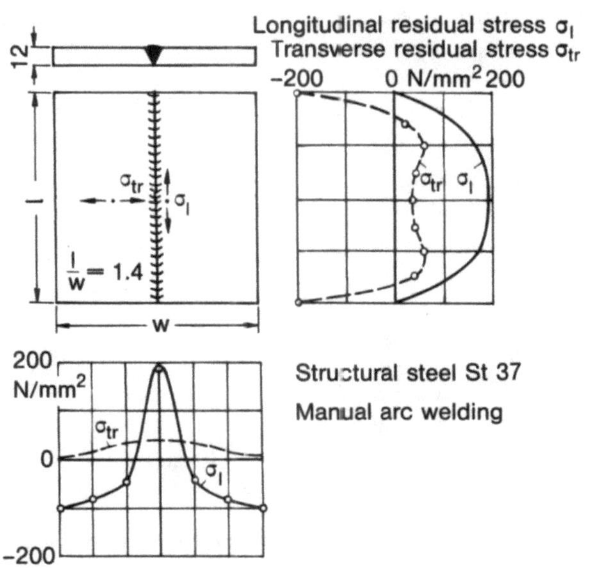

Fig. 8. Longitudinal and transverse residual stresses in centre cross-sections of rectangular plate with centric weld, analytical and experimental results; after Koch [274]

analysis). The longitudinal stresses show the tension maximum typical for mild steel, close to the yield limit in the centre of the weld and to the side of that the drop to compression values. The transverse stresses reveal high compression values at the weld ends and low tension values in the middle of the weld length. The symmetry of the distribution in the weld-longitudinal direction applies only to momentarily applied weld seams or to rapid welding relative to cooling down, i.e. to short welds. The equilibrium conditions can be examined by comparing the areas under the curves in suitable sections.They are obviously only inadequately met in this case. If the plate is extended in the weld longitudinal direction, the stress state at the plate ends is displaced accordingly. The dominant element in the case of this relatively short rectangular plate are the plate end effects. With a lower welding rate or longer weld, an asymmetrical distribution develops in the longitudinal direction of the weld. Transverse tension then occurs at the end of the weld.

The circular plate with plug weld on one side (Fig. 9) is typical in respect of residual stresses and also in spot and stud welded joints. The axisymmetrical residual stresses have been approximately calculated and compared with the results of measurements using the boring-out and turning-off method. High biaxial tension occurs in the weld zone, dropping to zero outside of this area; the radial stress directly, the tangential stress through a compression maximum. The radial tension in the inner zone is supported against tangential compression in the outer zone ("arch effect"). The plate bending effect, which in reality is superimposed, is not covered here. The type of residual stress field described also occurs in a semicircular form ahead of weld ends.

In view of the γα-transformation shifted to lower temperatures with the related transformation expansion, considerable deviations exist in the residual stress field of alloyed steels. The most striking is the compression zone at the former tension maximum of the thermal residual stresses. This is presented in

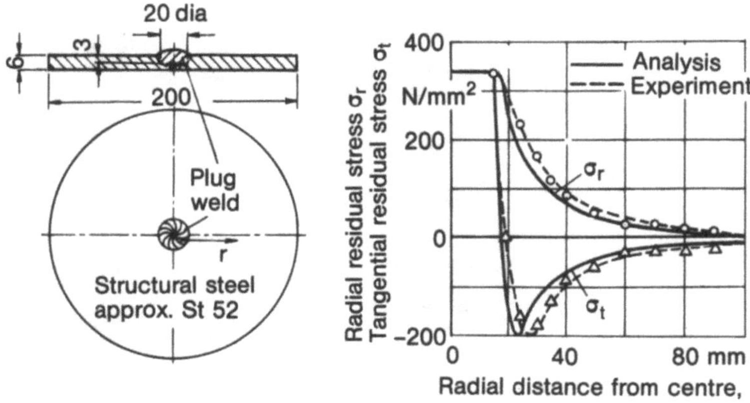

Fig. 9. Radial and tangential residual stresses in circular plate with plug weld, analytical and experimental results; after Watanabe and Satoh [213]

Fig. 10 for the longitudinal residual stresses in the top and bottom surface of a longitudinally welded plate of high-strength steel. The longitudinal residual stresses were determined applying the cutting method combined with strain gauges.

Particularly marked transformation stresses occur at flame cut edges (Fig. 11). Here, a tension maximum would be expected in view of the thermal contraction, just as in weld seams, but this is converted as a result of the transformation expansion into a compression maximum. The "compression skin" explains the high fatigue strength of flame cut surfaces and the decrease in strength as a result of subsequent machining. The residual stresses presented

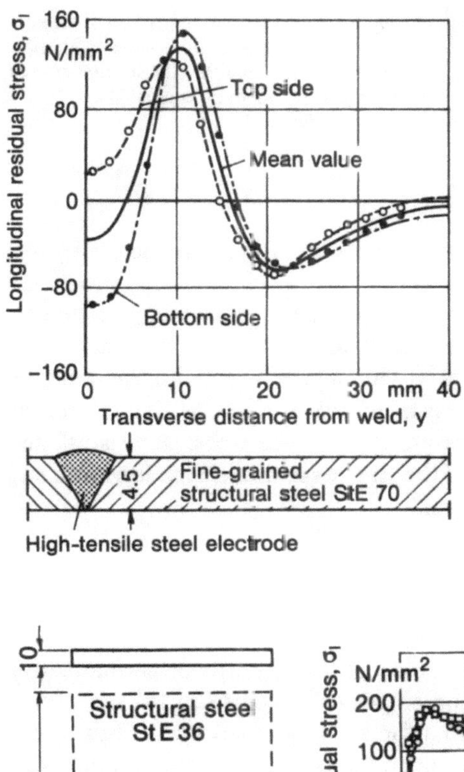

Fig. 10. Longitudinal residual stresses resulting from thermal and transformation strains, high-tensile fine-grained structural steel with electrode of matching material, experimental results; after Rappe [276]

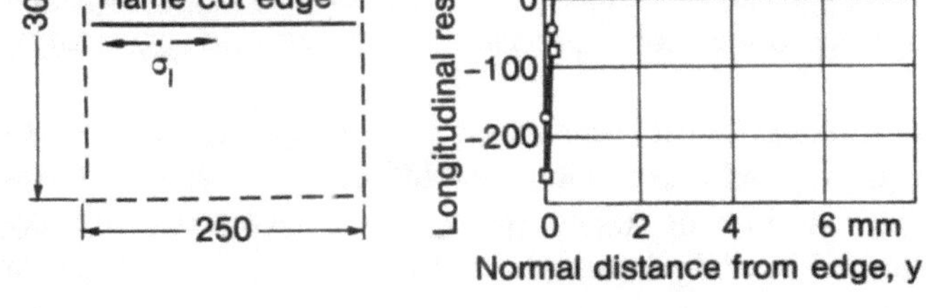

Fig. 11. Longitudinal residual stresses at flame cut edge of mild steel plate, experimental results; after Ruge and Schimmöller [285]

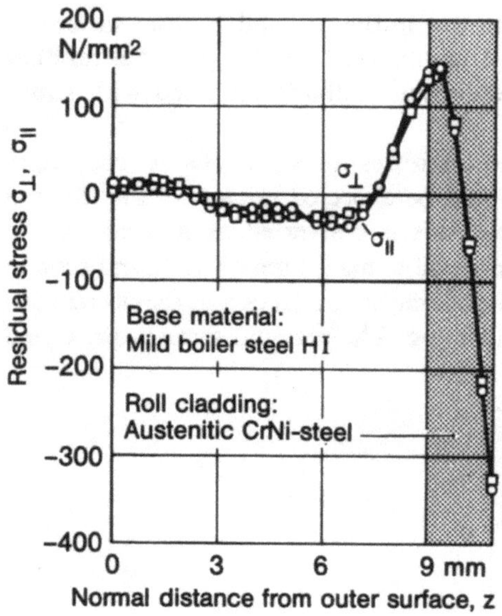

Fig. 12. Residual stresses in ferritic steel plate with austenitic roll-bonded cladding, stresses normal and parallel to direction of rolling, experimental results; after Schimmöller [281]

have been determined from the resilience deflections of the specimen with the flame cut edge being removed in layers.

The biaxial residual stresses caused by the local effect of (rolling) pressure in (cold) roll-bonded cladding on plate are presented in Fig. 12. They have been determined from the resilience deflections of the plate relieved with the removal of surface layers. High compression exists in the surface of the austenitic cladding while tension prevails in the bonding face. The stress state is (plane-hydrostatically) biaxial. The difference of the stresses in the rolling direction, $\sigma_{||}$, and transverse to it, $\sigma_{\perp}$, is negligibly small.

1.7 Welding deformations

With the welding residual stresses there also occur (temporarily or permanently) welding deformations (displacements and rotations), termed "shrinkage", "distortion" or "warpage". Stress and deformation are largely opposed. High stresses occur where the deformation is restrained (i.e. low), low stresses where the deformation is unrestrained (i.e. high). The task which exists in practice, though, is to achieve high shape and dimensional precision combined with low welding residual stresses.

The longitudinal shrinkage of the weld seams after welding caused by the longitudinal upsetting during welding results in a longitudinal shortening of the component, notably in the weld zone. Where the weld is arranged eccentrically, this produces the unwanted bending deformation of girders and plates (bending shrinkage). The transverse shrinkage of the weld seams, caused by transverse upsetting during welding, increased by the closing of originally open weld grooves, results in a transverse shortening of the component. Where the component is welded from one side, this results in additional angular shrinkage, or, where angular shrinkage is restrained, in bending deformations. Angular shrinkage is particularly marked in the case of single-side multi-pass welding.

The compressive residual stresses caused by shrinkage forces may produce an unstable transverse deformation of thin plates. Such "buckling" or "warpage" results in major deflections normal to the plate plane with a permanent reduction of the shrinkage forces.

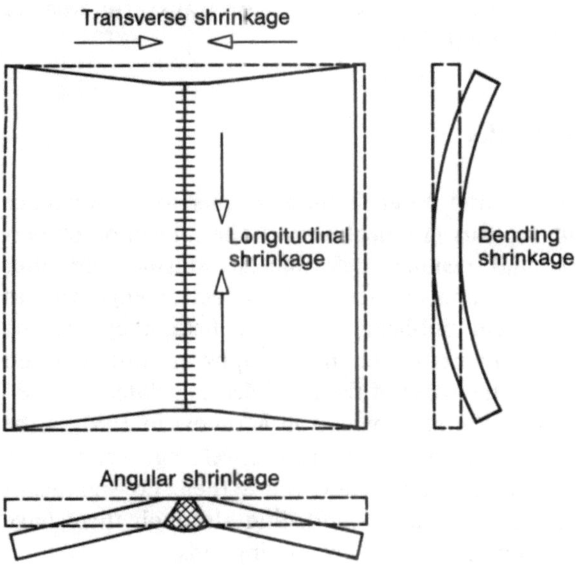

Fig. 13. Longitudinal and transverse shrinkage, angular and bending shrinkage of rectangular plate with single-side welded centric jointing seam

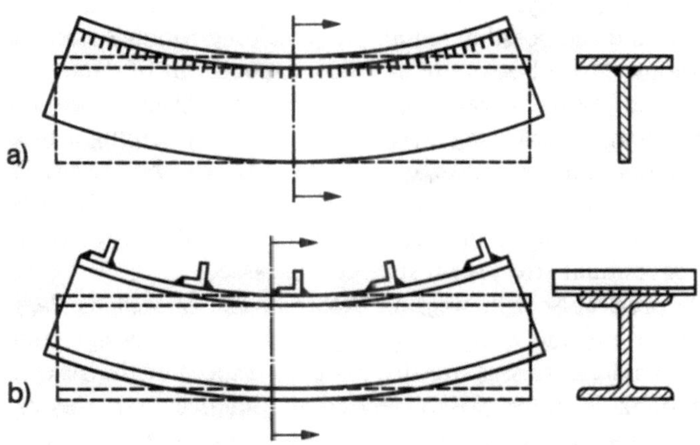

Fig. 14. Bending deformation of girders with eccentric longitudinal weld (**a**) and eccentric transverse weld (**b**)

Distortion is also observed during welding or flame cutting in the form of opening or closing of the weld groove or cutting faces.

The four basic types of welding deformations can be demonstrated for the rectangular plate with centric jointing weld, already used for presenting the welding residual stresses (Fig. 13). In contrast to the presentation of welding residual stresses, the two bending modes are also covered here, resulting from the unsymmetrical groove. The welding deformation can be subdivided into longitudinal and transverse shrinkage and angular and bending shrinkage. A further example for unwanted welding deformation in practice is the bending deformation of girders with an eccentric longitudinal or transverse weld (a similar effect occurs with stud welding) (Fig. 14).

1.8 References to related books

There are a number of German and English monographs and conference volumes (some translated from the Russian) dealing with the subject of temperature fields, residual stresses and residual deformations during and after welding, which are useful for design, analysis, test and production engineers. In addition to references to individual publications in the field, they provide a multitude of detail data. The contents of the monographs are outlined and assessed below, approximately in the order of their publication date.

The book by Rykalin [1], which first appeared in Moscow in 1951, is the basic work on the analysis of thermal processes during welding, which, even today, can claim to be the basic reference work. Since then, no significant advances have been noted in thermodynamic modelling although there have been improvements to numerical methods and computing aids.

The books by Okerblom [2, 3], first published in Moscow in 1947, contain reflections on the origin and classification of welding residual stresses and welding deformations, simple analyses and results of older measurements. Also presented are the effects and the possible reduction of welding residual stresses and welding deformations. These books are in many respects out of date.

The book by Gunnert [4] (his dissertation) focuses on a specific residual stress measurement method, the releasing of residual stresses by drilling a ring groove around the measuring marks arranged in a star pattern, the change in distance between the marks being detected by a detachable strain gauge. Results of measurements are reported, the verified reduction in residual stresses presented and the effects of residual stresses on strength discussed.

Malisius [5] deals preferably with welding deformations as well as their physical and technical origin and effect, including influences on design and manufacture. Welding residual stresses are discussed in respect of the risk of crack initiation. The book is more closely related to practice than theory. Qualitative statements dominate over quantitative statements. Individual cases are dealt with in greater depth, including a number of complete structures.

Hänsch and Krebs [6, 7] offer physical and illustrative basic facts in respect of the origin of welding residual stresses and welding deformations, methods of analysis in research and practice, longitudinal and transverse stress system effects on structural load bearing capacity, design and manufacture measures, the box girder of a railway bridge as an analysis example for welding deflections and an overview of related analysis programs. The book [7] offers an adequately systematic, thorough and condensed presentation in respect of rod-like parts, with detailed numerical data and an analysis example for practical applications.

Vinokurov [8] offers a comprehensive and theoretically substantiated presentation of the subject matter, while including numerous results of measurements, based in particular on the work of the Baumann Technical Institute in Moscow. The book includes fundamentals and applications, analysis and measurement methods, specimens and components as well as methods of reduction of stresses and deformations.

Neumann and Röbenack [9] compile results from literature together with the results of their own research in the form of a catalogue, organized systematically. The particular value of this book is its great variety of results and the quotation of the original results from measurement including approximative formulae.

Masubuchi [10] covers in a very extensive, manual-like book (624 pages, 1511 references) not only transient temperature fields, thermal stresses, welding residual stresses and welding deformations, but also the strength of welded structures (static strength, fatigue, corrosion, instability, weld cracking) on the basis of the above phenomena. Welding defects and fatigue fracture mechanics are also addressed. In view of its Japanese-American knowledge base, the book is a valuable supplement to the Russian and Western European oriented books [2 to 9]. It is not possible to deal here with several monographs in Russian and Japanese language relating to residual stresses and distortion during and after welding (references in [10]).

The above overview is rounded off finally by reference to the conference volumes [11 to 21]. Several existing books and conference volumes on residual stresses in general, including welding residual stresses, are not dealt with here.

1.9 Presentation aspects

Following these commentaries on older technical books available on the subject, what is now of interest are the presentation aspects of the present book, which, under the general subject of design, analysis and manufacture of welded structures, supplements the publication [436, 437] on their structural fatigue strength.

Welding residual stresses and welding distortion are kept within sustainable limits, proceeding from tediously acquired and frequently insufficiently verified empirical knowledge. Accordingly, the older publications on the subject at least are more a workshop manual than an instruction for action based on scientific-technical knowledge. On the other hand, considerable efforts have been made, initially in Russia, to shed light on the field also from the theoretical standpoint. In the West, the finite element method later opened up a wide-ranging, theoretical path, yet one closely related to practical applications. Proceeding from the finite element method, the subject of temperature fields, stresses, deformations and microstructural transformations arising by welding might be presented in concise mathematical form. This would be a specialized publication on non-linear and transient thermo-elastic-plastic field problems (supplemented by electromagnetic and fluid field problems), its possible application in no way being limited to welding engineering. On the other hand, relation to welding engineering practice could then only be presented in abbreviated form by way of examples. The numerous material and process parameters required for finite element simulation would not generally be sufficiently well known. It would not be possible to draw general conclusions from the numerical results relating to individual cases. The empirical knowledge from practice could not be included.

The theoretical basis which has been acquired so far is used in this book for convincingly structuring the subject matter, for integrating analysis and test results and for checking the logic of existing empirical knowledge. The physical and technical bases are presented, as are the results of analyses and measurements, but not the mathematical details of the analysis methods.

The uniqueness of welding phenomena necessitates the dominance of methodical aspects in structuring the subject matter. The phenomena are strongly non-linear, inhomogeneous and transient. They are determined by a multitude of geometry, material and process data. General statements and theorems, as exist, for instance, in the field of elastic structural behaviour, are not possible in this case for the reasons already stated. Each result applies only to the respective individual case of defined conditions. Each further case necessitates a detailed reanalysis. For this reason, the design, analysis, test and production engineer is best served with the presentation of the available methodical approaches. Account has also been taken, however, of application aspects in structuring the subject matter. The book is, therefore, only partially arranged according to methods. All in all, a compromise had to be adopted, as is inevitable with practical tasks. An elaborate subject index is available for readers requiring information on special issues.

The structuring employed here is of significance, beyond the didactic intent of the book, for the planning of expert systems for welding engineering, which are currently so topical.

2 Welding temperature fields

2.1 Fundamentals

2.1.1 Welding heat sources

2.1.1.1 Significance of welding temperature fields

The origin of welding residual stresses and welding distortion is the concentrated heat input, both locally and in terms of time, by which a fusion zone is produced at the welding spot (fusion welding). With the combined action of pressure producing local plastic deformations, heating to just below the melting temperature is also sufficient (pressure welding). Only in exceptional cases is welding performed solely under local pressure (cold welding). The high heat concentration is necessary because metallic materials rapidly diffuse the heat. The temperature fields during welding are consequently extremely inhomogeneous and transient. The basic temperature of the component in the most unfavourable case is $-40\,°C$ (severe frost); the local maximum temperature in the weld pool is at the evaporation temperature of the metal (approx. $3000\,°C$ in the case of steel). In this temperature range, base metal and filler metal fuse, metallurgical processes proceed in the weld pool, regions of the metal solidify and recrystallize, and microstructural transformations take place during heating and cooling. The temperature field, therefore, determines the welding residual stresses not only directly through the thermal strains, but also indirectly through the transformation strains which accompany the changes in state and microstructure (Fig. 3). The temperature field should be assessed in both respects. In addition, it is of interest in connection with questions of material and process engineering.

What is presented below in respect of welding temperature fields can basically be transferred to thermally comparable processes; for example, to thermal cutting and surface heating with the flame for the purpose of hardening, soldering, flame cleaning, stress relieving or flame straightening.

2.1.1.2 Types of welding heat sources

Local and time-related heat concentration is achieved by different types of welding heat sources.

In the case of arc welding, the heat is generated by means of the electric discharge at the anode and cathode spot as well as in the gas column (the thermal plasma). In welding processes using a direct arc, anode and cathode spot directly heat the base metal and the melting or non-melting electrode material, supported by indirect heating as a result of radiation and convection (the blow effect) proceeding from the arc column, as well as by radiation proceeding from the electrode spot. In welding processes using an indirect arc (plasma welding), only indirect heating is effective with a plasma jet velocity which can be increased up to flame cutting. In all cases, this involves surface heating. Arc welding is performed with a solid-coated melting electrode, with a melting or non-melting electrode shielded by active or inert gas or with a melting electrode shielded by slag-forming powder. Weld seams and weld spots are possible. An arc pressure welding method is stud welding.

In the case of welding (and spraying) with a gas flame, acetylene C_2H_2 is partially combusted with pure oxygen O_2 in the reduction zone surrounding the flame core (see Fig. 73) to form carbon monoxide CO and hydrogen H_2 and then completely combusted in the flame column with atmospheric oxygen to form carbon dioxide CO_2 and water vapour H_2O. The gas flow of the flame impacts the surface of the welding spot at high velocity. Heating is the result of convection and radiation. The method can be performed without and with filler metal. Seam and spot fusion welding are possible, as is gas pressure welding. Heating with a gas flame can, in addition to welding (and spraying), also serve other purposes (e.g. flame cleaning, soldering, heat treatment, preheating).

Electrical resistance heating is used in resistance spot welding (including also projection welding and roller seam welding), in resistance butt welding (pressure butt welding, flash butt welding, high frequency resistance welding of longitudinal and spiral-type seams) and in electric slag welding.

In resistance spot welding and resistance butt welding, the initially dominant contact resistance in the contact area of the parts being welded (and in the electrode face) results in heating of the surface. After local fusion, the contact resistance collapses (accelerated in flash butt welding by repeated separation). Therefore, it is the volume heating dependent on current density which predominates. In high-frequency resistance welding with conductive or inductive energy transfer, a thin surface layer is primarily heated as a result of skin effect and transmission resistance. In electric slag welding (of thick plates with vertical weld), the molten, electrically conducting slag layer is resistance-heated which, in turn, melts the base metal and then melts off the continuously fed electrodes.

In friction welding (of axisymmetrical parts), the contra-rotating surfaces are heated by friction, freed of foreign material layers and finally joined by axial compression at a temperature just below the melting point. In vibration welding (with ultrasound), high frequency friction effects are used; without, however, approaching the melting temperature.

In electron beam welding (performed in the vacuum), electrons (generated by hot cathode, focussed by electron optics) are absorbed in a surface layer of approximately 10 μm thickness, as a result of which heat is produced. If the power density is adequate, the surface may be fused and a deeply penetrating vapour capillary finally produced. The vapour capillary surrounded by the fused metal forms the welding heat source. It assumes the form of a "keyhole", if the workpiece is moved relative to the beam in order to produce a weld seam.

In laser beam welding, coherently focussed light is directed at the welding spot and (partially) absorbed here in a surface layer of approximately 0.5 μm thickness. If the power density is adequate, the surface is fused. Finally, in the same way as for electron beam welding, a vapour capillary occurs as the actual welding heat source. In addition, there is a (heat-inefficient) process variant in which the heat is generated only on the surface under a defocussed beam of low power density and penetrates into the interior of the welding spot solely as a result of heat conduction.

In alumino-thermal fusion welding (used for welding rails), the molten pool is formed by chemical conversion of aluminium powder with metal oxides. This results in aluminium oxide (slag), filler metal and heat. The heat is generated in the volume.

In flame cutting with oxygen beam, the (metallic) material combusts at the beam surface as soon as the ignition temperature is exceeded. Ignition occurs once the reaction heat released exceeds the heat which can be diffused.

From this description of the different welding heat sources, it is clear that heat is generated partially on the surface (and must then be propagated into the interior by conduction), and partially in the interior of the material. Combined with the different component and groove geometries as well as the variable behaviour of the welding heat sources, this results in a large number of possible variants in practice.

2.1.1.3 Output of welding heat sources

The principal parameter of the welding heat source for the temperature field is the heat input into the welding spot, in momentarily acting sources the heat quantity (or the heat energy) Q [J], and in continuously acting sources the heat flow (or heat output) q [J/s]. In both cases, what is considered are the net or effective values Q and q, respectively, which are related to the gross values $\bar{Q}$ and $\bar{q}$ by the heat efficiency η_h of the welding process. The gross values $\bar{Q}$ and $\bar{q}$, respectively, designate the total energy or total output generated at the heat source.

In arc welding, the total output is the product from amperage I [A] and voltage U [V] at the arc in the case of direct current. In the case of alternating current, effective values resulting from the momentary products have to be used (generally in the form RI_{eff}^2 with ohmic resistance R and effective amperage I_{eff}, [78]). In gas welding, acetylene consumption $\dot{V}_{\text{Ac}}$ [l/h] is taken as the basis [1]. The heat losses in welding which are taken into account by η_h include the heat dissipation to the surroundings as a result of convection and radiation, the spray losses and the losses caused by heating the electrode in the case of a non-melting electrode. Consequently, in continuous arc or gas welding, it holds:

$$q = \eta_h UI = \eta_h RI_{\text{eff}}^2, \tag{1}$$

$$q = \eta_h 3.2 \dot{V}_{\text{Ac}}. \tag{2}$$

The area-specific heat flow density q^* [J/mm^{2}s] and volume-specific heat source density q^{**} [J/mm^{3}s] transferred to the weld pool by the welding arc or welding flame, approximately follows a Gaussian distribution curve ("normal distribution source", see Section 2.1.3.3). With the same effective output of the source, the Gaussian distribution curve for the arc is narrow and high, for the flame wide and low (Fig. 16). The curve for plasma, electron and laser beam welding is once more substantially higher and narrower. A survey on power density of various welding processes is given in Fig. 15.

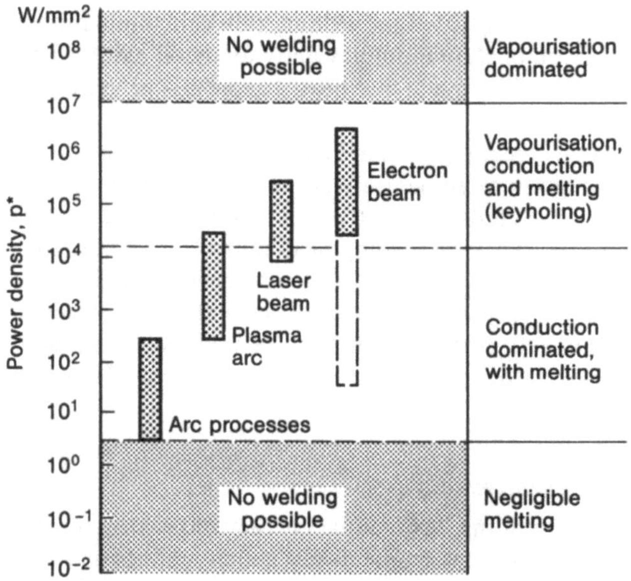

Fig. 15. Power densities of various arc and beam welding processes; after Quigley in Ref. [72]

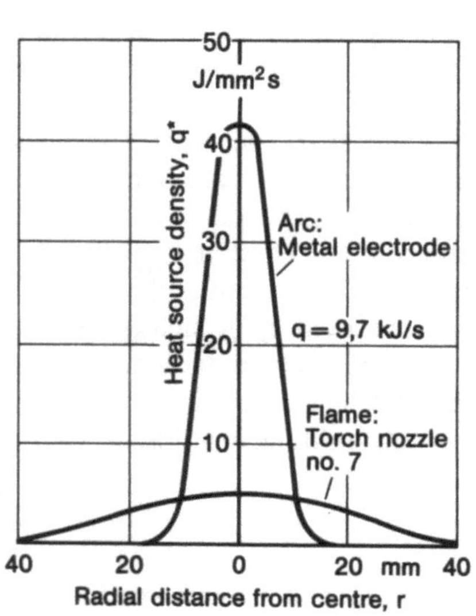

Fig. 16. Heat source density q^* at surface of plate during arc and flame welding with equal net heat output q, experimental results; after Rykalin [1]

Table 1. Output data of fusion welding methods used with steel and aluminium based on Ref. [1, 47, 72, 77, 156–158, 160, 169, 237]

Welding method	Heat output $\bar{q}$ [kJ/s]	Welding speed v [mm/s]	Output per unit length $\bar{q}_w$ [kJ/mm]	Efficiency η_h [–]
Covered electrode	1–20	< 5	< 3.5	0.65–0.90
Gas metal arc	5–100	< 15	< 2	0.65–0.90
Gas tungsten arc	1–15	< 15	< 1	0.20–0.50
Submerged arc	5–250	< 25	< 10	0.85–0.95
Electron beam	0.5–10	< 150	< 0.1	0.95–0.97
Laser beam	1–5	< 150	< 0.05	0.80–0.95
Acetylene flame	1–10	< 10	< 1	0.25–0.85

Instead of the heat input q per unit of time, the heat input per unit length of weld, q_w [J/mm], is used to advantage in seam welding (with speed v [mm/s]):

$$q_w = \frac{q}{v}. \tag{3}$$

The temperature field in seam welding is dominantly determined by the heat input per unit length of weld (and by the plate thickness in the case of through-welded seams).

Instead of q and q_w, respectively, the heat quantity q_m per unit mass of deposited metal is also specified, dependent on the process [6, 77].

In resistance spot and resistance pressure welding, respectively, the total energy is the product of ohmic resistance R, effective amperage I_{eff} and duration of current flow t_c:

$$Q = \eta_h R I_{eff}^2 t_c. \tag{4}$$

The heat output data of the welding process require detailed experimental and theoretical analyses in each individual case (see Ref. [78] and Section 2.3). The uncertainty of welding temperature field calculations is attributable to a high degree to uncertainties in respect of Q and q, respectively. A rough survey on the heat output of different welding processes is presented in Table 1. Further heat output or input data may be drawn from Ref. [79, 80, 77, 169].

2.1.2 Heat propagation laws

2.1.2.1 Law of heat conduction

The heat input in welding concentrated locally and in terms of time (the heat is required for fusing the welding spot) is, in metallic materials, propagated undesirably rapidly into more remote areas of the component. On the other hand, radiation and convection play an important role in most cases for heat input. Radiation and convection are also responsible for heat losses at the surface of the component. Firstly, the relation in terms of a law is stated between momentary local heat flow and momentary temperature field.

Fourier's law of heat conduction states that the heat flow density q^* [J/mm^{2}s] in isothermal faces of the body is proportional to the negative temperature gradient $\partial T/\partial n$ [K/mm] perpendicular to this area through the coefficient of thermal conductivity λ [J/mmsK] [22, 1]:

$$q^* = -\lambda \frac{\partial T}{\partial n}. \tag{5}$$

The geometric location of all points of a solid which have the same temperature is designated the isothermal face (by analogy to the isotherms in the area).

The coefficient of thermal conductivity of metals depends on the chemical composition, the microstructural state and the temperature (see Section 2.1.2.6). As heat conduction in metals, just as current conduction, is based on the movement of free electrons, a simple relation exists between thermal and electrical conductivity.

2.1.2.2 Law of heat transfer by convection

Heat is primarily propagated in gases and liquids by means of the movement of particles. Natural convection occurs if this movement is produced solely by the differences in density caused by the temperature differences. Forced convection occurs if external (flow) forces maintain the movement (as, for example, the blow effect of the arc or flame).

For a surface element of a solid body contacted by flowing gas or liquid, the heat flow density q_c^* is, according to Newton's law, proportional to the difference between surface temperature T and gas or liquid temperature T_0 through the coefficient of convective heat transfer α_c [J/mm^{2}sK] [1]:

$$q_c^* = \alpha_c(T - T_0). \tag{6}$$

The coefficient of convection heat transfer α_c depends on the flow conditions on the surface (specifically, the structure of the boundary layer), on the surface properties, on the properties of the flowing medium and (unintentionally) on the temperature difference $T - T_0$. The dependence is presented by means of empirical formulae using dimensionless parameters.

2.1.2.3 Law of heat transfer by radiation

Heat radiation of heated bodies propagates as an electromagnetic wave in space, penetrates transparent bodies and is absorbed by bodies which are impermeable to light and converted back into heat. The bodies are thus in a state of mutual heat exchange.

The heat radiated per unit of area and time by a heated body, the heat flow density q_r^*, is, according to Stefan-Boltzmann's law, proportional to the fourth power of the surface temperature T [K] through the radiation coefficient εC_0 [J/mm^{2}sK^4] [1]:

$$q_r^* = \varepsilon C_0 T^4. \tag{7}$$

The radiation coefficient $C_0 = 5.67 \times 10^{-14}$ [J/mm^2sK^4] applies to the "absolutely black body". The "grey body" is characterized by the blackness degree $\varepsilon < 1.0$. For polished metallic surfaces, $\varepsilon = 0.2$–0.4. For rough, oxidized surfaces of steel, $\varepsilon = 0.6$–0.9. The blackness degree increases with the temperature, $\varepsilon = 0.90$–0.95, in the range of the melting temperature.

Heat dissipation for the important case, in welding technology, of the cooling of a relatively small body (body temperature T) in relatively extensive surroundings (body temperature T_0) occurs by means of radiation (the latter predominating compared to convection at high temperatures) in accordance with

$$q_r^* = \varepsilon C_0 (T^4 - T_0^4). \tag{8}$$

The corresponding boundary value problem of heat conduction is linearized by also using the (only apparently) linearized form of equation (8) in which the coefficient of heat radiation α_r [J/mm^2sK] heavily depends on T and T_0:

$$q_r^* = \alpha_r (T - T_0). \tag{9}$$

2.1.2.4 Field equation of heat conduction

The dependence of the local temperature change on the temperature field in its surroundings is now considered. Locally supplied heat increases the local temperature according to the local volume-specific heat capacity $c\varrho$ [J/mm^3K]. The parameter c [J/gK] is the mass-specific heat capacity and the parameter ϱ [g/mm^3] is the density. The more unevenly the temperature is distributed at a particular time, the more rapidly the temperature changes. For the homogeneous and isotropic continuum with temperature-independent material characteristic values, the following field equation of heat conduction applies (e.g. Ref. [22, 1]), which is based on the principle of conservation of energy:

$$\frac{\partial T}{\partial t} = \frac{\lambda}{c\varrho}\left(\frac{\partial^2 T}{\partial x^2} + \frac{\partial^2 T}{\partial y^2} + \frac{\partial^2 T}{\partial z^2}\right) + \frac{1}{c\varrho}\frac{\partial Q_v}{\partial t}. \tag{10}$$

The parameter Q_v [J/mm^3] is the heat energy released or consumed per unit of volume. The material- and temperature-dependent term $\lambda/c\varrho$ is abbreviated to the coefficient of thermal diffusivity a [mm^2/s]:

$$a = \frac{\lambda}{c\varrho}. \tag{11}$$

The differential term in parentheses is abbreviated introducing the Laplacian operator ∇^2 to give $\nabla^2 T$.

In tensor notation, equation (10) reads:

$$\lambda T_{,ii} + \dot{Q}_v = c\varrho \dot{T}. \tag{12}$$

Whereas ϱ is only weakly temperature-dependent, λ, c and a are markedly temperature-dependent (Figs. 17 to 19). The specific heat capacity, a measure for the temperature increase caused by the input of heat, is moreover infinitely large at the transformation temperatures of the material (and correspondingly the thermal diffusivity infinitely small) because, at these temperature points, heat is consumed or released without changing the temperature. Transformation temperatures in the case of steel are the melting point (1528 °C for pure iron), the $\delta\gamma$-transformation temperature (1401 °C for pure iron) and the $\gamma\alpha$-transformation temperature (906 °C for pure iron). The temperature field analysis for welding is often simplified by the fact that the thermal diffusivity is introduced, location- and temperature-independent, as an averaged constant value in the area and range under study.

The field equation of heat conduction is reduced for the plate to two, for the rod to one coordinate. The temperature in the stationary (i.e. steady-state) temperature field is time-constant in all points, i.e. $\partial T/\partial t = 0$. Equation (10) is then simplified to the material-independent Laplacian differential equation:

$$\nabla^2 T = 0. \tag{13}$$

The instationary (i.e. transient) temperature field, by contrast, is determined on the material side by the thermal diffusivity. Welding temperature fields are highly instationary. The term "heat diffusion" is used in the following text instead of "heat conduction" if the instationary character of heat flow is to be stressed.

The quasi-stationary temperature field of the continuously acting heat source moving with constant speed v along the x-axis (see Section 2.2.2) is given related to the moving ξyz-coordinate system [29, 30]:

$$-v\frac{\partial T}{\partial \xi} = \frac{\lambda}{c\varrho}\left(\frac{\partial^2 T}{\partial \xi} + \frac{\partial^2 T}{\partial y^2} + \frac{\partial^2 T}{\partial z^2}\right) - \frac{v}{c\varrho}\frac{\partial Q_v}{\partial \xi} \tag{14}$$

2.1.2.5 Initial and boundary conditions

The question which is generally posed relates to the temperature field which results from the effect of the welding heat source in a dimensionally limited structure, proceeding from a defined initial temperature state. The task which is thus set is to solve the field equation of heat conduction for given initial and boundary conditions. The initial condition is the location-constant temperature of the surrounding medium or of preheating; in exceptional cases also a certain temperature distribution, e.g. the temperature field of the preceding pass in the case of multi-pass welding. The boundary conditions are the conditions of heat dissipation at the boundary surfaces of the structure.

Three types of boundary conditions occur in the case of application analyses [1]. Firstly, the temperature may be prescribed, in the simplest case, as a constant temperature (isothermal boundary condition). Secondly, the heat flow density on the surface may be prescribed, in the simplest case, as vanishing heat flow (adiabatic boundary condition). According to equation (5), the temperature profiles perpendicular to the boundary have the gradient zero in this case. Thirdly, heat transfer to the surrounding medium in accordance with equations

(6) and (9) may occur. In the simplest case of a constant ambient temperature, the temperature gradient perpendicular to the boundary is then all the greater, the greater the local temperature difference to the surrounding medium. Equating q^* from equation (5) with q_c^* and q_r^* from equations (6) and (9) results in

$$(\alpha_c + \alpha_r)(T - T_0) = -\lambda \frac{\partial T}{\partial n}. \tag{15}$$

The isothermal boundary condition follows with $(\alpha_c + \alpha_r)/\lambda = \infty$, i.e. strong heat transfer with low heat conduction, the adiabatic boundary condition with $(\alpha_c + \alpha_r)/\lambda \to 0$, i.e. low heat transfer with strong heat conduction. The temperature field during welding can be calculated with the adiabatic boundary condition as far as short-time processes in quiet air outside of the welding spot are considered.

2.1.2.6 Thermal material characteristic values

Considering temperature field calculations based on the basic equation of heat conduction, the following thermal material characteristic values are required:

- thermal conductivity λ [J/mmsK],
- specific heat capacity c [J/gK],
- density ϱ [g/mm^3] and
- thermal diffusivity a [mm^2/s].

For function-analytical solutions, the basic equation is linearized, i.e. the material characteristic values λ, c, ϱ or their combination in the thermal diffusivity a are introduced as constants. As the characteristic values are in reality temperature-dependent, the constants selected are the mean values in the temperature range under study. In the case of residual stress analyses, this is the range of relatively low temperatures in which the yield limit displays relatively high values (e.g. 20–500 °C in the case of mild steel). For finite element solutions, by contrast, the temperature dependence of the material characteristic values can be taken into account.

Knowledge relating to the thermal material characteristic values [198, 199] and their temperature dependence is not always adequate, with the result that the numerical possibilities are limited solely as a result of this. Pertinent information in the textbooks relating to welding residual stresses [1–10] is rather scanty. By contrast, reference can be made to an excellent general compilation by Richter [23] of values for ferrous materials.

A distinction requires to be made in connection with the material characteristic values between temperature dependent momentary values and temperature range dependent mean values. The former are more suitable for finite element analyses, the latter for (linearized) function-analytical solutions.

A first survey on the values of λ, $c\varrho$ and a (and the thermal expansion coefficient α) for steel, aluminium and titanium is provided in Table 2 on the basis of a Russian source [8]. For λ, $c\varrho$ and a, obviously mean values between

Table 2. Thermal material characteristic values of base metals, the values of λ, $c\varrho$, a averaged between 0 °C and T_{max}; after Vinokurov [8]

Base metal	α [K^{-1}]	λ [J/mms K]	$c\varrho$ [J/mm^{3}K]	a [mm^2/s]	T_{max} [°C]
Mild and low-alloy steels	$12–16 \times 10^{-6}$	0.038–0.042	$4.9–5.2 \times 10^{-3}$	7.5–9.0	500–600
Austenitic CrNi-steels	$16–20 \times 10^{-6}$	0.025–0.033	$4.4–4.8 \times 10^{-3}$	5.3–7.0	600
Aluminium alloys	$23–27 \times 10^{-6}$	0.27	2.7×10^{-3}	100	300
Titanium alloys	8.5×10^{-6}	0.017	2.8×10^{-3}	6	700

ambient temperature and expected maximum temperature T_{max} are stated (T_{max} is not clearly defined in Ref. [8]). The particularly high coefficients of thermal conductivity and diffusivity of the aluminium alloys are remarkable.

The characteristic values λ, c, ϱ for a fine-grained structural steel with transformation at 700 °C (according to Ref. [234]), for an aluminium alloy (according to Ref. [10]) and for a titanium alloy (according to Ref. [215]) are presented in Figs. 114 to 116 together with the mechanical material characteristic values. The temperature dependence of the characteristic values of the nickel-chrome-iron alloy, Inconel 600, is presented in Ref. [249].

In Figs. 17 to 19, derived on the basis of Ref. [23] the characteristic values λ, c and ϱ are presented as ranges of curve profiles over temperature, on the one hand for mild (ferritic or eutectoid) and low-alloy steels, on the other hand for high-alloy (austenitic) steels, which cover the individual curves of Ref. [23]. The discontinuity of the curves for mild and low-alloy steel at 700 °C is caused by the transformation starting here (A_{c1} temperature).

Thermal conductivity λ is presented in Fig. 17 together with thermal diffusivity a. The conductivity is calculated from the specific electrical resistance, which is determined by measuring amperage and voltage. In accordance with Lorenz's law, a definite quantitative correlation exists between thermal and electrical conductivity. Both are based on the mobility of free electrons in the metallic lattice. Thermal conductivity and diffusivity decrease at a given temperature as the alloying content increases. With mild and low-alloy steels, the values drop over temperature; with high-alloy steels, they rise to a maximum and then drop.

The momentary specific heat capacity c according to Fig. 18, measured as a function of temperature in the mixing calorimeter, displays an infinitely large value, with mild and low-alloy steels in the transformation point, caused by the transformation heat released or consumed with constant temperature. With high-alloy steels, the heat capacity rises steadily with temperature. The alloying elements have only a slight influence on the curve position.

The density ϱ, measured at ambient temperature in accordance with Archimedes' principle and converted through the mean, linear thermal expansion coefficient to increased temperature (Fig. 19), drops with the temperature. The discontinuity in the transformation point of mild and low-alloy steels is due to the denser atom packing of the face-centred austenite compared to the body-centred ferrite.

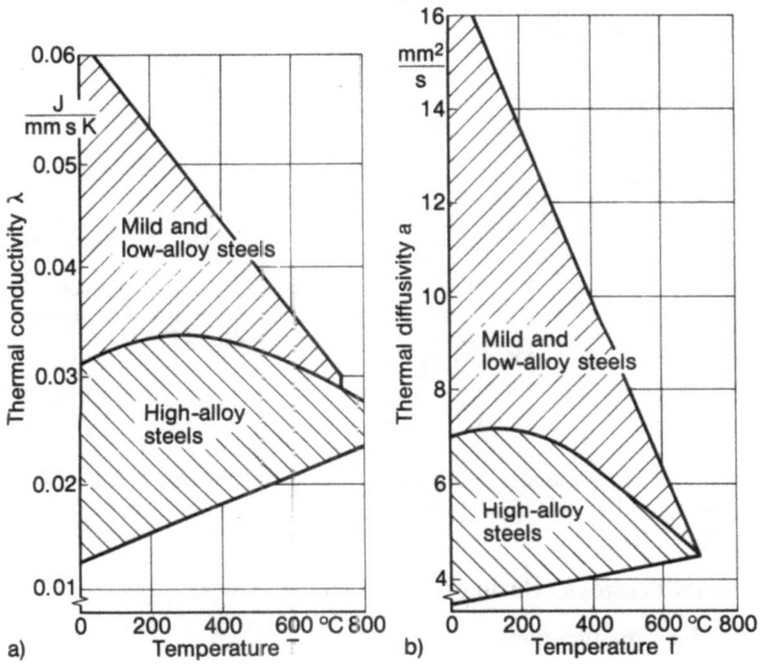

Fig. 17. Thermal conductivity λ (**a**) and thermal diffusivity a (**b**) dependent on temperature; areas of individual curves according to Richter [23]

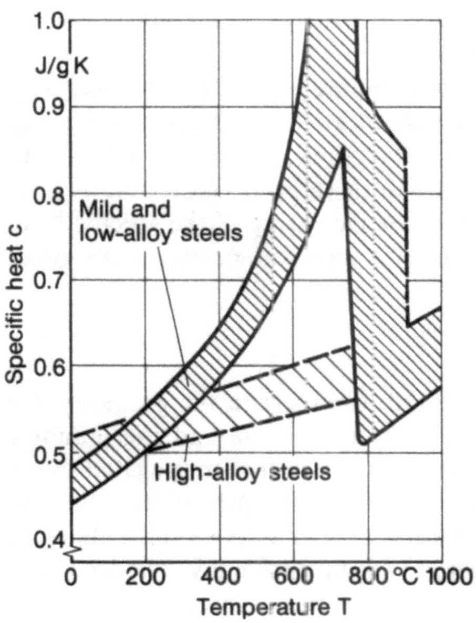

Fig. 18. Mass-specific heat capacity c of steels dependent on temperature, transformation heat at 700 °C covered by c; areas of individual curves according to Richter [23]

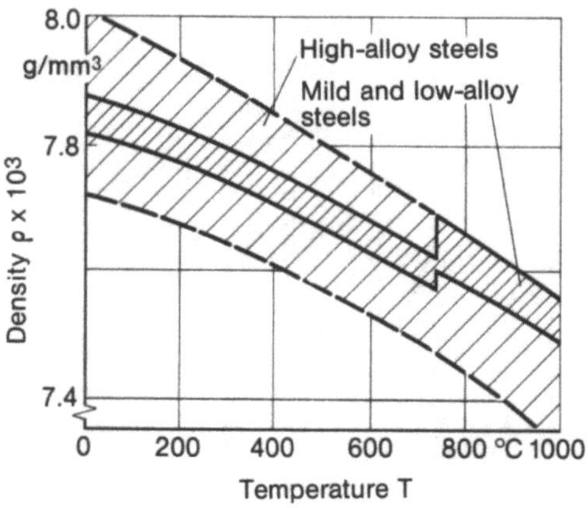

Fig. 19. Density ρ of steels dependent on temperature; areas of individual curves according to Richter [23]

The coefficients of heat transfer α_c and α_r for convection and radiation on steel plates are presented in Ref. [248] as a function of temperature.

2.1.3 Model simplifications relating to geometry and heat input

2.1.3.1 Necessity for simplifications

Simplifications relating to geometry and heat input as part of the analysis model are absolutely necessary for functional analysis solutions resulting in compact final formulae (additionally to linearizing the analysis model in the form of location- and temperature-independent diffusivity and of adiabatic boundary conditions). Finite element solutions, on the other hand, in principle permit nearly any complexity to be taken into account although, in practical terms, economic requirements set a limit here as well. The following simplifications are tailored to functional analysis solutions.

Common to all models is the fact that the complex processes in the weld pool are ignored, in particular the melting and solidification processes with melting movement and heat transfer by means of convection and radiation. Only the melting or solidification heat which appears in these processes is occasionally taken into account as an apparent lowering or increasing of the specific heat. The complex thermal process in the weld pool is approximated by heat sources in the heat-conducting continuum.

2.1.3.2 Simplifications of the geometry

Three basic geometrical bodies are introduced in respect to the geometry, the semi-infinitely extended solid, the infinitely extended plate and the infinitely extended rod (Fig. 20). In the semi-infinite solid, the heat is propagated three-dimensionally. The heat source acts in the centre of the surface of the solid. The plane layer of this solid is of interest in application as a model for thick plates. In the infinite (plane) plate, heat is propagated two-dimensionally. The heat flow density is constant over the plate thickness. The heat source, likewise with

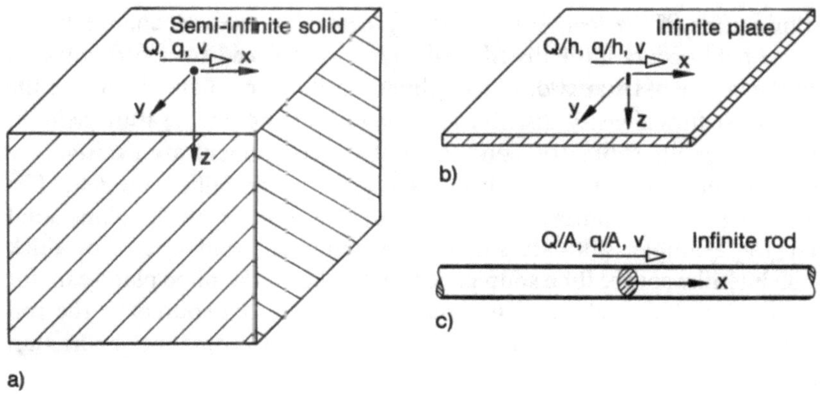

Fig. 20. Basic geometrical bodies introduced for temperature field analysis on the basis of functional analysis: semi-infinite solid with point heat source in the surface (**a**), infinite plate with line source perpendicular to the surface (**b**) and infinite rod with area source perpendicular to the axis (**c**); these bodies infinitely extended perpendicular to the hatched surfaces of cut

constant output over the thickness, acts in the centre of the plate. In the infinite (straight) rod, the heat is propagated one-dimensionally. The heat flow density is constant over the rod cross-section. The heat source, likewise with constant output over the cross-section, acts in the centre section of the rod. One-dimensional heat flow is also possible as a special case in the semi-infinite solid and in the infinite plate with a planar heat source.

The simplification of an infinite extension instead of finite dimensions is all the more justified, the larger the dimension of the component in the appropriate direction, the shorter the period of heat propagation under study (heating and cooling), the lower the thermal diffusivity, the less distant from the heat source the area under study, and the greater the coefficient of heat transfer.

The simplification of two- and one-dimensional heat flow, on the other hand, is all the more justified, the smaller the plate thickness or the rod cross-section, the more centric to it the effect of the heat source, the longer the period of heat propagation under study, the greater the thermal diffusivity, the more distant from the heat source the area under study, and the smaller is the coefficient of heat transfer. As these requirements relating to plate and rod are largely the opposite of those stated in respect of infinite extension, there then exists an unresolvable modelling contradiction for infinite plate and infinite rod. Only the first-mentioned two criteria speak in favour of plate and rod including their infinite extension.

2.1.3.3 Spatial simplifications of the heat source

The local concentration of the welding heat input suggests corresponding simplifications of the heat source for the purpose of analysing the temperature field. A significant dependence of the temperature field on the distribution of the heat flow density in the heat source exists only at distances of equal magnitude as the dimensions of the heat source. At a greater distance, no change in the temperature field occurs if the distributed source is replaced by a concentrated source in the centre of its area or of its volume. In the immediate vicinity of the source, the heat flow density distribution determines the temperature field; at a greater distance the geometric conditions of the component are decisive.

The following concentrated sources are used in simplified models. The point-like source (point source) on the semi-infinite solid or on the plane solid layer can, for example, simulate surfacing welding on solid bodies or thick plates. The line-like source (line source) perpendicular to the plate plane can, for example, simulate butt welding. The area-like source (area source) in the rod cross-section can, for example, simulate heating of an electrode face or friction welding.

By restricting the analysis to concentrated sources, the temperature distribution in close vicinity of the welding heat source is not adequately simulated. This applies especially to the centre of the source where temperature rises to infinity. For more realistic investigations in this area in respect of welding with arc, beam or flame, surface sources with a normal distribution of heat source density q^* (normal distribution sources) have proven satisfactory in practice (what is meant is the Gaussian normal distribution of probability analysis, the Gaussian distribution curve):

$$q^* = q^*_{\max} e^{-kr^2}, \tag{16}$$

$$q^*_{\max} = \frac{k}{\pi} q. \tag{17}$$

In this connection, q [J/s] is the effective output of the heat source and the factor k [1/mm^2] is a factor disignating the heat source concentration (width of the Gaussian distribution curve) (Fig. 21). In general, the parameter r [mm] designates the distance from the centre within a circular source. The parameter r can also be used, however, to designate the distance from the centre line in the transverse direction within a strip-shaped source. The circular and strip source described is illustrated in Fig. 22.

As the Gaussian distribution curve described by equation (16) runs toward zero only at infinity, an agreement is required regarding which small q^* values of

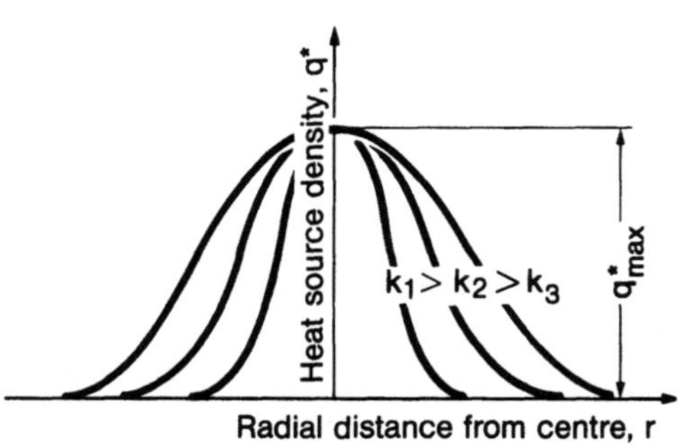

Fig. 21. Heat source density q^* over radial distance from centre r, normal distribution sources, equal maximum value $q^*_{\max}$, different concentration factors k_1, k_2, k_3

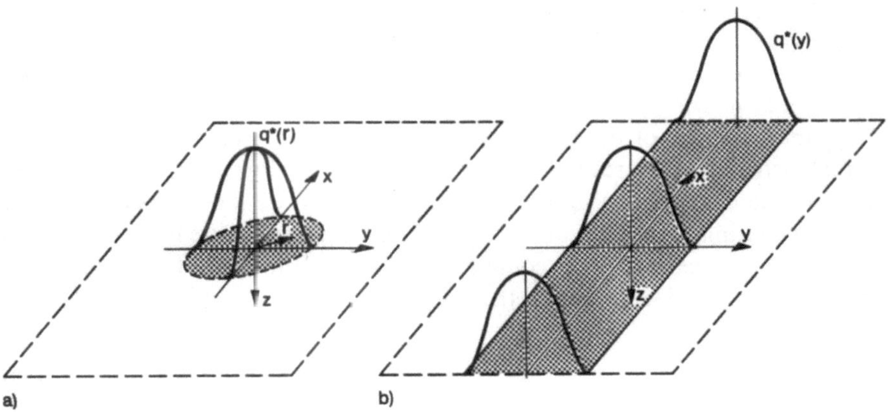

Fig. 22. Circular source (**a**) and strip source (**b**) with normal distribution of heat source density q^*

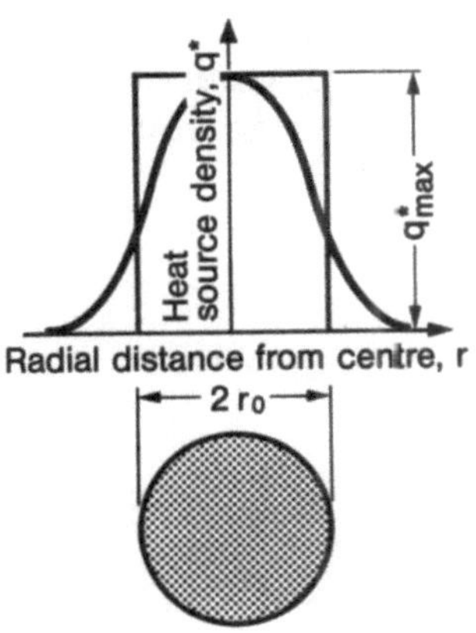

Fig. 23. Circular source with normal distribution of heat source density q^* and equivalent circular source of equal output with constant heat source density q^*_{max}

the Gaussian distribution curve can be regarded as negligible. In accordance with Ref. [1] (ibid., p. 70), $q^*_{min} = 0.05\, q^*_{max}$ is introduced. From this follows the outer diameter d_n of the normal distribution source ("heating spot"):

$$d_n = \frac{2\sqrt{3}}{\sqrt{k}} \tag{18}$$

On the other hand, the output of the normal distribution source can be numerically converted to an equivalent source with a constant heat source density equal to the maximum value of the normal distribution source (Fig. 23). The diameter $d_0 = 2r_0$ of this equivalent heat spot results from:

$$d_0 = \frac{2}{\sqrt{k}}. \tag{19}$$

According to figures in Ref. [1] (ibid., pp. 71 and 145), $d_n = 14\text{–}35$ mm was measured for arcs with an electrode spot diameter of about 5 mm. Weld torches with gas flame, on the other hand, resulted in $d_n = 55\text{–}84$ mm depending on the size of torch nozzle.

As the structure of the temperature field equation relating to the circular normal distribution source in the plate (assuming a uniform source distribution over the plate thickness) agrees with the structure of the temperature field equation relating to the line source in the plate, the temperature field of a normal distribution source can be set equal to the temperature field of the line source which starts operation earlier by the time interval Δt_0, [1] (ibid., pp. 164–166):

$$\Delta t_0 = \frac{1}{4ak}. \tag{20}$$

The time interval Δt_0 indicates how much earlier the line source must start operation in order to produce the temperature field of a normal distribution source (Fig. 24). It is dependent on the heat source concentration of the normal distribution source and (through a) on the material. According to the figures for steel in Ref. [1] (ibid., pp. 71 and 145), $\Delta t_0 = 0.6\text{–}3.1$ s was determined for arc welding and $\Delta t_0 = 8\text{–}19$ s for gas welding. With an increasing heat source concentration factor k, Δt_0 tends toward zero.

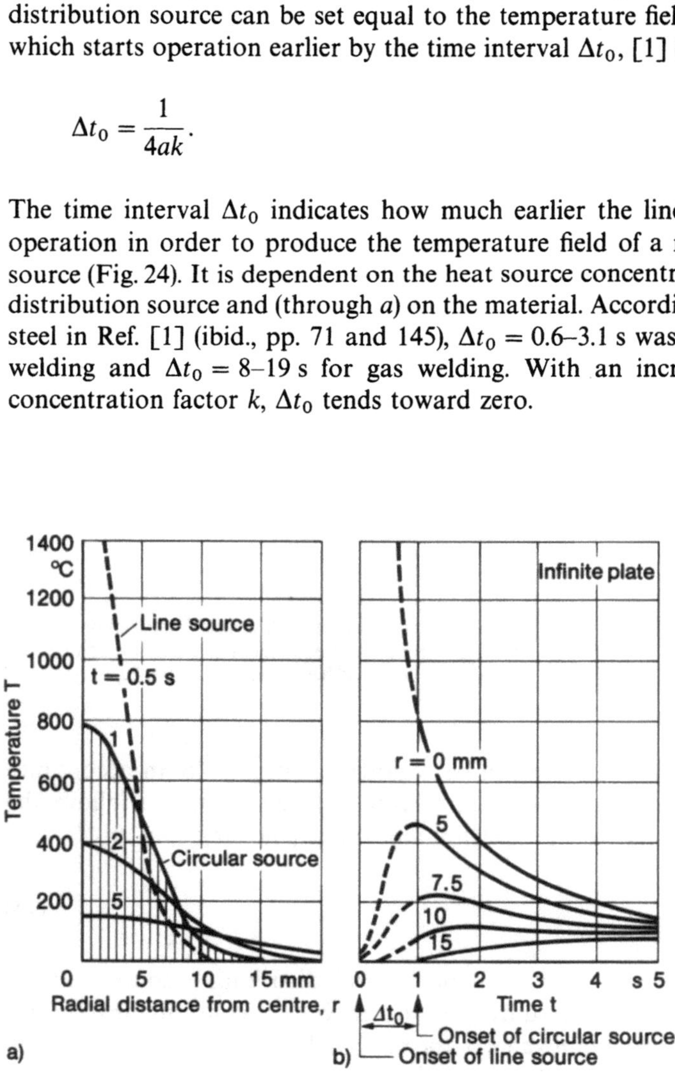

Fig. 24. Temperature field around momentary circular heat source in plate (full curves) considered as part of the temperature field of a momentary line source starting operation earlier by the time interval Δt_0 (broken curves); temperature T over radial distance from centre r (**a**) and over time t (**b**); after Rykalin [1]

The heat quantity $q\mathrm{d}t$ to be introduced through the normal distribution source is advanced by Δt_0 as heat quantity $\mathrm{d}Q$ of the line source:

$$\mathrm{d}Q = q\mathrm{d}t \tag{21}$$

The principle explained above of the time-advanced, concentrated source in the plate applies in a restricted form in the case of the semi-infinite solid (see Section 2.2.3.1).

An alternative to the time-advanced concentrated source avoiding the infinitely high temperature in the source point or source line in the case of the semi-infinite solid (or the related thick plate model) is an apparent concentrated source located at a small definite distance above the surface [150].

A semi-ovaloid arrangement of volumetric heat sources, approximately in the shape and size of the weld pool, is used for the moving welding heat source of deep-penetrating surfacing or butt welds [32, 38, 28] (Fig. 25). In the ovaloid, the volume-specific heat source density q^{**} is assumed to be (Gaussian-)normally distributed (a corresponding area-specific heat source density distribution in the surface is also possible [243]). The heat source density drops from a maximum value in the centre of the ovaloid exponentially toward the edges. The dimensions of the ovaloid should be selected approximately 10% smaller than the dimensions of the weld pool. The total heat output should be equal to the effective output of the welding process. A concluding parameter correction is made on the basis of a comparison of calculated and measured weld pool and temperature field, respectively.

The longitudinal dimensions of rapidly moving high power sources may be reduced to zero without major loss of accuracy of results. A plane surface source distribution may be substituted by a line source of finite width in the surface perpendicular to the moving direction. A volumetric source distribution may be replaced by a corresponding plane source perpendicular to this direction.

It is not always possible to find realistic source distributions for actual welding situations. On the other hand, temperature measurement in the weld

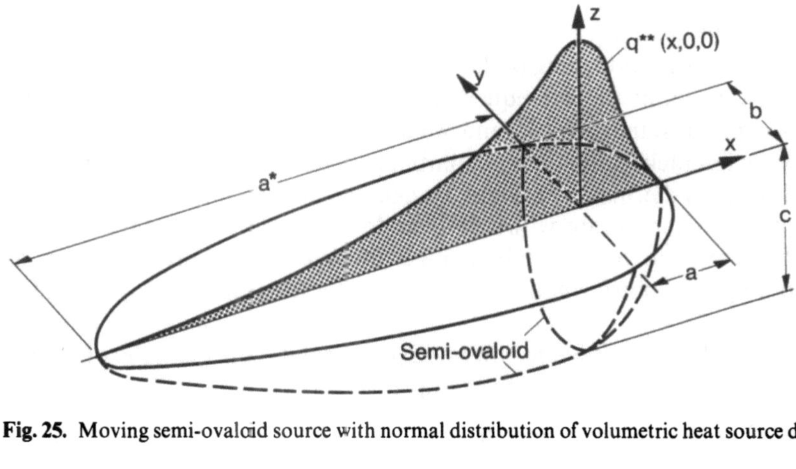

Fig. 25. Moving semi-ovaloid source with normal distribution of volumetric heat source density q^{**}

pool region is highly developed. So it may be advantageous to prescribe the temperatures in that region instead of the heat sources as the basis of analysis.

2.1.3.4 Time simplifications of heat source

A further model simplification relates to the duration of the heat source operation. The momentarily acting source is characterized by the heat quantity Q [J], which is introduced at that moment. This assumption all the better reflects reality, the shorter the time for which the heat source acts there. The continuously acting source is characterized by the heat flow (or by the heat output) q[J/s], which is introduced with constant intensity during the period the heat source is acting. This assumption corresponds in general to the reality of seam welding.

In the case of plate models with line source, Q or q is related to the plate thickness h, as it is in rod models with area source to the cross-section area A.

Finally, model simplifications are introduced in respect of the movement of the heat source. A distinction is made between stationary, moving and rapidly moving sources, the movement being assumed as linear and uniform (constant speed). The rapidity is related to the heat propagation rate in the direction of the source movement. If movement is rapid, it is assumed that heat propagates only perpendicular to the weld. Consequently, the temperature field equations are greatly simplified. This simplification is acceptable if the welding heat source is in fact moved rapidly (mechanized welding) or if only areas close to the weld are considered.

In finite element solutions, these simplifications are less necessary. In particular, complex geometry and temperature-dependent material characteristic values may be taken into account without restrictions. The result, however, also does not contain any general statement.

2.1.3.5 User questions addressing welding temperature fields

The appraisal of the various model simplifications is also oriented to the question the respective user will ask, and to which an answer is expected from temperature field analysis.

A relatively rough modelling of the thermal process, which is matched to the degree of resolution of the subsequent continuum-mechanical model, is sufficient for welding residual stresses and welding deformations (including plastic strains and cold cracks which may be initiated) caused directly by the temperature field through thermal expansion. No detailed information at all is required in areas of high temperature with relatively low or even vanishing yield limit because the internal forces vanish here. Finer modelling is required, however, as soon as cooling and transformation stresses at lower temperature play a significant role.

A finer modelling is needed for thermodynamic and thermokinetic processes in the area of the welding pool; for example:

- melting-off of the electrode and surface fusing of the base metal with forma-

tion of the fusion zone (metal deposition rate, shape and size of the fusion zone, bead surfacing height and fusion zone depth);
- crystallization of the fusion zone and subsequent microstructural transformations;
- microstructural changes in the transition zone of the base metal during heating and cooling in the high temperature range (e.g. formation of rough-grained structure and hardening), and layer formation in the heat-affected zone;
- formation of hot cracks in the weld metal and transition zone as a result of large deformation strains in the high temperature range of reduced ductility.

The necessary coupling of the energy relation, equation (10), with weld pool physics has not yet been accomplished but progress has been made in respect of the heat-affected zone.

2.1.3.6 Numerical solution and comparison with experiments

The functional analysis solution (and also the formerly frequent solution based on the finite difference method) of the field equation of heat conduction assumes simplifications of the model described above. Introducing a constant thermal diffusivity (constant as averaged quantity in the temperature range under study) results in a linearized field problem. With moving sources, only the build-up, quasi-stationary state of uniform and linear source movement is considered. Differential equation solutions for such simplified problems are offered, in particular by Rykalin [1]. Significant further contributions are available from Greenwood [48], Rosenthal [29, 30], Christensen, Davies and Gjermundsen [31], Myers, Uyehara and Borman [160] and Carslaw and Jaeger [22].

Functional analysis solutions to heat conduction processes in welding are presented in Sections 2.2.1 to 2.2.5 (see also Ref. [52]).

The finite element solution of the heat conduction problem proceeds from an integral equation which corresponds to differential equations (10) and (12). The field is discretized into simply shaped elements. This leads to a system of non-linear, ordinary differential equations of the first order for the unknown nodal temperatures T (see Ref. [26, 27, 193, 194]), the matrix equation of heat conduction:

$$[C_{\mathrm{T}}]\{\dot{T}\} + [K_{\mathrm{T}}]\{T\} = \{\dot{Q}\}. \tag{22}$$

where $[C_{\mathrm{T}}]$ is the heat capacity matrix, $[K_{\mathrm{T}}]$ the heat conduction matrix, $\{T\}$ and $\{\dot{T}\}$ the column vector of the nodal temperatures and their time derivatives, respectively, and $[\dot{Q}]$ the column vector of nodal heat sources. $[C_{\mathrm{T}}]$ and $[K_{\mathrm{T}}]$ are (non-linearly) temperature-dependent. The transient temperature field is solved by means of special (usually implicit) time step methods. Such finite element solutions are presented in Section 2.2.7. A more sophisticated analysis of transient temperature fields taking into account coupling effects in respect of microstructural transformation may be performed according to the flow chart in Fig. 90, originally developed in respect of quenching and tempering processes.

A boundary element solution for transient heat-conduction problems of welding comprising temperature-dependent thermal coefficients is found in Ref. [197].

Temperature fields calculated for welding processes require to be checked by means of temperature measurements [42]. The uncertainties of the analysis result from the approximation assumptions in respect of the extent and distribution of the effective welding heat, the heat losses by convection and radiation, the magnitude and temperature dependence of the material characteristic values as well as the geometrical circumstances. Temperature measurements are usually conducted with thermocouples which are attached to the surface of the part and also in the interior through blind holes. Measurements are also conducted in the area of the fusion zone, the thermocouples being adjusted in the weld pool by means of special methods. Infrared thermography is used for obtaining the field information on temperatures before attaching the thermocouples. In addition to temperature measurements, the total heat absorbed by the component as a result of welding can also be measured. This is done by using calorimeters which ensure contact over a large area between the component and a heat-dissipating liquid.

2.2 Global temperature fields

2.2.1 Momentary stationary sources

2.2.1.1 Momentary point source on the semi-infinite solid

The momentary stationary source is a possible model simplification for welding processes with short-time heating and subsequent cooling (e.g. spot-welding). The corresponding mathematical solutions are additionally significant as a starting basis for welding processes with a continuous and moving heat source.

The heat quantity Q is considered as acting at the time $t = 0$ momentarily in the centre of the surface of the semi-infinitely extended solid (Fig. 20a); the heat propagates in three dimensions. The temperature increase, $T - T_0$, in the (arbitrarily directed) distance R from the point source at the time t (from operation of source) is (equation (60) in Ref. [1]):

$$T - T_0 = \frac{2Q}{c\varrho(4\pi at)^{3/2}} e^{-R^2/4at}. \tag{23}$$

In the source point itself ($R = 0$) the temperature increase is

$$T - T_0 = \frac{2Q}{c\varrho(4\pi at)^{3/2}}. \tag{24}$$

At the start of the process ($t = 0$) the temperature in $R = 0$ is infinitely high and then drops hyperbolically with $1/t^{3/2}$. The altitude of the hyperbolic curve is proportional to Q. The infinitely high temperature does not in fact occur during welding (arc welding: $T_{max} \approx 2500\,°C$). It is a consequence of the simplification as point source (see Section 2.1.3.3). In the distance R from the centre, the

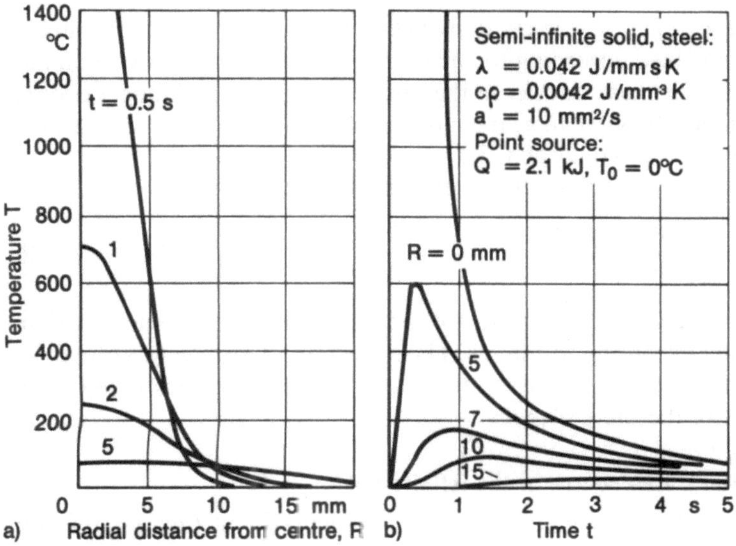

Fig. 26. Temperature field around momentary point source on semi-infinite solid, temperature T over radial distance from centre R (**a**) and over time t (**b**); after Rykalin [1]

temperature follows an inverse exponential function over time, the maximum value of which in $R = 0$ is identical to the values of the hyperbolical curve mentioned above (Fig. 26a). In points outside the centre the temperature rises initially over time, reaches a maximum and drops to zero with $R \rightarrow \infty$ (Fig. 26b). The rise is steeper and the maximum higher, the closer to the source is the considered point. The sequence of temperature maxima, the height of which is rapidly decreasing with R, characterizes the tapering-off of the heat source effect.

2.2.1.2 Momentary line source in the infinite plate

The heat quantity Q is considered as acting at time $t = 0$ in the central line element of the infinitely extended plate of thickness h (Fig. 20b). Q is uniformly distributed over h forming the thickness related heat intensity Q/h. The heat propagates in two dimensions. If the top and bottom surface of the plate are assumed to be heat-impermeable, the temperature in all points of the plate is constant over the plate thickness during the heat propagation process. Where heat exchange exists through the ambient air, the temperature in the middle of the plate thickness is slightly increased. The difference to the surface temperature, however, is normally negligibly slight. The temperature increase in the (arbitrarily directed) distance r from the line source at time t (from operation of source) is (equation (62) in Ref. [1]):

$$T - T_0 = \frac{Q}{hc\varrho(4\pi at)} \, e^{(-r^2/4at - bt)}. \tag{25}$$

The parameter $b = 2(\alpha_c + \alpha_r)/c\varrho h$ designates the coefficient of heat transfer which can be explained according to equations (6) and (9) with convection and radiation of heat. The heat transfer in the case of thin plates and longer cooling time is not negligible (relative to heat conduction). In the case of the semi-infinite solid it is negligible (see equation (23)).

In the source line itself ($r = 0$) the temperature increase is:

$$T - T_0 = \frac{Q}{hc\varrho(4\pi at)} e^{-bt}. \tag{26}$$

The temperature profiles in the plate are similar to those in the surface of the semi-infinite solid according to Fig. 26. The hyperbolical drop with $1/t$ in $r = 0$ is weaker than in the semi-infinite solid because the heat flow is restricted to the two dimensions of the plate plane.

2.2.1.3 Momentary area source in the infinite rod

The heat quantity Q is considered as acting momentarily at time $t = 0$ in the central cross-section $x = 0$ of the infinitely extended rod with cross-section area A (Fig. 20c). Q is evenly distributed over A forming the area-related heat intensity Q/A. The heat propagates in one dimension. A plate model with momentary area source, as is occasionally used for very rapidly applied weld seams, is identical with the rod model. The result for the temperature distribution in the rod (equation (64) in Ref. [1]) is:

$$T - T_0 = \frac{Q}{Ac\varrho(4\pi at)^{1/2}} e^{(-x^2/4at - b^*t)}. \tag{27}$$

The heat transfer coefficient introduced here is $b^* = (\alpha_c + \alpha_r)P/c\varrho A$, with P designating the perimeter and A the area of the rod cross-section.

In the source area itself ($x = 0$) the temperature increase is:

$$T - T_0 = \frac{Q}{Ac\varrho(4\pi at)^{1/2}} e^{-b^*t}. \tag{28}$$

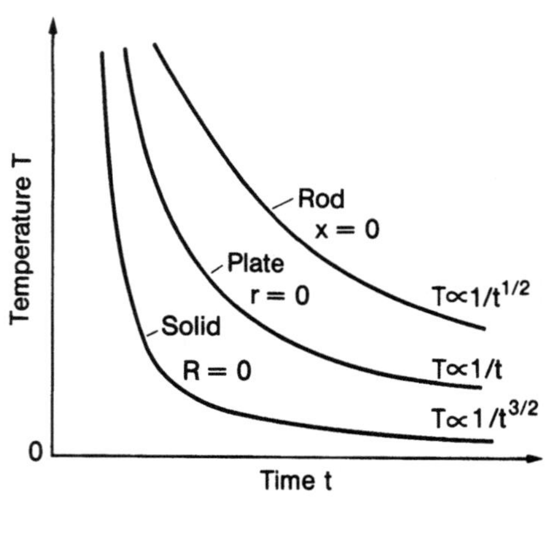

Fig. 27. Temperature profile in the source point, source line and source plane, respectively, of momentary heat sources; body with point source (three-dimensional heat diffusion), plate with line source (two-dimensional heat diffusion, $b = 0$) and rod with area source (one-dimensional heat diffusion, $b^* = 0$); after Rykalin [1]

The hyperbolical drop with $1/t^{1/2}$ in $x = 0$ is once again weaker than in the case of the plate because the heat flow is restricted to a single dimension.

The temperature drop of varying gradient in the centre of solid, plate and rod is compared in Fig. 27. The more the heat flow is spatially restricted, the more the temperature gradient is reduced. The rapidity of heat propagation thus diminishes from solid to plate and from plate to rod.

2.2.2 Continuous stationary and moving sources

2.2.2.1 Moving point source on the semi-infinite solid

The mathematical expression for the temperature field around a continuously acting and moving heat source is obtained proceeding from the superposition principle. This principle is valid only in the case of a linear differential equation, which is based on the assumption that the material characteristic values are independent of temperature. This linearization is in many cases an acceptable assumption.

A sequence of differential momentary sources is obtained by a differential subdivision of the operative time of the continuous source. Consequently, with a moving source, the operative distance is also subdivided; in other words, the differential momentary sources are arranged in succession in the direction of movement. The effect of the differential momentary sources in individual points of the heat-conducting solid now has to be added up, taking into account time and location of the occurrence of the momentary sources, i.e., in mathematical terms, it has to be integrated.

The temperature field around a source moving uniformly and linearly in an infinite solid, in an infinite plate or in an infinite rod is quasi-stationary if the processes at the start and end of operation are ignored. The field appears as stationary, i.e. with fixed field parameters in a coordinate system moving with the source (hereinafter designated by x, y, z and R, r).

The point-type heat source is considered as moving uniformly and linearly (velocity v and heat output q) on the surface of the semi-infinite solid. The temperature increase in the (arbitrarily directed) distance r from the moving source (coordinate x in the direction of movement) is derived as (equation (74) in Ref. [1], see also Ref. [29, 30]):

$$T - T_0 = \frac{q}{2\pi\lambda R}\, e^{-v(x+R)/2a}. \tag{29}$$

In the special case of the stationary ($v = 0$) continuous point source, what applies in the limit state of prolonged heating ($t \to \infty$) is:

$$T - T_0 = \frac{q}{2\pi\lambda R}. \tag{30}$$

The isothermal faces which result are of concentric semi-spherical shape. The temperature drops hyperbolically with $1/R$. The highly heated zone is all the greater the smaller the coefficient of thermal conductivity, $\lambda = ac\varrho$.

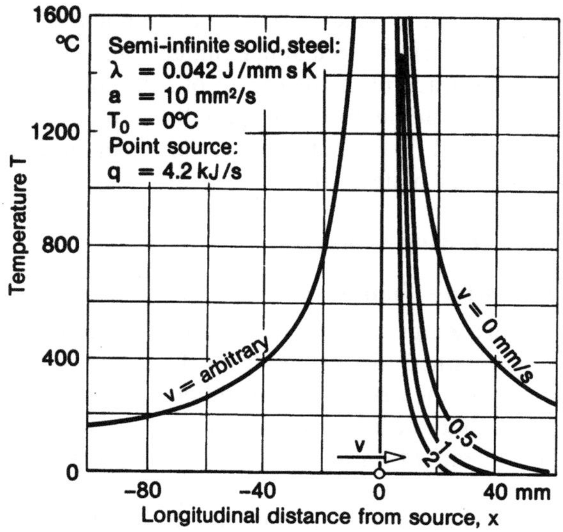

Fig. 28. Temperature profiles ahead of and after moving point source on semi-infinite solid, quasi-stationary limit state; moving coordinate x, different velocities v; after Rykalin [1]

What applies for the moving source in the line of motion after the source ($x = -R$) is the temperature distribution identical with equation (30) irrespective of the velocity v. By contrast, what applies for points in the line of movement ahead of the source ($x = R$) is:

$$T - T_0 = \frac{q}{2\pi\lambda R}\,e^{-vR/a}. \tag{31}$$

The temperature drop ahead of the source is all the steeper the larger the velocity v. In the case of a very large v, the heat propagates almost exclusively in transverse direction (high power sources, see Section 2.2.4.1). Figure 28 serves to illustrate this.

The temperature drop transverse to the moving source ($x = 0$) is given by

$$T - T_0 = \frac{q}{2\pi\lambda R}\,e^{-vR/2a} \tag{32}$$

i.e. by the drop after the source in accordance with equation (30) multiplied by the value of the inverse exponential function (factor less than 1.0).

As the distance from the source increases, the temperature ahead of the source drops most sharply, after the source most weakly and alongside the source with medium gradient. The complete temperature field is presented in Fig. 29 (as the model of a surfacing weld). The isotherms on the surface form ovally closed curves with large density ahead of and lower density after the source. The elongation of the isotherms is determined by the parameter vR/a. The more rapidly the source is moving (relative to the thermal diffusivity), the greater is the elongation. The isotherms in the cross-section form concentric circles. The isothermal faces are, consequently, axisymmetrical with the line of

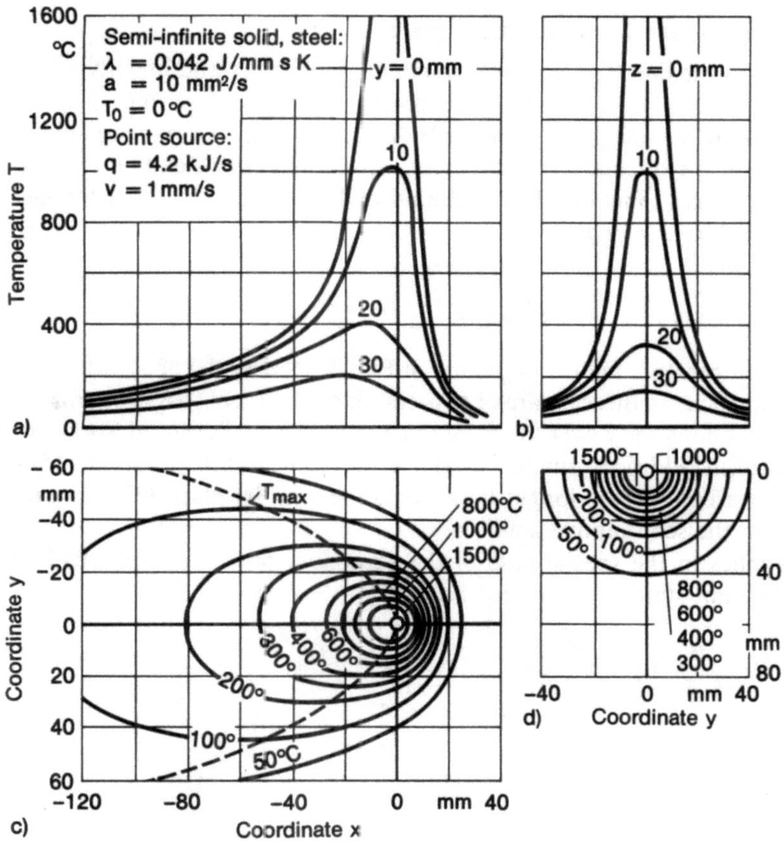

Fig. 29. Temperature field around moving point source on semi-infinite solid, quasi-stationary limit state in moving coordinate system x, y, z; temperature T over x and y (**a**, **b**), isotherms in surface and cross section (**c**, **d**); after Rykalin [1]

motion of the source as the axis of symmetry. The temperature at the location of the point source is infinitely large.

In points alongside (and below) the line of motion of the source, the local temperature maximum is achieved with a time lag compared to the passage of the source through the relevant cross-section plane. The time lag is all the greater and the maximum all the smaller, the greater the distance is from the line of motion of the source. The location of the temperature maxima is identified in Fig. 29 by a broken curve.

Reference is made to Section 2.2.2.2 with respect to the influence of the working parameters during welding as well as to the influence of the thermal material characteristics on the extent of the areas heated beyond certain temperatures.

2.2.2.2 Moving line source in the infinite plate

What applies to the line-type heat source moving uniformly and linearly in the infinitely extended plate (velocity v, thickness-related heat output q/h) for temperature T in the (arbitrarily directed) distance r from the moving source (coordinate x in the direction of movement, equation (80) in Ref. [1]; see also Ref. [29, 30]). Bessel function of second kind and zero order Y_0, heat transfer coefficient $b = 2(\alpha_c + \alpha_r)/c\varrho h)$ is:

$$T - T_0 = \frac{q}{h2\pi\lambda} e^{-vx/2a} Y_0\left(r\sqrt{\frac{v^2}{4a^2} + \frac{b}{a}}\right). \tag{33}$$

The related temperature field is presented in Fig. 30 (the model of a butt weld). The isothermal faces in this case are of non-circular cylindrical shape, because of the temperature constancy over the plate thickness. The ovally closed isotherms presented are, in addition, similar to those of the semi-infinite solid. They enclose larger areas and are longer because the heat in the plate diffuses more slowly than in the semi-infinite solid.

What applies in the special case of the stationary ($v = 0$) continuous line source in the limit state of prolonged heating ($t \to \infty$) is:

$$T - T_0 = \frac{q}{h2\pi\lambda} Y_0\left(r\sqrt{\frac{b}{a}}\right). \tag{34}$$

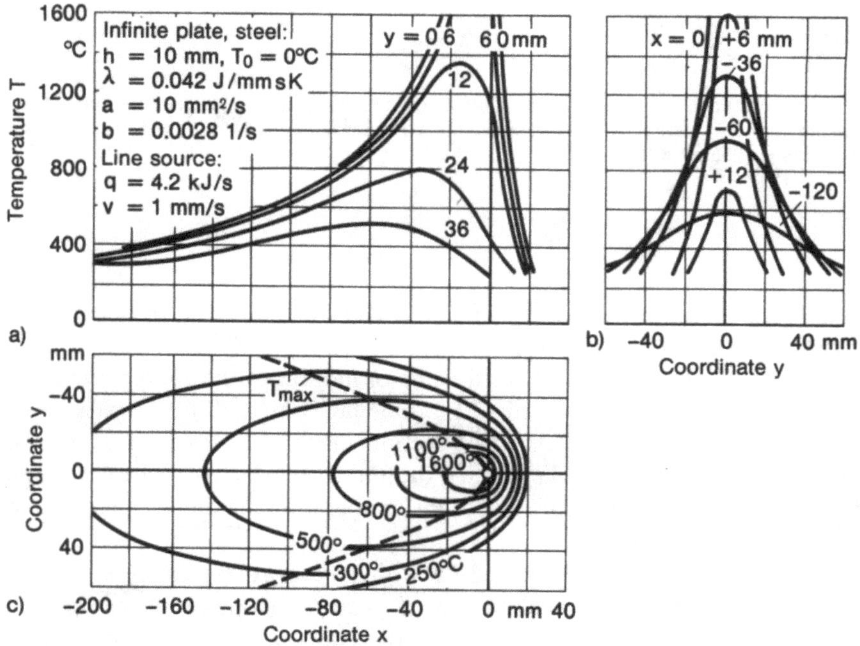

Fig. 30. Temperature field around moving line source in plate, quasi-stationary limit state in moving coordinate system x, y; temperature T over x and y (**a**, **b**), isotherms in plate plane (**c**); after Rykalin [1]

Concentric circular cylinders result as isothermal faces. The temperature drop over r is less steep than in the case of the semi-infinite solid. It depends on the parameter $b/a = 2(\alpha_c + \alpha_r)/h\lambda$, i.e. on the ratio of heat transfer to heat diffusion.

With the moving line source in the plate, the temperature drop after the source is not independent of the velocity v, contrary to the point source on the semi-infinite solid.

As regards the influence of the working parameters during welding on the extent of the areas heated above certain temperatures (e.g. 600 °C), the following statements can be made, proceeding from equation (33) (see Ref. [1]):

- Influence of rising welding speed v at constant output q: the areas heated beyond certain temperatures are reduced, the isotherms contract transverse to the weld and are shortened in weld direction.
- Influence of rising heat output q at constant speed v: the areas heated beyond certain temperatures enlarge over-proportionally, the isotherms bulge transverse to the weld and elongate in weld direction.
- Influence of rising output q and speed v at constant heat input per unit length of weld, $q_w = q/v$: the areas heated above certain temperatures enlarge approximately proportional to output and speed, respectively.
- Influence of rising preheating temperature T_0 at constant output q and speed v: the areas heated above certain temperatures enlarge.

As regards the influence of the thermal material properties, it can be stated:

- The thermal conductivity $\lambda = ac\varrho$ is decisive for the extent of the areas heated above certain temperatures (Fig. 31). With a small λ, a small q_w is sufficient for welding; with a large λ, a large q_w is required. For this reason, austenitic CrNi

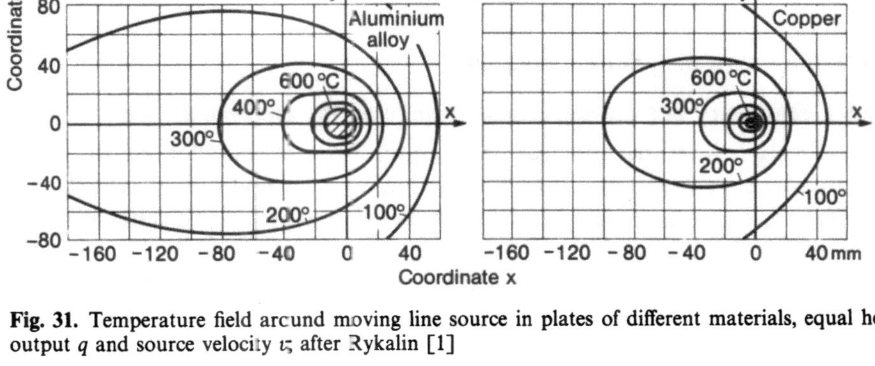

Fig. 31. Temperature field around moving line source in plates of different materials, equal heat output q and source velocity v; after Rykalin [1]

steel (small λ) can be welded with a small heat input per unit length of weld whereas aluminium and copper (large λ) require a high heat input per unit length of weld.

2.2.2.3 Moving area source in the infinite rod

For the area-type heat source moving uniformly in the infinitely extended rod (velocity v, area-related heat output q/A), what holds for the temperature increase at the distance x from the moving source ($x > 0$ ahead of the source, $x < 0$ after the source; according to Ref. [30], with perimeter P and area A of the cross-section) is:

$$T - T_0 = \frac{q}{Ac\varrho v}\, e^{-\left(\sqrt{\left(\frac{v}{2a}\right)^2 + \frac{P}{A}\frac{\alpha_c + \alpha_r}{\lambda}} + \frac{v}{2a}\right)x} \qquad (x > 0), \tag{35}$$

$$T - T_0 = \frac{q}{Ac\varrho v}\, e^{\left(\sqrt{\left(\frac{v}{2a}\right)^2 + \frac{P}{A}\frac{\alpha_c + \alpha_r}{\lambda}} + \frac{v}{2a}\right)x} \qquad (x < 0). \tag{36}$$

The term ahead of the exponential function is identical to the maximum temperature, $T_{max} - T_0$, occurring at the location of the source ($x = 0$):

$$T_{max} - T_0 = \frac{q}{Ac\varrho v} \qquad (x = 0). \tag{37}$$

This maximum temperature is retained after the source at a constant level if heat transfer is negligibly small, $\alpha_c + \alpha_r = 0$. If heat transfer is not negligible, the temperature decreases depending on the parameter $P(\alpha_c + \alpha_r)/A$. Ahead of the source, an exponential temperature drop occurs, which, if heat transfer is negligible, has the following simple form:

$$T - T_0 = (T_{max} - T_0)e^{-vx/a} \qquad (x > 0). \tag{38}$$

It is not clarified whether equation (72) stated in Ref. [1] can be transformed to the above-mentioned equations (35) and (36) (although in the special case $\alpha_c + \alpha_r = 0$, equation (184) in Ref. [1], agrees with equation (38) here).

The relations of this section can be applied to the melting-off of electrodes (see Section 2.3.1.3).

2.2.3 Gaussian distribution sources

2.2.3.1 Stationary and moving circular source on the semi-infinite solid

In this section is treated the heat propagation from the circular-shaped Gaussian distribution source introduced in Section 2.1.3.3, equations (16) to (21). The circular source is considered as stationary or moving in the surface of the semi-infinite solid.

What holds for the stationary circular surface source with momentary heat output $q\mathrm{d}t$ (for correlation with q^* see equations (16) and (17)) is (equation (146) in Ref. [1]):

$$dT = \frac{2q\,dt}{c\varrho}\,\frac{e^{-z^2/4at}}{(4\pi at)^{1/2}}\,\frac{e^{-r^2/4a(t+\Delta t_0)}}{4\pi a(t+\Delta t_0)}. \tag{39}$$

The term in the middle of the right-hand side of equation (39) characterizes the heat diffusion from the surface into the interior of the solid in accordance with the relation for the momentary area source (equation (27)). The subsequent term characterizes the heat diffusion in the surface of the solid (distance r from the source) in accordance with the relation for the momentary line source in the plate (equation (25)). In the latter case, the principle of the concentrated source advanced in time by Δt_0 is obviously valid.

For the uniformly moving circular source with heat output q, the mathematical expression for the temperature T in the likewise moving coordinate system cannot be presented by elementary functions (equation (148) in Ref. [1]). Equation (47) is approximately valid, particularly in points close to the source.

In the special case of the stationary ($v = 0$) continuous circular source, an expression applies in the limit state of prolonged heating ($t \to \infty$) (equation (153) in Ref. [1]), which has been evaluated in Fig. 32 (with $T_0 = 0$). What results for the maximum temperature T_{max} in the centre of the source (equation (150) in Ref. [1]) is:

$$T_{max} - T_0 = \frac{q}{2\lambda}\sqrt{\frac{k}{\pi}}. \tag{40}$$

With a moving circular source, the maximum temperature is reached all the less, the greater the velocity of the moving source. In addition, the location of the maximum temperature is displaced to after the centre of the circular area. A steep temperature drop occurs ahead of the source, and a flatter drop after the source (Fig. 109 in Ref. [1]). The isotherms are similar to those of the moving point source. The infinitely high temperature in the centre of the source is avoided. The temperatures at a greater distance from the source are nearly identical.

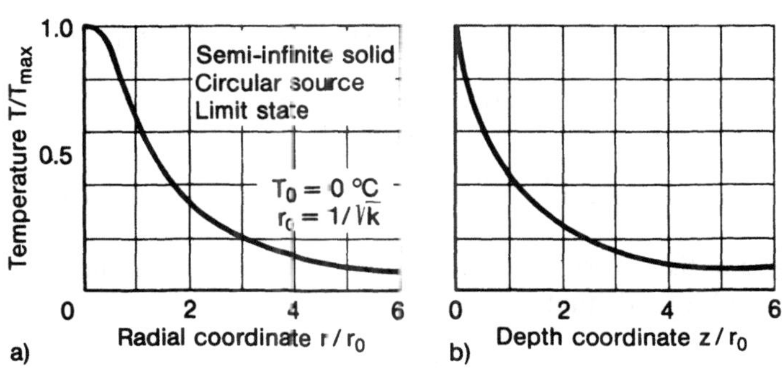

Fig. 32. Temperature profile over radial coordinate (**a**) and depth coordinate (**b**), stationary circular source on semi-infinite solid, stationary limit state; after Rykalin [1]

2.2.3.2 Stationary and moving circular source in the infinite plate

Heat propagation from the Gaussian distribution circular source introduced in Section 2.1.3.3, equations (16) to (21) is now considered. The circular source is assumed to act evenly distributed over the thickness h of the plate.

What applies for the stationary circular source with momentary heat output $q\mathrm{d}t$ (for correlation with q^*, see equations (16) and (17)), according to the principle of the line source advanced in time by Δt_0 (equation (124) in Ref. [1]), is:

$$\mathrm{d}T = \frac{q\mathrm{d}t}{hc\varrho 4\pi a(t + \Delta t_0)} \mathrm{e}^{-r^2/4a(t+\Delta t_0)}. \tag{41}$$

No solution in functions which can directly be evaluated is known for the uniformly moving circular source with heat output q. Equation (50) is approximately valid, particularly in points close to the source. A solution without the simplification related to high power sources is stated only for the line source fictitiously advanced in time Δt_0 (equation (132) in Ref. [1]); except for $v = 0$, however, this is not the temperature maximum:

$$T - T_0 = \frac{q}{h4\pi\lambda} \mathrm{e}^{b\Delta t_0} \left\{ -\mathrm{E_i}\left[-\left(b + \frac{v^2}{4a} \right) \Delta t_0 \right] \right\}. \tag{42}$$

In this connection, $\mathrm{E_i}(-u) = \int_u^\infty (\mathrm{e}^{-u}/u)\mathrm{d}u$ for $u > 0$ is the integral exponential function and $b = 2(\alpha_c + \alpha_r)/c\varrho h$ the heat transfer coefficient.

2.2.3.3 Stationary strip source in the infinite plate

Heat propagation from the strip-shaped Gaussian distribution source introduced in Section 2.1.3.3, equations (16) through (21) (heat output q_1 per unit of length in direction of the strip) is now considered. The strip source is assumed to act evenly distributed over the thickness h of the plate. The strip is located on the x axis (see Fig. 22); heat conduction thus occurs in the direction of the y axis.

A simple solution is available only for the centre line of the strip source ($y = 0$) in the limit state of prolonged heating (equation (141) in Ref. [1]):

$$T_{\max} - T_0 = \frac{q_1}{2[2(\alpha_c + \alpha_r)\lambda h]^{1/2}} \mathrm{e}^{b\Delta t_0}[1 - \Phi(b\Delta t_0)^{1/2}]. \tag{43}$$

In this connection, $\Phi(u) = (2/\sqrt{\pi}) \int_0^u \mathrm{e}^{-u^2} du$ is the Gaussian probability integral.

In the special case of the source concentrated on the centre line of the strip ($y = 0$, $k \to \infty$, $\Delta t_0 \to 0$) it holds that:

$$T_{\max} - T_0 = \frac{q_1}{2[2(\alpha_c + \alpha_r)\lambda h]^{1/2}}. \tag{44}$$

2.2.4 Rapidly moving high-power sources

2.2.4.1 Rapidly moving high-power source on the semi-infinite solid

Rapidly moving high-power sources are characterized by high heat output q at high source velocity v. The process parameters q and v are increased with a possibly constant heat input per unit length of weld, $q_w = q/v$. Rapidly moving high-power sources reduce the welding time and are, therefore, of considerable practical significance.

The theoretician is interested in rapidly moving high-power sources because of the possibility of simplifying the equations. With a constant q_w, the limit values $q \to \infty$ and $v \to \infty$ are introduced. The error which occurs as the consequence of the introduction of the limit values is relatively small directly next to the source. The simplified formulae are, therefore, particularly suited for the areas close to the weld.

Heat propagation around rapidly moving high-power sources can also be discussed. With increasing source velocity v and proportionally enlarged heat output q, the areas heated beyond certain temperatures are enlarged (see Section 2.2.2.2). Their length grows proportionally to the velocity and their width tends toward a limit value. With a very high velocity, the heat propagates mainly perpendicular to the direction of movement of the source. The propagation in the direction of movement is negligibly small. The semi-infinite solid or plate can be considered to be subdivided into a large number of thin plane layers perpendicular to the direction of movement of the source. The heat introduced as the heat source passes through the respective layer is then diffused solely in this layer, irrespective of the state of the adjacent layers. This has the effect of simplifying the equations.

For the rapidly moving high-power point source on the semi-infinite solid, the following relation holds (with distance r from the point source in the layer (equation (82) in Ref. [1]):

$$T - T_0 = \frac{q}{v 2\pi\lambda t} \, e^{-r^2/4at}. \tag{45}$$

Hence, there follows the local maximum temperature in the layer positioned perpendicular to the movement of the source (equation (87) in Ref. [1]):

$$T_{\max} - T_0 = 0.234 \, \frac{q}{vc\varrho r^2}. \tag{46}$$

The rapidly moving high-power circular source can be converted to an equivalent line source advanced in time by Δt_0 (Gaussian distribution source on semi-infinite solid over line perpendicular to direction of movement), the heat of which propagates only perpendicular to the direction of movement. What results for the temperature field (with coordinate z in direction of depth and coordinate y in direction transverse to movement in the surface, equation (157) in Ref. [1]) is:

$$T - T_0 = \frac{2q}{vc\varrho} \frac{e^{-z^2/4at}}{(4\pi at)^{1/2}} \frac{e^{-y^2/4a(t+\Delta t_0)}}{[4\pi a(t + \Delta t_0)]^{1/2}}. \tag{47}$$

The commentary on equation (39) applies here by analogy.

A solution for the line source of finite width in the surface of a semi-infinite solid perpendicular to the moving direction of the source is derived in Ref. [150]. In order to avoid infinitely high surface temperatures in the source line, an apparent source is introduced, located at a small definite distance above the surface, replacing the original source.

2.2.4.2 Rapidly moving high-power source in the infinite plate

For the rapidly moving high-power line source in the plate, the following relation holds (coordinate y in direction transverse to movement, equation (83) in Ref. [1]):

$$T - T_0 = \frac{q}{vh(4\pi\lambda c\varrho t)^{1/2}} e^{-[y^2/4at + bt]}. \tag{48}$$

Hence, there follows the local maximum temperature in the layer positioned perpendicular to the movement of the source (equation (89) in Ref. [1]):

$$T_{max} - T_0 = 0.242 \frac{q}{vc\varrho hy}. \tag{49}$$

The rapidly moving high-power circular source can be converted to an equivalent strip source advanced in time by Δt_0 (Gaussian distribution source over strip positioned perpendicular to the direction of movement), the heat of which propagates only perpendicular to the direction of movement. What results for the temperature field (with coordinate y in direction transverse to movement, equation (135) in Ref. [1]) is:

$$T - T_0 = \frac{q}{vh[4\pi\lambda c\varrho(t + \Delta t_0)]^{1/2}} e^{-[y^2/4a(t+\Delta t_0) + bt]}. \tag{50}$$

2.2.5 Heat saturation and temperature equalization

What has been considered in the preceding Sections 2.2.2 to 2.2.4 was the limit state which results after a prolonged period during which the stationary or moving source is acting. In the case of the stationary source, the relevant temperature field is likewise stationary, i.e. the local temperatures are independent of time. In the case of the moving source, the relevant temperature field is quasi-stationary, in other words the local temperatures in a likewise moving coordinate system are independent of time.

The limit state cannot appear instantaneously. Before the heat source starts operation, the solid under study has in all points the constant ambient or preheating temperature. As a result of the heat input, a heated zone is formed locally, which is increasingly enlarged and tends to a limit state. The limit state is achieved later if the point under study is more distant from the source.

The period of time between start of heat input and (practical) achievement of the local temperature limit state T_{li} is designated as the heat saturation time. To simplify the analyses for moving sources, the local temperature transition is described in accordance with Ref. [1] by a generally applicable heat saturation function $\psi(\varrho_i, \tau_i)$:

$$T(t) - T_0 = \psi(\varrho_i, \tau_i) T_{li}. \tag{51}$$

In this connection, τ_i is a dimensionless parameter which is proportional to time t and ϱ_i a dimensionless parameter which is proportional to the distance r of the considered point from the heat source (i = 1, 2, 3).

For the three-dimensional heat diffusion (in space) from the moving point source in the surface of the semi-infinite solid, the following relation holds:

$$\varrho_3 = \frac{v}{2a} R, \quad \tau_3 = \frac{v^2}{4a} t. \tag{52}$$

The relevant heat saturation function ψ_3 is shown in Fig. 33.

For the two-dimensional (plane) heat diffusion from the moving line source in the plate, the following relation holds:

$$\varrho_2 = \left(\sqrt{\frac{v^2}{4a^2} + \frac{b}{a}} \right) r, \quad \tau_2 = \left(\frac{v^2}{4a} + b \right) t. \tag{53}$$

The relevant heat saturation function ψ_2 is shown in Fig. 34.

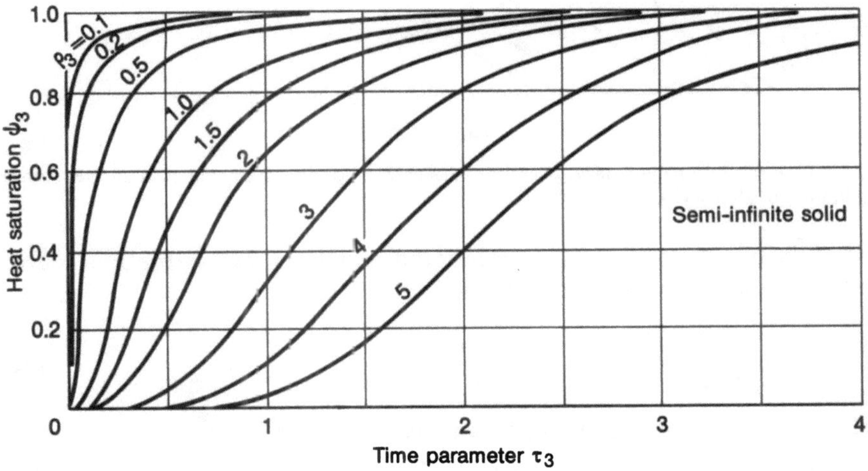

Fig. 33. Heat saturation function ψ_3 over time parameter τ_3 for distance parameters ρ_3, point source on semi-infinite solid; after Rykalin [1]

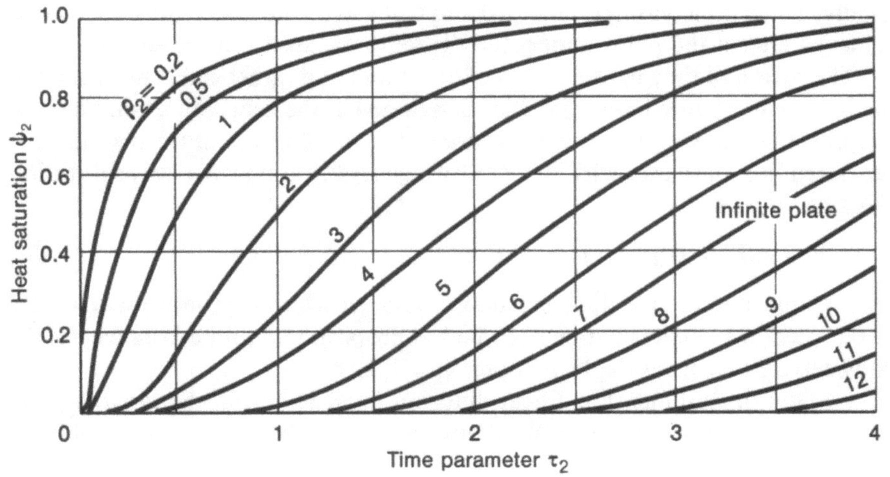

Fig. 34. Heat saturation function ψ_2 over time parameter τ_2 for distance parameters ρ_2, line source in infinite plate; after Rykalin [1]

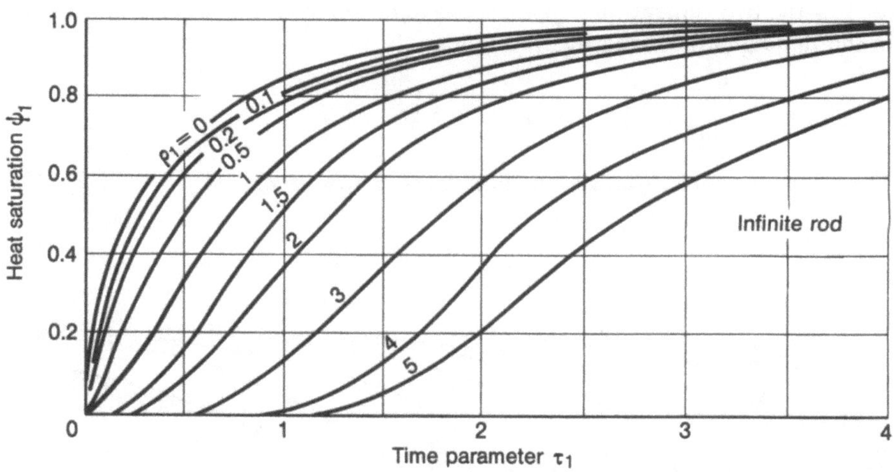

Fig. 35. Heat saturation function ψ_1 over time parameter τ_1 for distance parameters ρ_1, area source in infinite rod; after Rykalin [1]

For the one-dimensional (linear) heat diffusion from the moving area source in the rod, the following relation holds:

$$\varrho_1 = \left(\sqrt{\frac{v^2}{4a^2} + \frac{b^*}{a}}\right)|x|, \quad \tau_1 = \left(\frac{v^2}{4a} + b^*\right)t. \tag{54}$$

The relevant heat saturation function ψ_1 is shown in Fig. 35.

Heat saturation proceeds more slowly if the heat flow in space is restricted to plane or linear conditions. It proceeds more rapidly if the point under study is closer to the heat source.

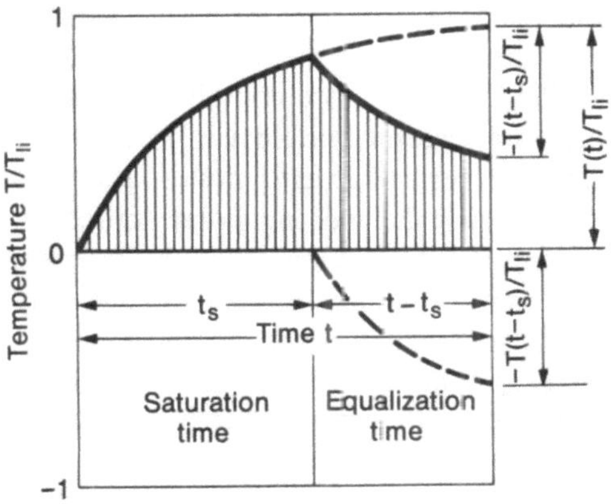

Fig. 36. Modelling of temperature equalization as superimposition of positive and negative heat saturation, temperature T overtime t; after Rykalin [1]

A process opposed to heat saturation commences when a stationary or moving source of constant output is switched off. The irregularities of the temperature field maintained by the source then begin to balance out until the solid has achieved a constant temperature, increased slightly by the former effect of the source. The related time span is designated as the temperature equalization time.

Switching-off the heat source is simulated in the temperature field analysis as introducing an equivalent heat sink (with negative heat output) compared to the continuing, i.e. not switched-off, heat source (with positive heat output) in accordance with Ref. [1].

This is presented schematically in Fig. 36 for an arbitrary point of the solid. The negative heat saturation curve is subtracted from the positive heat saturation curve starting at the moment of switch-off, t_s. It should be noted that the process proceeds in the coordinate system which is stationary or moving with the source and that the superposed sink is also stationary or moving with the source. The temperature in the equalization time is thus calculated as follows:

$$T(t) - T_0 = T_{li}[\psi(t) - \psi(t - t_s)]. \tag{55}$$

2.2.6 Effect of finite dimensions

The expressions for the temperature field stated in Section 2.2.1 to 2.2.5 presume unlimited extension of the body under consideration. What is described by functional analysis is the momentary and continuous, stationary and moving source on the semi-infinitely extended solid, in the infinitely extended plate and in the infinitely extended rod. In reality, the dimensions of the bodies are finite. The influence of the boundary faces of the bodies on the temperature field is great if their distance to the source is small. If the boundary faces are assumed to be heat-impermeable, the temperatures calculated for the infinitely extended bodies are to be subsequently increased.

Boundary faces parallel to the source movement can in many cases be allowed for in functional analysis in accordance with the method of (periodical) mirror-image source arrangement (see Ref. [1]). In more complex geometry, only a finite element solution of the heat propagation problem is possible.

In view of its significance for application to surfacing welds, multi-pass welds and fillet welds, the solution from Ref. [1] for the moving point source on a (thick) plane layer is illustrated below. Upper and lower surface of the layer are assumed to be heat-impermeable. The solution is gained through the infinitely extended solid with periodically reiterated arrangement of the moving source at distance $2h$ with the output doubled ($2q$) compared with the single layer (Fig. 37). As seen graphically, the single layer with surface source is correctly presented in this arrangement.

Higher temperatures are determined in the upper surface of the layer relative to the lower surface. The difference is particularly large close to the source. The difference at distance $4h$ from the source is already as little as 5%. The temperature in the lower surface of the layer is a multiple of the temperature in the semi-infinite solid in the comparable plane. The result of the analysis is presented in Fig. 38.

A solution for the stationary or moving line source of finite width perpendicular to the surface of a thicker plate is derived in Ref. [39].

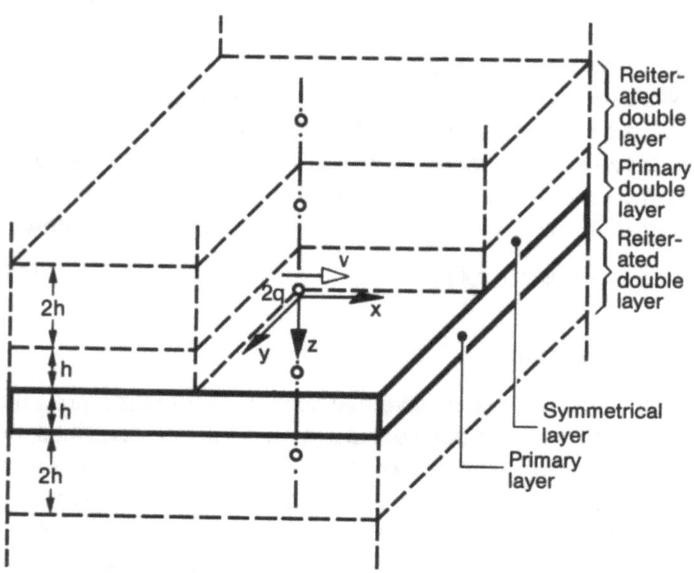

Fig. 37. Point sources (heat output $2q$) periodically reiterated in infinite solid, model for temperature field in plane layer (thickness h) with point source (heat output q) single-sided on surface

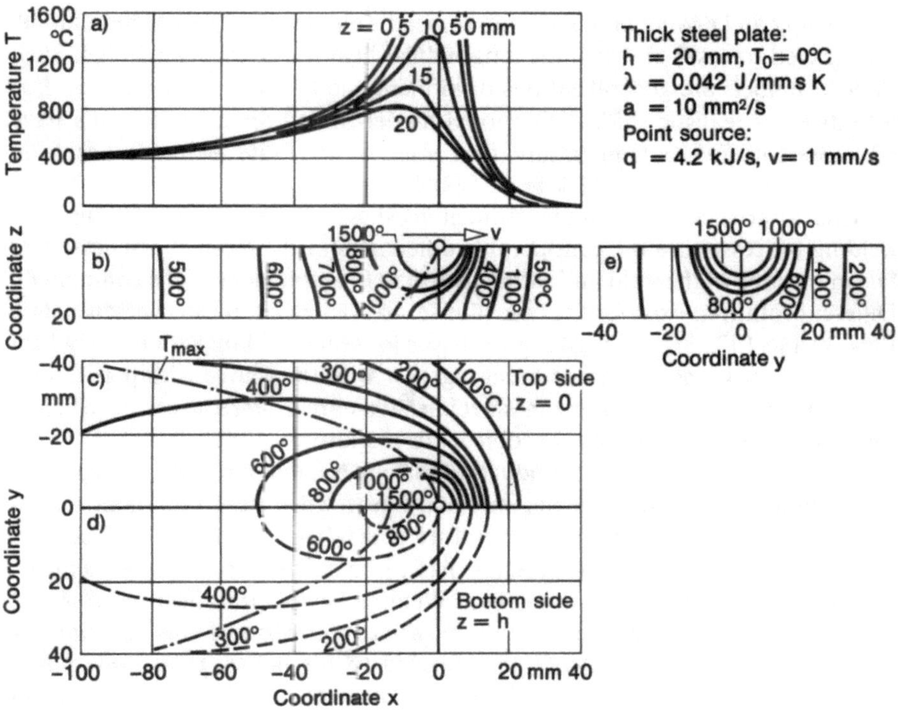

Fig. 38. Temperature field around point source moving on one side of a thick plate, quasi-stationary limit state in moving coordinate system *x*, *y*, *z*; temperature *T* over coordinate *x* (**a**), isotherms in longitudinal section (**b**), on topside (**c**), on bottomside (**d**) and in cross-section (**e**); after Rykalin [1]

2.2.7 Finite element solution

2.2.7.1 Fundamentals

Temperature-independent (mean) material characteristic values are assumed in the functional analysis solutions evaluated above. Heat transfer and heat radiation are often neglected. The boundary faces are generally assumed as being infinitely distant. Only in special cases are heat transfer or heat radiation and finitely distant faces considered, the latter by using a periodic solution (method of mirror-image source arrangement).

In more complex applications, temperature field solutions based on the finite element method [26, 27] in simpler (plane and linearized) cases also based on the finite difference method, are suitable. In accordance with the present state of development of welding residual stress analysis by means of the finite element method, two-dimensional problem solutions are of primary interest. These include axisymmetrical and plane models, the latter subdivided according to temperature fields in the plate plane and those in the plate cross-section perpendicular to the weld. The in-plane models are primarily used to simulate the starting and ending phase of seam or bead welding. The cross-section model is restricted to cases in which the longitudinal heat flow can be neglected. Shell element models are generalizations of in-plane models to curved surfaces. For several years realistic three-dimensional models based on solid elements have been presented. The temperature field analysis by finite elements has been reviewed especially by Goldak et al. [28, 38, 41, 44, 45].

Characteristics of the finite element meshes required for modelling such welding processes are the considerable differences in the mesh coarseness, very fine in the area of the weld and rather coarse in the remainder of the component. Different solutions exist for the task thus set of grading the mesh coarseness (see Figs. 39 and 147). Special meshing strategies for seam welding with moving heat source consist in moving a fine mesh together with the source within a stationary coarse mesh. Such moving mesh refinements are a prerequisite for efficient solutions of three-dimensional problems in the future.

The field equation of heat conduction in finite element formulation (equation (22)), limited to two dimensions, but non-linear when heat radiation, microstructural transformation and temperature dependence of the material characteristic values are taken into account, has to be solved for the nodal temperatures dependent on time and position. Welding heat source, heat transfer and heat radiation are taken into account by surface sources, and also occasionally interior sources at the elements, transformed to the nodes. Heat radiation is introduced in accordance with equation (8). The heat capacity matrix $[C_T]$ and heat conduction matrix $[K_T]$ depend on the temperature field. The non-linear field equation of heat conduction is solved in time steps by Euler's (explicit or implicit) forward or backward integration [26, 27, 248].

Transformation processes in the solid-solid and also in the solid–liquid phases are, if they take place at a definite temperature (i.e. without a temperature range), connected with moving transformation fronts in which the heat content changes abruptly. The abrupt transition in heat content is caused by the latent

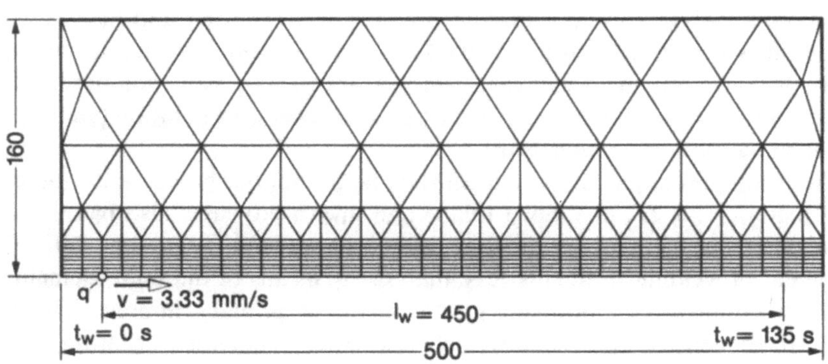

Fig. 39. Finite element mesh (symmetry half) for numerical analysis of the temperature field in the plate plane during seam welding (weld length l_w, weld beginning and ending at a small distance from the plate edge); after Argyris, Szimmat and Willam [248]

heat released or consumed during transformation. The specific heat displays a discontinuity at this (temperature) point. Problems involving moving fronts therefore require adaptive meshes for finite element modelling. In the case of real materials, transformation usually takes place prolonged over a temperature range so that the discontinuity is blurred. Accordingly, the transformation heat is reflected in an apparent increase in specific heat and can be considered in this form without adaptive meshes, solely with an appropriately refined mesh coarseness [36, 221, 248].

Finite element solutions (including finite difference method) for the temperature field during welding are contained in the majority of publications relating to finite element analysis of welding residual stresses, but are also presented separately [28, 34, 35, 38, 41, 44, 45, 54–56]. The finite element analysis of the electrical (potential) field may precede in models relating to resistance spot welding.

2.2.7.2 Ring element model

The axisymmetrical finite element model (primarily in the simpler form of the finite difference method) is applied for the temperature field analysis relating to spot welding [48–54, 57–67, 216, 217, 220]. Besides rendering the basis for subsequent residual stress analyses (see Section 3.2.3), this enables the prediction of weld nugget formation dependent on weld current, weld time, plate thickness, electrode shape, coatings and other influence parameters under the simplifying condition of neglected mechanical behaviour (models including mechanical behaviour in Section 3.2.3). The process of heat generation is treated in more detail in Section 2.3.3.

Before reviewing the various references on this topic, three modelling and result examples are presented The finite element mesh of a typical analysis model for spot welding is shown in Fig. 40. The model consists of solid ring elements simulating at first thermoelectrical and later thermomechanical behaviour supplemented by surface ring elements simulating the respective interface phenomena (e.g. contact resistance, lifting off, slipping) between electrode and plate and between the two plates (surface element thickness equal to oxide layer thickness, i.e. approx. 0.05 mm). The fully developed weld nugget in stainless steel, plate thickness 1 mm, determined with this model is shown in Fig. 41. The heat source and temperature distribution in plate and electrode according to another combined current and heat conduction analysis for aluminium without considering melting and contact resistance is shown in Fig. 77.

The heat source and temperature distribution in weld spots is analysed by Greenwood, Williamson, Ruge et al. [48, 50] proceeding from Ref. [24, 25, 65] using a linearized model without considering the contact resistance. The latter was considered in an improved model [49] from the results of which it was concluded that the contact resistance played a major role only in the early stages of heat production and became less influential in the later stages of the weld nugget formation. Preceding investigations on contact resistance aiming at

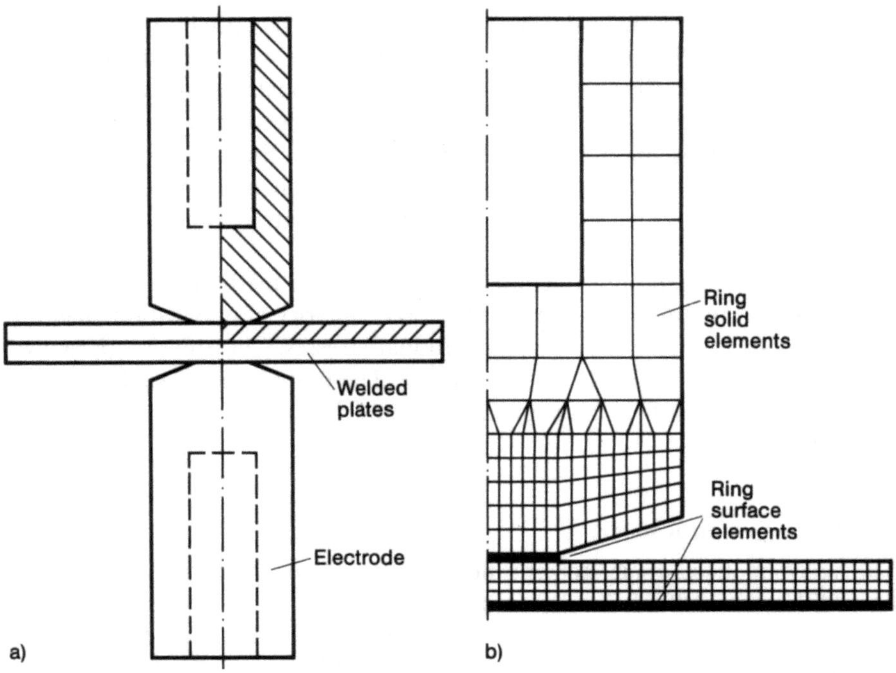

Fig. 40. Axisymmetric finite element model simulating spot welding: geometry of electrode and sheet metal (**a**), discretization into solid ring elements; after Dickinson, Tsai and Jammal [66]

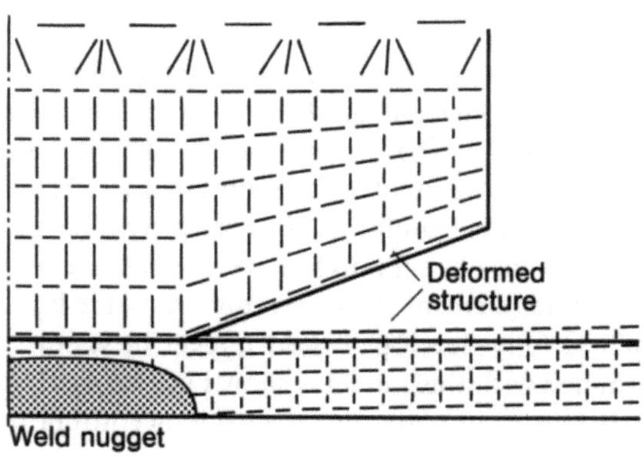

Fig. 41. Weld nugget formation in stainless steel after six current cycles or 0.06 s ($I = 8000$ A, $U = 1$ V); after Dickinson, Tsai and Jammal [66]

a dynamic control of welding parameters such as welding current, electrode pressure and hold time based on actual nugget formation are reported on in Ref. [62–64].

A one-dimensional heat transfer multilayer model consisting of different materials and subjected to current pulses of different shape and length is investigated by Rice and Funk [51] in respect of resistance welding in semiconductor manufacture.

The temperature field in a circular plate made of aluminium during spot melting using the gas or tungsten arc is analysed by Pelli et al. [52] based on a ring element model.

The formation of the spot weld nugget is presented by Kaiser et al. [53] for low- and high-tensile steel in comparison dependent on the ratio of material resistance to contact resistance as well as on contact force and current. The transient electrical and thermal field during spot welding is analysed by Nied [54] and Dickinson et al. [65–67] proceeding from the thermoelastically modelled contact pressure distribution and from the shape and size of the weld nugget derived therefrom. The finite element model takes into account lifting-off, slip and change in resistance in the contact area, but not yielding and fusing. The above investigations, from which Figs. 40 and 41 are taken, concentrate on weld nugget formation without any statement on residual stresses (the thermo-elastic model does not comprise this effect). On the other hand, valuable information related to process control have been derived, e.g. on weld nugget formation and penetration as fraction of plate thickness dependent on weld time (considering squeeze, weld and hold cycle) and weld current, on unequal plate thickness and material in one joint.

The suitability of the axisymmetrical model for the geometrically and physically more complex process of commutator welding is presented by Schwab [55, 56].

The temperature fields during laser beam spot welding are analysed by Öhlschläger [59].

2.2.7.3 Plate element models

The example of a finite element analysis (by Argyris et al. [248] using high-order triangular and rectangular elements) for the temperature field in the plane of a rectangular plate made from mild steel St37, with butt weld passes deposited simultaneously from both sides in an double-V groove, is shown in Figs. 39 and 42. Finite element analysis in this plane is necessary in place of functional analysis because the effect of plate edges in finite distance to the weld have to be taken into account, and the temperature field can no longer be assumed to be quasi-stationary. The heat flow density of the moving source is modelled according to the Gaussian distribution within an area of oval shape. Interestingly enough, the temperature maximum during cooling occurs within the weld length near the weld end. It should be noted that the weld end in turn is shifted from the transverse edge of the plate in the considered case.

The temperature field (one symmetry half) in a fine-grained structural steel plate of $1 \times 1 \times 0.01$ m^3 with a butt weld in the x-axis shortly before the end of welding is presented in Fig. 43 as an example of finite element analysis (by Jonsson et al. [238] using low-order triangular elements) of heat diffusion restricted to planes transverse to the rapidly moving high-power line source (cross-section models). The local temperature gradients are greater in the transverse direction than in the longitudinal direction, with the exception of a small area around the heat source. This is an important aspect for finite element mesh

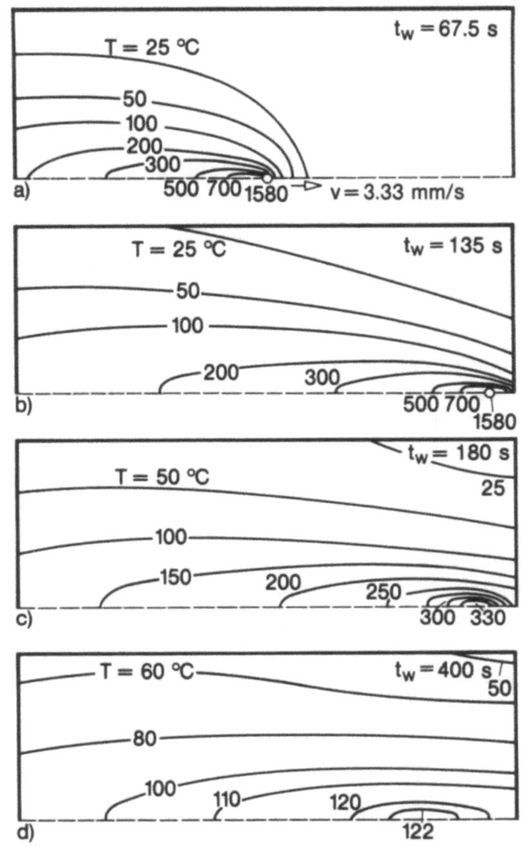

Fig. 42. Temperature field (symmetry half) in rectangular plate with centric weld seam, during welding (**a**), at end of welding (**b**) and during cooling (**c**, **d**), time interval t_w after start of welding; after Argyris, Szimmat and Willam [248]

structuring in the subsequent mechanical model (Fig. 147). The temperature profile over time for two points at a smaller and larger distance from the weld centre line is shown in Fig. 44. The very steep cooling gradient over time close to the weld determines the transformation and hardening behaviour of the base metal (i.e. of the heat-affected zone). Other temperature field analyses in planes transverse to the weld proceed from heat source profiles in the weld surface in accordance with a moving area source [249, 250] or specify the temperature profile at the outer contour or in the weld centre plane in accordance with the temperature field calculation in the plane as a control variable [248, 251].

2.3 Local heat effect on the fusion zone

2.3.1 Electric arc as a welding heat source

2.3.1.1 Physical-technical fundamentals

The most frequently used welding heat source is the electric arc. It converts electrical energy into heat energy. It produces the melting of the filler metal and

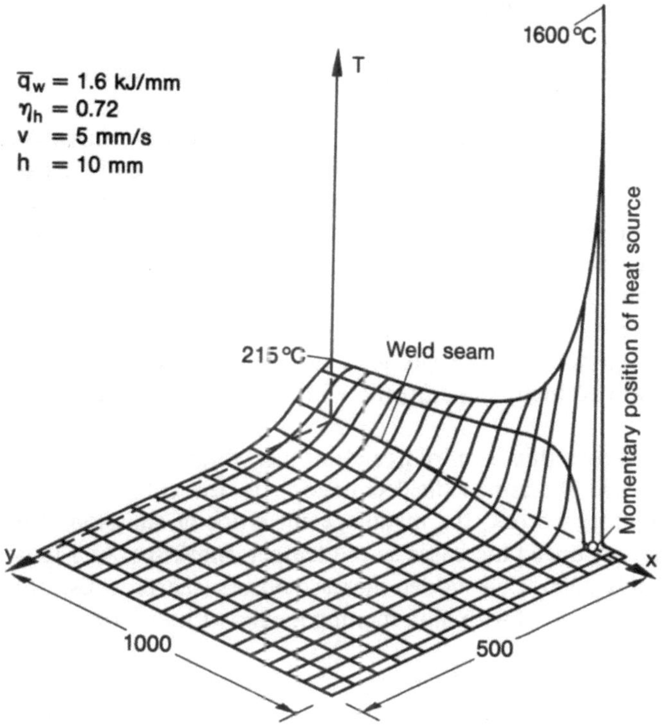

Fig. 43. Temperature field shortly before end of welding in steel plate with centric gas metal arc weld seam; after Jonsson, Karlsson and Lindgren [238]

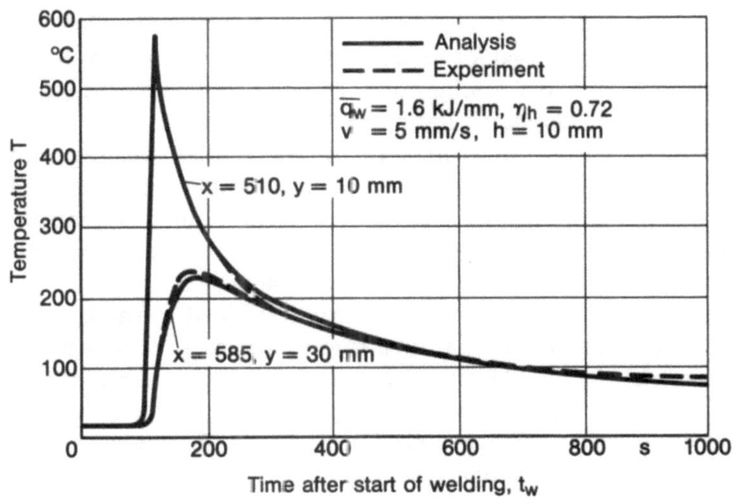

Fig. 44. Temperature profile over time in steel plate (size $1 \times 1 \times 0.01\ m^3$) alongside centric weld, smaller and larger transverse distance from weld y, longitudinal coordinate x according to Fig. 43; after Jonsson, Karlsson and Lindgren [238]

the fusing of the base metal. It maintains the endothermic and exothermic chemical processes which occur in the arc. The arc is a particular type of electric discharge in gas. It is based on free-moving charged particles (electrons and ions) in the arc gap.

A precondition for current flow in the gas between two fixed poles (negative cathode and positive anode) when voltage is applied is the ionization of the gas. As a result of the voltage of the electrical field between the poles the electrons are accelerated toward the anode. In so doing, they impact on gas molecules, which, as a result of the impact, are decomposed into positively charged atoms (ions), releasing further electrons. The gas ions are accelerated toward the cathode. The process of impact ionization snowballs when the arc is ignited. When the electrons impact on the anode or the gas ions on the cathode, anode and cathode are heated locally to high temperatures, with the result that metal ions evaporate.

Metals are enveloped by a thin layer of free electrons, which are accelerated in the electric field toward the anode (field emission). In the cold state, this requires performing minimum work, the metal-specific work of emission. The electron current density produced increases as the temperature rises. At a temperature of approximately 3500 °C, a sudden increase occurs (thermoemission), although this can only be achieved with high-melting metals without prior metal evaporation.

The arc is ignited by brief contact of the two poles. During contact, a short-circuit current flows, which is significantly higher than the normal welding current. If contact ignition is not desired, ignition aids in the form of high-voltage pulses may be used.

The voltage drop over the arc gap consists of the steep "cathode drop" and "anode drop" in a very thin layer and the flatter drop over the plasma gap. The voltage-current characteristic curve comprises an initially steep drop in voltage over the amperage followed by a moderate rise at larger amperage. Only the latter ohmic resistance range is used for welding and, together with the voltage-current characteristic curve of the welding current source, constitutes the working range. Each arc has its own particular characteristic curve.

Welding with direct current presents fewer problems than welding with alternating current. With direct current, both types of polarity of electrode and workpiece are possible, although with slightly different results in practice. Polarity does not play any role in the case of alternating current. The arc is reignited in the 50 cps rhythm of the alternating current.

The arc as a (moving) electric conductor is surrounded by its own magnetic field, which accelerates the charged particles toward the axis of the arc. As a result, the arc constricts itself and forms small attachment spots on anode and cathode ("anode spot" and "cathode spot"). The spot on the electrode is always smaller than that in the weld puddle, irrespective of polarity. The anode spot is relatively fixed, while the cathode spot is relatively mobile. The arc may easily be deflected by external magnetic forces (the blow effect).

Anode spot and cathode spot are heated to high temperatures (approximately 3000 to 4000 °C depending on the material) as a result of the impact of the electrons or ions, the fixed anode spot to higher temperature than the mobile cathode spot. The maximum arc temperatures occur in the arc plasma (up to 30 000 °C depending on nature and composition). The heat generated is proportional to the local voltage gradient. As the voltage gradient is particularly large at the anode and cathode, this is where a major part of the arc heat is generated.

The arc generates radiation in the visible as well as in the ultraviolet and infrared range. This increases the heat effect and, on the other hand, necessitates protective clothing and goggles.

The most widespread welding processes operate with a melting electrode. The filler metal passes in the form of droplets from the electrode to the workpiece. The very different forces which occur in this process are illustrated in Fig. 45, i.e. viscosity and surface tension, gravity force, force of inertia, suction force of the plasma flow, and electromagnetic and electrostatic force. Electromagnetic constriction (the pinch effect) is the essential factor for the formation of droplets, while expanding gases effect the detachment of the droplets (intensified by crater formation in the case of covered electrodes).

In the case of plasma arc welding, the arc is burning between the non-melting tungsten electrode and the nozzle mouth. Argon gas which is jetted through the nozzle and arc is forming a plasma beam of high ionization and temperatures. The pressure of this impinging beam may be increased to the extent that it exceeds the pressure caused by vapourization in the weld pool. A vapour capillary is then generated which takes the shape of a "keyhole" if the workpiece

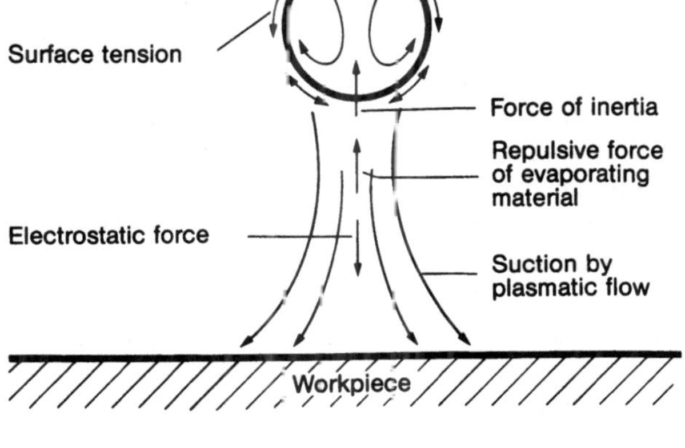

Fig. 45. Forces effecting drop-formation at melting electrode; after Killing [75]

is moved relative to the beam. The metal is melted in front of the advancing keyhole, moving round the sides of the hole and solidifying to form the weld bead at the rear. This process is the same for electron and laser beam welding. All three welding processes are characterized by the extremely high power density. They achieve very narrow and deep welds.

Reference is made in respect of further details to Killing [75], Ruge [74] and Schellhase [73]. An excellent survey on the physics of arc welding has been given by Lancaster [72].

2.3.1.2 Heat balance and heat source density

Statements relating to the heat balance of the welding arc enable its effective power to be assessed without the need for analysing the complex physical processes in more detail. With the additional statement of the heat source density distribution, relatively accurate calculations of heat propagation including the fusion zone can be performed.

The correlation between electrical gross power at the arc and effective heat input at the component (measured in a calorimeter) is expressed by the heat efficiency η_h, equation (1) and Table 1. The higher values which belong to welding with melting electrode can be explained by the fact that part of the heat consumed for melting the electrode passes with the droplets into the weld pool and contributes to heating the base metal. On the other hand, with a given welding process, η_h depends only to a slight extent on the type, polarity and intensity of the welding current. The efficiency decreases with increasing arc length (or arc voltage). It attains particularly high levels when the arc is immersed in the weld pool. Efficiency is additionally improved by covering the weld pool with flux.

The heat balance of three thermal process variants is presented in the form of graphs in Fig. 46, i.e. for welding with carbon electrode, melting metal electrode and melting metal electrode under flux.

The current density is concentrated on the anode and cathode spot respectively. The heat source density to be introduced numerically in this zone should be defined with a larger base area (the heating spot). It is approximated by the Gaussian normal distribution according to equation (16). In the centre of the heating spot, heating occurs mainly as a result of the impacting charged particles. Heating by means of convection and radiation dominates in the surrounding ring zone. The diameter of the anode or cathode spot is in the millimetre range, that of the heat spot in the centimetre range [1].

With rising amperage, maximum value and extension of the heat source density are increased (Fig. 47). With rising voltage, the maximum value is reduced while the extension is increased. The metal electrode (with open arc) displays a higher heat source density with equal extension compared to the carbon electrode (Fig. 48). Covering with flux considerably concentrates the heat source density.

The question relating to the parameter dependence of heat balance and heat source density is posed indirectly in connection with the heat input per unit

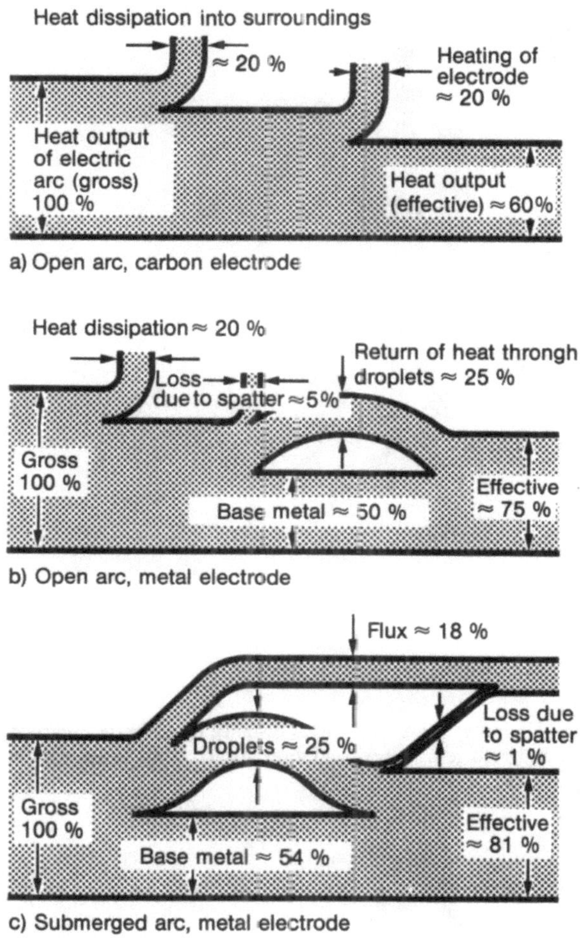

Fig. 46. Heat balance of welding arcs after Rykalin [1]; non-melting carbon electrode ($I = 1000$ A, $U = 40$ V) (**a**), melting uncovered metal electrode ($I \leq 250$ A, $U \leq 25$ V) (**b**), metal electrode melting under flux ($I = 1000$ A, $U = 36$ V, $v = 6.7$ mm/s) (**c**)

length of weld, q_w, according to equation (3). The numerically determined temperature field of a moving source is dependent on the effective heat input per unit length of weld. It is unimportant whether welding is performed slowly with low power or rapidly with high power. It is also unimportant whether the power is produced in the combination of small amperage with high voltage or vice versa (assuming that η_h remains the same).

To what extent this view is correct has been examined by Eichhorn and Niederhoff [156] by means of temperature and heat quantity measurements (especially austenitizing time Δt_a above 800 °C and cooling time $\Delta t_{8/5}$ from 800 to 500 °C) for surface arc welding with melting electrode protected by active gas (GMA welding). On the one hand amperage and voltage were varied in opposite directions (with constant power and welding speed) and, on the other hand, power and welding speed were varied in the same direction (with constant ratio of amperage and voltage). In both cases, the (gross) heat input per unit length of

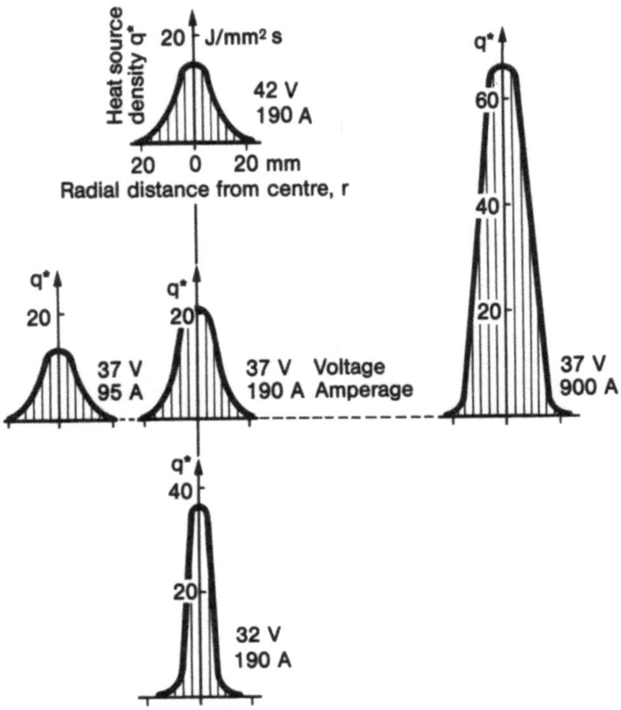

Fig. 47. Heat source density q^* of rapidly moving arc over radial distance from centre r, carbon electrode, dependency on amperage I and voltage U; experimental results according to **Kulagin** in Ref. [1]

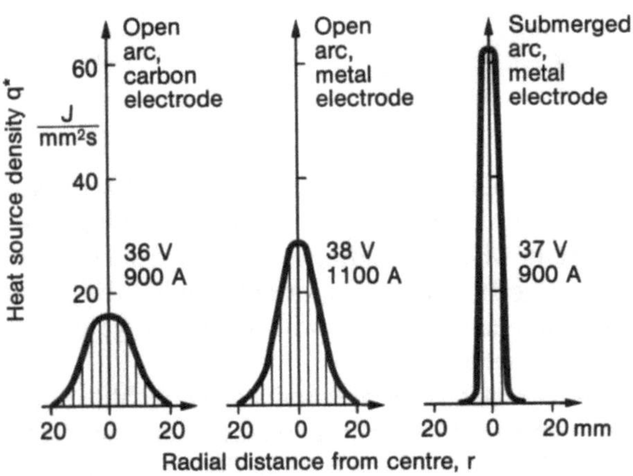

Fig. 48. Heat source density q^* of rapidly moving arc over radial distance from centre r, non-melting carbon electrode, melting uncovered metal electrode and metal electrode melting under flux; experimental results according to Kulagin in Ref. [1]

weld was maintained at a constant level. With the heat efficiency varying relatively little ($\eta_h = 0.8$–0.9), changes in width of the fusion zone and in cooling time up to a factor of two were determined. The different settings, particularly those of amperage and voltage, also resulted in significantly different shape and

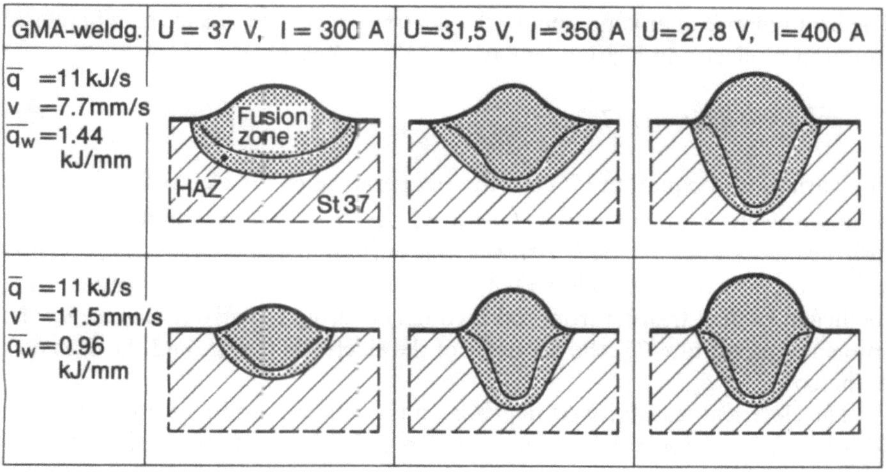

Fig. 49. Cross-section of fusion zone and heat affected zone (HAZ) after gas metal arc (GMA) surface welding (with active gas CO_2), different settings of voltage U and amperage I with equal electrical gross input $\bar{q}$ or input per unit of weld length $\bar{q}_w$; after Eichhorn and Niederhoff [156]

size of the cross-section of the surfacing bead (Fig. 49). Penetration and reinforcement are low with a low amperage and high voltage, and are enlarged in the reverse case. The size of the bead cross-section obviously increases with the amperage, which is attributable to an improvement of η_h and η_t. The parameter reduction achieved by the heat input per unit length of weld may, therefore, require correction in individual cases (see Ref. [157, 158]).

The effective heat input of various conventional welding processes and welding conditions has been determined dependent on the process parameters in Ref. [169, 77]. The heat source density of plasma, electron and laser beam welding is much higher and more concentrated than that of the conventional arc welding processes (see Section 2.1.1.3).

2.3.1.3 Heat conduction modelling of fusion welding

2.3.1.3.1 Melting of the electrode

The melting of the electrode (and thus the deposition of the filler metal) is an important sub-function of the welding arc, in addition to the main function of fusing the base metal.

The electrode is resistance-heated in the entire volume through which the current flows. In the section close to the end of the electrode, the heat spot produces an additional steep rise in temperature, which results in the melting of the electrode. The resistance heating, which is analysed first, increases with the current density and the duration of the current flow. The relevant temperature increase is equally large at any moment in the entire volume of the electrode. By contrast, a radial temperature gradient occurs in the covering of the electrode.

The heat generated as a result of the current flow is distributed on electrode rod and electrode covering, a minor part being dissipated to the surroundings.

The heat balance at any moment of the heating process results in the differential equation of electrode heating (equation (167) in Ref. [1]):

$$\overline{c\varrho}\frac{\mathrm{d}T}{\mathrm{d}t} = \varrho^* j^2 - \alpha_t (T - T_0)\frac{4d_c}{d_e^2}. \tag{56}$$

The variables in equation (56) are the temperature of the electrode rod T, the volume-specific heat capacity $\overline{c\varrho}$ (averaged over rod and covering), the specific electrical resistance ϱ^* ($R = \varrho^* l/A_e$), the current density j ($j = I/A_e$), cross-section area of the electrode rod A_e (without covering), the rod length l, the coefficient of heat transfer from rod through covering to surrounding air α_t, the ambient temperature T_0, the diameter of the electrode rod d_e, and the diameter of the covering d_c.

The electrical resistivity ϱ^* depends on the material and its composition. In addition, it increases with temperature (Fig. 50). The result of an evaluation of equation (56) applying the finite difference method shows the temperature rise over current flow time for different current densities for uncovered and covered electrodes (Fig. 51). The heating rate increases with current density. Covered electrodes heat up slightly more slowly than uncovered electrodes because the covering stores part of the heat.

In addition to resistance heating, the heat resulting from the arc acts in the immediate vicinity of the melting-off electrode end, i.e. within approximately 10 mm of electrode length. The steep temperature drop in this area can be approximated by the temperature field of the area source moving in a rod (equation (38), the latent melting heat is neglected):

$$T = T_r + (T_m^{**} - T_r)\mathrm{e}^{-v_e x/a}. \tag{57}$$

The variables in equation (57) are the temperature rise at the end of the electrode rod T, the temperature increase as a result of resistance heating T_r, the temperature of the molten droplet T_m^{**} ($T_m^{**} > T_m$), the melting-off or feed rate of the electrode v_e, the thermal diffusivity a, and the rod-longitudinal coordinate x.

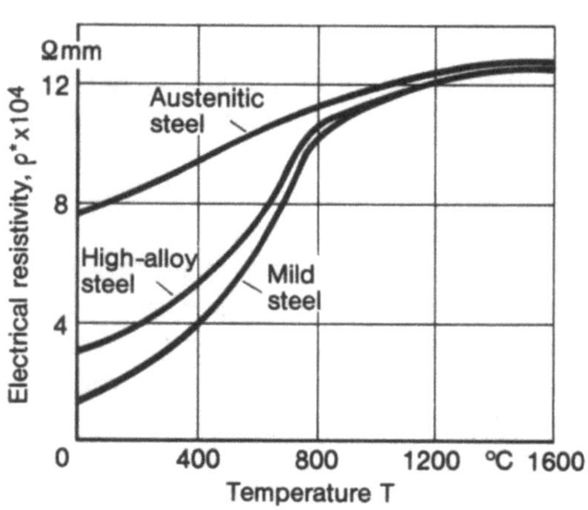

Fig. 50. Electrical resistivity ϱ^* of different electrode steels; after Rykalin [1]

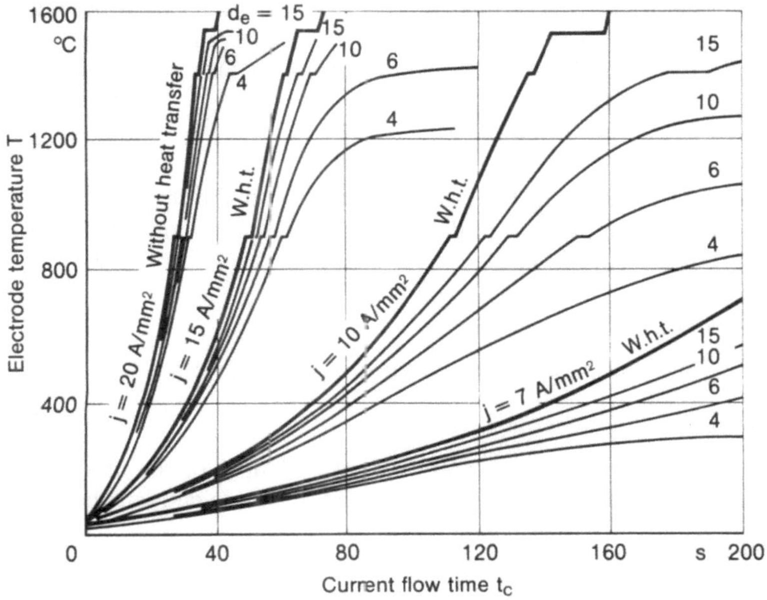

Fig. 51. Temperature rise in uncovered mild steel electrodes, resistance heating without and with heat transfer, different diameters d_e and current densities j; after Rykalin [1]

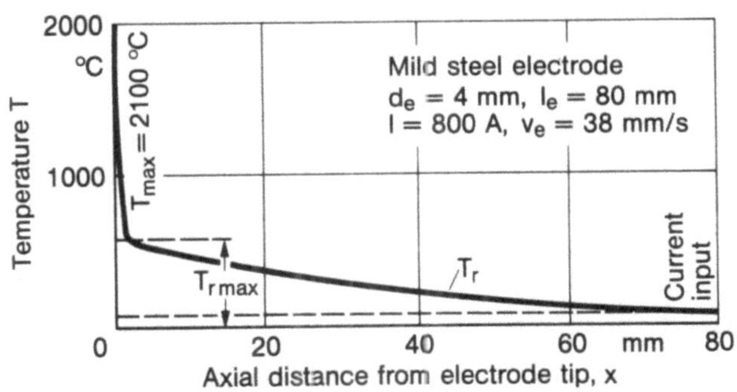

Fig. 52. Temperature profile in steadily fed wire electrodes of mild steel, resistance and arc heating; after Rykalin [1]

Arc and resistance heating are superimposed. In rod electrodes, the temperature T_r increases steadily in the whole rod with welding time t_w. In wire electrodes with automatic steady feed, the temperature T_r increases nearly linearly from the current input point to the end of the wire (Fig. 52). In both cases, the exponential pattern of the temperature increase resulting from the arc is superimposed. In the case of automatic welding, the maximum temperature T_{rmax} increases with the exposed length of the welding wire with the result that a higher melting rate is achieved.

The melting rate of electrodes can be described in general terms, proceeding from heat balance and temperature distribution. The effective heat output q_e of the arc at the electrode is determined from voltage U (at the arc), amperage I and heat efficiency η_e ($\eta_e \approx 0.1$):

$$q_e = \eta_e U I. \tag{58}$$

The effective heat output rises the temperature of the end of the electrode, which is moving on and melting off with feed rate v_e, to the temperature $T_m^{**} > T_m$ (melting temperature T_m):

$$q_c = v_c A_e \overline{c\varrho}(T_m^{**} - T_{rmax}). \tag{59}$$

The melting rate $\dot{m}_e$ [g/s] may be introduced in equation (59):

$$\dot{m}_e = v_e A_e \varrho \tag{60}$$

The mass-specific change in heat content (i.e. in enthalpy) of the arc-melted droplets compared to that of the resistance-heated electrode end is given by:

$$\Delta i = c(T_m^{**} - T_{rmax}) \tag{61}$$

The, in reality temperature-dependent, material characteristic values, ϱ and c, are introduced into equation (59) as independent of temperature, but averaged in the temperature range studied and modified in respect of the latent melting heat (cross-bar above $c\varrho$).

By equating the formulae for q_e from equations (58) and (59), taking into account equation (60), it follows for the melting rate

$$\dot{m}_e = \frac{\eta_e U I}{c(T_m^{**} - T_{rmax})} \tag{62}$$

and for the melting and feed speed

$$v_e = \frac{\eta_e U I}{A_e \overline{c\varrho}(T_m^{**} - T_{rmax})} \tag{63}$$

In practice, melting rate $\dot{m}_e$ and melting speed v_e are controlled, above all, by the amperage I. The possible amperage is limited, however, dependent on the type of electrode (melting instabilities, spatter losses). The melting rate of rod electrodes rises with increasing welding duration as a consequence of the temperature increase T_r. The temperature increase T_{rmax} after completion of the welding operation with a given electrode, on the other hand, is all the greater the slower the welding operation is performed, and in addition higher with an uncovered electrode than with a covered electrode.

The ratio α_e of melting rate $\dot{m}_e$ per unit of amperage I,

$$\alpha_e = \frac{\dot{m}_e}{I} \tag{64}$$

is used as a process- and electrode-typical melting index which, according to equation (62), is approximately constant if the deposition rate is not extremely high. In accordance with Ref. [1], $\alpha_e = 5$–14 g/Ah for hand welding and $\alpha_e = 13$–23 g/Ah for mechanized submerged-arc welding.

The molten-off mass m_e reappears in the deposited mass m_d (deposited with welding speed v) after subtraction of the mass losses in the arc and from the molten material (spatter and evaporation):

$$\dot{m}_d = v A_d \varrho, \tag{65}$$

$$\dot{m}_d = \dot{m}_e(1 - \psi_d). \tag{66}$$

The loss factor ψ_d is stated with $\psi_d = 0.05$–0.2 for common welding methods and with $\psi_d = 0.01$–0.02 for submerged-arc welding.

The ratio α_d of deposition rate $\dot{m}_d$ per unit of amperage I,

$$\alpha_d = \frac{\dot{m}_d}{I} \tag{67}$$

is used as a process- and electrode-typical deposition index. It is obvious that

$$\alpha_d = \alpha_e(1 - \psi_d). \tag{68}$$

Calculation examples relating to the statements above in this section are given in Ref. [1]. Further information on electrode heating in terms of welding parameters is given in Ref. [72] (ibid. pp. 330–336).

The statements can be transferred by analogy to welding methods with direct arc and non-melting electrode as well as to methods with indirect arc (plasma welding). There is no resistance heating of the filler metal in these cases. The heat is transferred from the arc to the filler and base metal solely by means of convection and radiation. Deposition rate and deposition index are reduced as a consequence.

The temperature distribution in gas shielded electrodes has been analysed in Ref. [43].

2.3.1.3.2 Fusion of the base metal

Fusion of the base metal, with or without filler metal, is decisive for firmly connecting the parts to be joined. Theoretically, an extremely thin fused layer might be sufficient for this. In fact, the aim is to achieve a larger layer thickness of approximately 1 mm in order to bridge geometrical, material and process deviations from the specified values without causing poor fusion. On the other hand, however, the fusion layer should also not be thicker than necessary in order to avoid waste of energy, edge burn-off, sagging of the weld pool and deep weld end craters.

The statements below refer throughout to arc welding, but they may be applied analogously to flame welding.

Fusion and overheating of the weld pool occur momentarily at the heat spot of the welding arc on the surface of the base metal. The expanding fusion zone forms the weld pool. The surface of the weld pool is impressed in the shape of

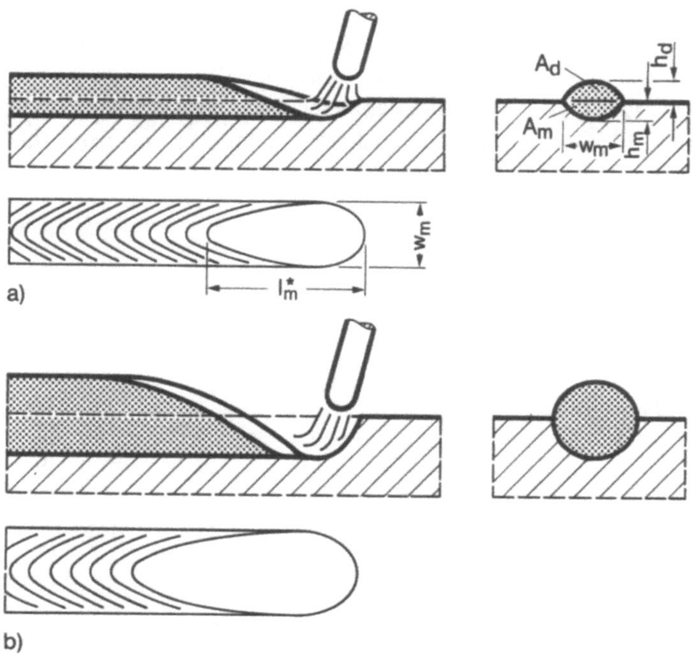

Fig. 53. Weld pool with surfacing arc (a) and immersing arc (b); after Rykalin [1]

a crater by the blow effect of the arc. Consequently, in addition to the longitudinal and transverse movability (oscillation), the arc acquires a further movability in the direction of the depth.

A distinction is made between surfacing and immersing arc (Fig. 53). In the surfacing arc, which is promoted by low amperage, the heat spot is located a little below the surface of the workpiece, a molten layer is always retained below the heat spot and the fusion depth is relatively small. In the immersing arc, which is promoted by high amperage, the heat spot (and thus also a part of the arc) is located deep below the surface of the workpiece, the molten layer is pushed away towards the solidifying weld and the fusion depth is large; intensive heat transfer also favourably occurs at the crater walls. In submerged-arc welding, the immersing arc is additionally covered upwards by slag powder, which further reduces the heat loss.

The dimensions of the weld pool or of the fusion zone are, in the case of surfacing and butt welds, characterized by the weld pool length l_m^*, the weld pool width w_m, the weld bead height h_d, the weld pool depth h_m, the weld bead cross-section area A_d, and the weld pool cross-section area A_m (Fig. 53). The parameters w_m, h_d, h_m, A_d and A_m can be measured in the polished transverse section, fusion zone and weld pool being set equal in this case. In reality, the fusion zone is given by the enveloping surface of the moving weld pool. The difference in the transverse section is negligibly small. The area A_d identifies the quantity of deposited filler metal; in contrast, the area A_m identifies the quantity of fused base metal. The length l_w^* can be measured on the solidified weld end crater. The depth of the end crater, however, is not a measure for the weld pool depth because the crater fills up when the arc is switched off.

Additionally, the relative fusion depth h_m/w_m and the relative fusion area, the fullness ratio $\mu = A_m/(h_m w_m)$, are introduced as dimensionless shape parameters. The relative fusion depth differs in size dependent on welding method, electrode and base metal. In contrast, the relative fusion area, $\mu = 0.6$–0.8, is relatively constant.

The temperature field equations derived for heat conduction in the semi-infinite solid and infinite plate with point and line source respectively (neglecting heat transfer at the surface, $b = 0$) can be used for determining the geometrical parameters of the fusion zone by numerical analysis (they are particularly influential, among other things, on residual stresses and distortion). The area within the melting temperature isotherm is regarded as the fusion zone. In this way, the influence of the principal parameters of geometry, material and welding method are covered. The deviations of the analysis results from reality are large, however, because the fusion process is inadequately modelled.

The model takes account solely of heat conduction, proceeding from the point or line source. In reality, the heat input is effected through the planar heat spot with the contribution of convection and radiation at the crater walls. Fusion heat is consumed at the crater front wall and released again during solidification at the crater back wall. The filler metal supplied by the electrode is melting above the weld pool with heat consumption and solidifies after dropping off with heat release.

When the weld pool forms, energy is absorbed as latent heat of melting. Although it is released later, this heat is temporarily removed from the input energy and stored in the weld pool. If the weld pool dimensions are calculated on the basis of the temperature fields derived for heat conduction, the latent heat of melting has to be subtracted from the heat input to give the dimension-effective heat (alternatively, the heat content of the pool has to be increased).

In conformity with Ref. [1], the numerical assessment of the geometrical weld pool parameters on the basis of the temperature field equations derived for heat conduction is performed introducing the concept of thermal efficiency of the welding process.

The heat input q for melting the base metal (per unit of time) is determined from the fusion volume vA_m [mm^3/s] and its volume-specific heat content ϱi_m [J/mm^3] (including latent fusion heat, but without overheating) and equated through the fusion efficiency η_m to the heat output of the electric arc:

$$vA_m \varrho i_m = \eta_m UI. \tag{69}$$

This statement on heat input and heat output relates to fusion without filler metal (weld bead cross-section area $A_d = 0$). Otherwise, it would be more correct to introduce $A_d + A_m$ on the left-hand side of the equation in place of A_m. If this is not done, the melting heat of the filler metal then occurs in η_m incorrectly as a loss factor. This has been ignored in Ref. [1].

The fusion efficiency η_m is split up into the heat efficiency η_h of heat generation and heat transfer in the arc gap, into the thermal efficiency η_t of base metal fusing determined by calculation (a significant part of the effective heat output of the arc is lost as a result of heat diffusion into the base metal) and into the correction factor p for the deviations of the calculation results from reality:

$$\eta_m = p\eta_t\eta_h. \tag{70}$$

The heat efficiency may be assumed according to Table 1 dependent on the welding method or determined more accurately by measurement. It can be assumed for the surfacing arc that $p < 1.0$ ($p \approx 0.5$–1.0), and for the immersing arc that $p > 1.0$ ($p \approx 1.0$–1.2). With improved analysis methods for the temperature field (more accurately taking into account the thermodynamics of the weld puddle), it is possible to achieve $p \approx 1.0$. The thermal efficiency η_t is determined from the (calculated) fusion area A_m of the zone lying within the isotherm T_m, converted to heat input and related to the effective heat input q.

In a semi-infinite solid with rapidly moving high power point source, a half cylinder around the motion line of the source (radius r_m) is heated to T_m, in the analysis:

$$\eta_t = \frac{vc\varrho(T_m - T_0)\pi r_m^2}{2q}. \tag{71}$$

It follows with equation (46) that $\eta_t = 0.234 \times \pi/2 = 0.368$. Only 36.8% of the effective heat input is available for fusing.

In a plate with rapidly moving high power line source, a strip on both sides of the source motion plane (width w_m, thickness h) is heated to T_m in the analysis:

$$\eta_t = \frac{vc\varrho b_m h(T_m - T_0)}{q}. \tag{72}$$

It follows with equation (49) that $\eta_t = 0.242 \times 2 = 0.484$. Only 48.4% of the effective heat input is available for fusing.

The values above apply to rapidly moving high power sources, without taking filler metal into account. They are based on heat diffusion exclusively perpendicular to the direction of the moving source. With lower source speed, as occurs in reality, the heat loss as a result of heat conduction in the direction of the moving source is increased. The thermal efficiency η_t for non-rapidly moving sources is presented in Ref. [1] in the form of a graph, depending on the dimensionless parameters ξ_3, for the semi-infinite solid, index 3 for three-dimensional heat diffusion, and ξ_2, for the plate, index 2 for two-dimensional heat diffusion (Figs. 54 and 55):

$$\xi_3 = \frac{qv}{a^2\varrho i_m}, \tag{73}$$

$$\xi_2 = \frac{q}{ah\varrho i_m}. \tag{74}$$

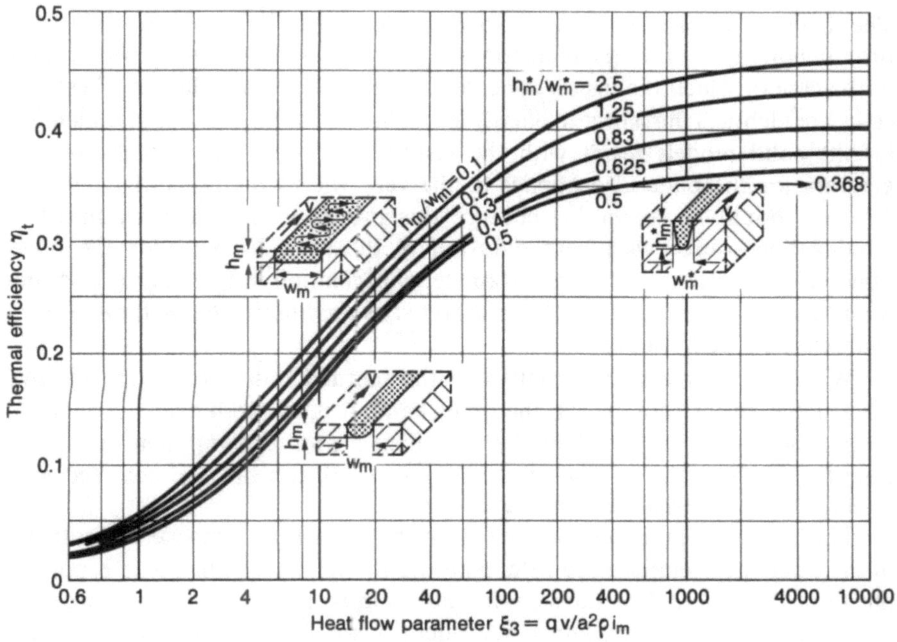

Fig. 54. Thermal efficiency η_t of surface welding with three-dimensional heat diffusion dependent on heat flow q, welding speed v, thermal diffusivity a and melting heat content ϱi_m for different ratios h_m/w_m or h^*_m/w^*_m; after Rykalin [1]

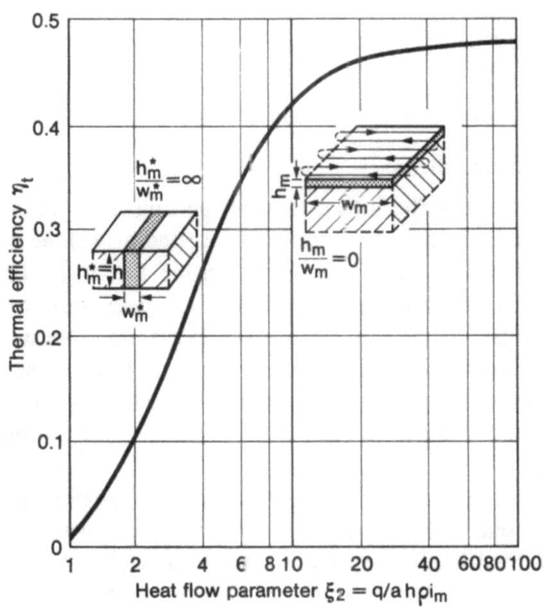

Fig. 55. Thermal efficiency η_t of butt welding or large-width surface welding with two-dimensional heat diffusion dependent on heat flow q/h, thermal diffusivity a and melting heat content ϱi_m; after Rykalin [1]

The diagram for the semi-infinite solid is enlarged by the weaving weld (with progression rate v of the weaving loops) and by the weld with deep penetration, in the case of which the fusion zone deviates from the half cylinder. These two cases are identical in respect to energy for $h_m^*/w_m^* = 1/(4h_m/w_m)$ because identical infinite solid models result with the respective symmetrical extensions. The diagram for the plate also applies to the very wide weaving (surfacing) weld ($h = w_m$ should be set in ξ_2) because this process can be approximated as moving line source with heat diffusion of the same kind as for the plate. Obviously, despite the different parameter combination in ξ_3 and ξ_2 (v does not occur in ξ_2), η_t over ξ_2 should be interpreted as the limiting curve for η_t over ξ_3 characterized by $h_m/w_m = 0$.

With low source speed, relative to the thermal diffusivity a, the thermal efficiency η_t drops rapidly. On the other hand, it is all the higher the wider or deeper the line source acts. The thermal efficiency for the point source is lower than for the line source. The ideal value, $\eta_t \approx 1.0$, can only be achieved with volume-effective resistance heating of rods.

The length l_m^* of the weld pool results from the melting temperature isotherm of the calculated temperature field, ignoring the relatively small length portion ahead of the source. For the semi-infinite solid with moving point source, the length l_m^* results explicitly and independent of speed in accordance with equation (30):

$$l_m^* = \frac{q}{2\pi\lambda(T_m - T_0)}. \tag{75}$$

For the infinite plate with moving line source, the length l_m^* results implicitly and dependent on speed in accordance with equation (33) setting $b = 0$:

$$\mathrm{Y}_0\left(l_m^* \frac{v}{2a}\right) \mathrm{e}^{-l_m^* v/2a} = \frac{h 2\pi\lambda(T_m - T_0)}{q}. \tag{76}$$

The effective heat input by the moving source is designated by q.

2.3.1.3.3 Interaction of melting-off and fusion

The welding process is characterized by the interaction of melting filler metal and fusing base metal in the area of the arc. The coupling is particularly close in the case of the processes with melting electrode. The decisive factor is the distribution of the arc heat to electrode and base metal. With a given welding method, it depends on base and filler metal, composition of electrode covering or welding flux, electrode polarity and arc length. Whereas the total heat output of the arc can be simply controlled through the amperage in combination with a suitable electrode diameter, the possibilities for controlling the division of the arc heat are limited.

The problem of heat division is illustrated by the following equations for the cross-section areas A_d and A_m produced by deposition and fusion, respectively. In this case, in conformity with the critical remarks in respect of equation (69), the fusion efficiency η_m is defined initially more practically in terms of simultaneous deposition (or melting) and fusion. In order to keep definition and equations as simple as possible, the specific heat contents i_m for fusion of base metal and Δi for melting of filler metal are set equal ($\Delta i = i_m$). Equation (69) is thus replaced by:

$$v(A_m + A_d)\varrho i_m = \eta_m UI. \tag{77}$$

This corresponds to the model conception that the filler metal is deposited in the weld groove and is heated and fused there together with the base metal by the arc.

The deposition area follows from equations (62), (65) and (66) with $\Delta i = i_m$:

$$A_d = \frac{\eta_e UI(1 - \psi_d)}{v\varrho i_m}. \tag{78}$$

The combined fusion and and deposition area, on the other hand, follows from equation (77):

$$A_m + A_d = \frac{\eta_m UI}{v\varrho i_m} \tag{79}$$

After rearrangement, it follows from equations (70), (78) and (79) that:

$$\frac{A_m}{A_d} = \frac{p\eta_t\eta_h}{\eta_e(1 - \psi_d)} - 1 \tag{80}$$

In practice, greatly different values for A_m/A_d are aimed at depending on the type of weld. The surfacing welds are at one end of the scale of values with $A_m/A_d \ll 1$, the butt welds without filler metal at the other end with $A_m/A_d \gg 1$.

Of the quantities occurring in equation (80), only η_t can be set in a greatly variable manner. If A_m is to be enlarged compared to A_d, a welding method with a high η_t is required. Only rapidly moving high power sources satisfy this requirement. If A_m is to be small compared to A_d, a method with a low η_t can be considered. With the same heat input per unit length of weld, $q_w = q/v = \eta_h UI/v$, η_t differs in value depending on the choice of q and v, respectively. However, in view of the fact that the impairment of the economies of the welding process resulting from small values v and η_t is not acceptable, preference should be given to the measures acting on η_e, mentioned at the beginning of this section.

The decisive factor for the productivity of welding methods is their welding speed, v. The question of how the welding speed can be increased is answered on the basis of the derived relations. In welds with $A_m/A_d \ll 1$, the speed v with given A_d is increased mainly by increasing the melting index α_e and the amperage I, for, in accordance with equations (64), (65) and (66):

$$v = \frac{\alpha_e(1 - \psi_d)I}{\varrho A_d}. \tag{81}$$

In welds with $A_m/A_d \gg 1$, the speed v with given h_m is increased mainly by increasing the (effective) heat input q (by means of the amperage I) and the relative fusion depth h_m/w_m (by means of deep penetration measures), for, in accordance with equations (1) and (69), with the fullness ratio $\mu = A_m/(h_m w_m)$:

$$v = \frac{1}{\mu h_m^2 \varrho i_m} \frac{h_m}{w_m} q. \tag{82}$$

A simple formula for the total (effective) heat input q per unit plate thickness h in single pass arc butt welding has been derived by Wells [47] proceeding from the melting temperature isotherm of the moving source, equation (33) with $b = 0$. The heat input is expressed in terms of that necessary to melt the fused zone (width w_m), multiplied by a factor, which depends on the non-dimensional parameter $vw_m/4a$:

$$\frac{q}{h} = 8\lambda(T_m - T_0)\left(\frac{1}{5} + \frac{vw_m}{4a}\right) \tag{83}$$

Another formula is given proceeding from the molten filler metal in the gap between the plate edges (width w_g) and assuming that 15% of the (effective) heat is used for melting the electrode and providing the filler metal:

$$\frac{q}{h} = \frac{80}{3}\lambda(T_m - T_0)\frac{vw_g}{4a} \tag{84}$$

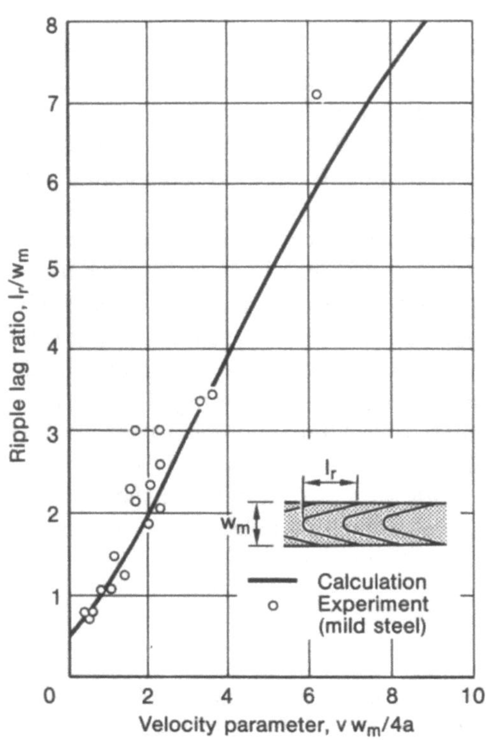

Fig. 56. Ripple lag ratio derived from calculated melting temperature isotherm and from experiment; after Wells [47]

It appears that for an initial separation w_g of the plate edges at a given heat input q/h, there is only one welding speed, which will just allow a flush weld bead to be deposited. At other speeds the bead is either convex or concave.

Evaluating the melting temperature isotherm additionally after the moving source, a formula for the ripple lag ratio l_r/w_m is derived and plotted in Fig. 56 together with the experimental results. For low heat inputs the ripples appear to be semi-circular, while for high inputs corresponding with maximum economy of energy, the ripple lag is greater than three times the molten width.

2.3.1.4 Weld pool modelling

2.3.1.4.1 Weld pool physics

The heat conduction model, on which temperature field, residual stress, distortion and microstructural change analysis for welded joints is mainly based, gives reliable results outside the heat-affected zone and under certain conditions also for this zone itself. One of these conditions is that the actual fused zone shape and dimensions are correlated to the results of the heat conduction model, which is achievable to some extent by adjusting the heat source distribution therein. The model is unsuited for treating weldability problems of the fusion zone itself. Such problems comprise a well designed weld shape such as adequate penetration, smooth transitions, not too large a reinforcement, the absence of weld imperfections such as large undercut, burnthrough, pores and other defects, the generation of a strength and ductility enhancing grain structure. These problems occur especially when extending the proven range of application of a process, material or design or when optimizing the latter within a proven range. Even under conditions of normal use appropriate machine setting is not a trivial task despite the well-founded empirical formulae which are available in certain cases [102–104].

In recent years, remarkable progress has been made in modelling the complex process of arc and beam welding [72, 81–101]. The following survey comprises those models which are directly related to the weld pool behaviour. Such processes are considered first, in which the conventional, surface-related weld pool is formed, i.e. mainly arc welding, and beam welding only as far as a defocused beam is applied. Beam welding with focused beam and keyhole formation is treated subsequently (see Section 2.3.1.4.6).

First the physical phenomena characterizing the weld pool behaviour are described [72]. The processes within the pool and the conditions on its surface are mainly considered without covering further details on arc or beam formation, on drop formation at the electrode tip and on mass transport from the electrode to the weld pool.

In contrast to conventional heat conduction models, heat flow in weld pools is mainly convectional [72]. Convection is based on fluid motion in the pool. The temperature gradients within the pool decrease with increasing flow velocity. The fluid motion is generated by asymmetric magnetic fields (spinning flow), by surface tension gradients (toroidal flow), by drag forces of the plasma and gas jet, by its stagnation pressure, by impinging metal drops and by buoyancy forces.

Surface tension and its temperature dependence is often the decisive driving force determining the dominating pattern of fluid motion (thermocapillary flow, Marangoni effect). If surface tension σ^* decreases with increasing temperature T ($d\sigma^*/dT$ negative), an outward flow is generated on the surface transporting heat mainly horizontally so that a wide and shallow weld pool is achieved (Fig. 57b). If surface tension σ^* increases with increasing temperature T ($d\sigma^*/dT$ positive), an inward flow is generated on the surface transporting heat mainly vertically so that a deep and narrow weld pool is achieved (Fig. 57d). The deep penetration narrow weld behaviour is enhanced by certain "surface active elements" (i.e. elements segregating preferentially on the surface of the liquid metal) such as sulphur, oxygen and aluminium in the case of steels. The term "dopant" is used if these elements are intentionally applied on filler metals. It has been found [93, 82], that the penetration behaviour of welds (and therefore its weldability) may depend on the delivered melting charge, especially on its sulphur content, the reason being varying surface tension behaviour. Surface tension is changed by a slag covering as applied in submerged arc welding.

The surface of the weld pool does not remain plane as it is incorrectly assumed in oversimplified models [81]. The surface is deformed primarily by the

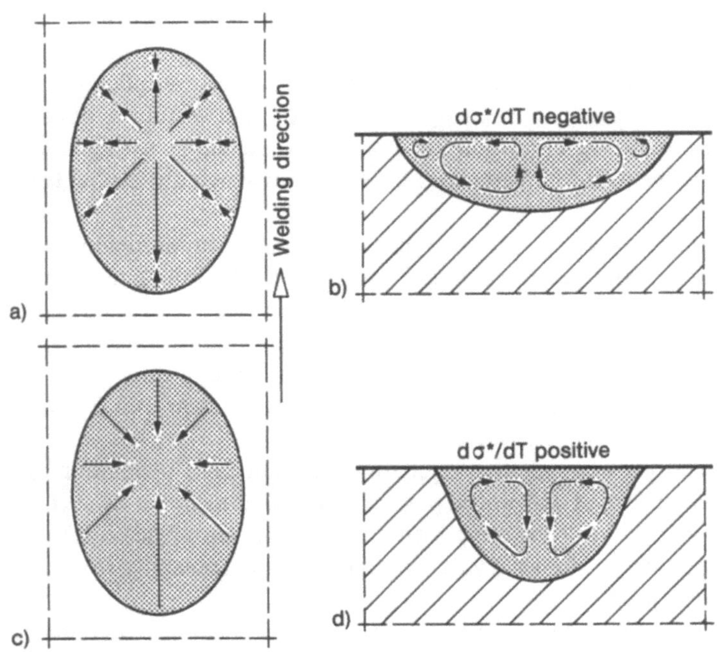

Fig. 57. Pattern of fluid motion in weld pool: wide and shallow pool with negative surface tension versus temperature gradient (**a, b**), narrow and deep with positive surface tension versus temperature gradient (**c, d**); top view (**a, c**) and cross-section (**b, d**); after Heiple et al. [88]

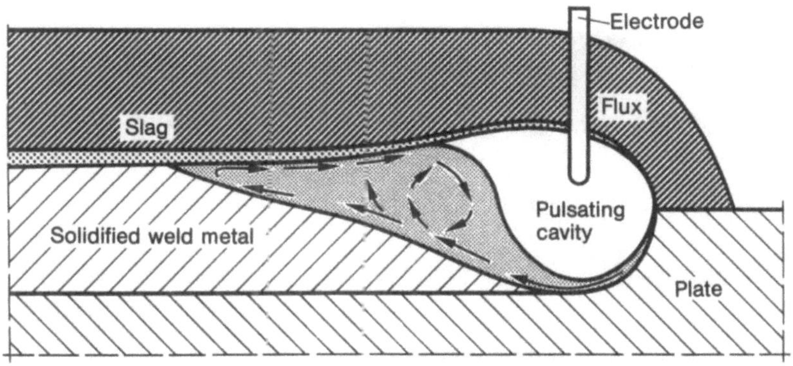

Fig. 58. Surface depression and flow pattern in submerged arc welding pool; after Lancaster [72]

arc and gas stagnation pressure which increases with higher current levels. Significant surface deformations occur especially in submerged arc welding (Fig. 58) (enlarged by cavity formation) and in high power gas metal arc processes. They affect the arc and modify the heat and current flow. They may give rise to surface waves generating the ripples and to local shape instabilities such as void formation by surface projection and subsequent collapse.

The simplifying assumption of a single melting temperature isotherm disregarding the latent heat effects of melting and solidification is not acceptable in fusion zone analysis [84, 130]. A partially molten, i.e. two-phase, zone surrounds the moving weld pool (Fig. 59a). The solid line bounding the weld pool designates the liquidus temperature of the alloy; the dashed line the effective solidus temperature. The effective solidus temperature is always lower than that predicted from microstructural equilibrium. The volume of the molten phase present varies in a continuous fashion from 0 at the dashed line to 100% at the solid line. The partially molten zone is of high practical significance because welding defects originate in this zone, especially hot and cold cracks.

At the leading edge of the weld pool, ABC, melting occurs. This requires that the latent heat of fusion is supplied to cause liquidation. At the trailing edge, CDA, the latent heat must be liberated to cause solidification. On average, the growth of the solid occurs parallel to the maximum temperature gradient, which is perpendicular to the advancing solid–liquid interface. The growth velocity must be zero at the sides of the weld pool (points A and C) and largest at the centre line of the weld (point D). Unfortunately, the minimum temperature gradient and therefore the minimum capability to conduct the latent heat to the cooler surroundings, is found at this location.

Consequently, when welding velocity reaches a critical value, such that the rate of growth parallel to the welding direction can no longer be sustained, the weld pool will assume a teardrop shape (Fig. 59b). This critical velocity is inversely proportional to the welding current, whereas the welding voltage has only a minor influence on the size and shape of the weld pool. The grain

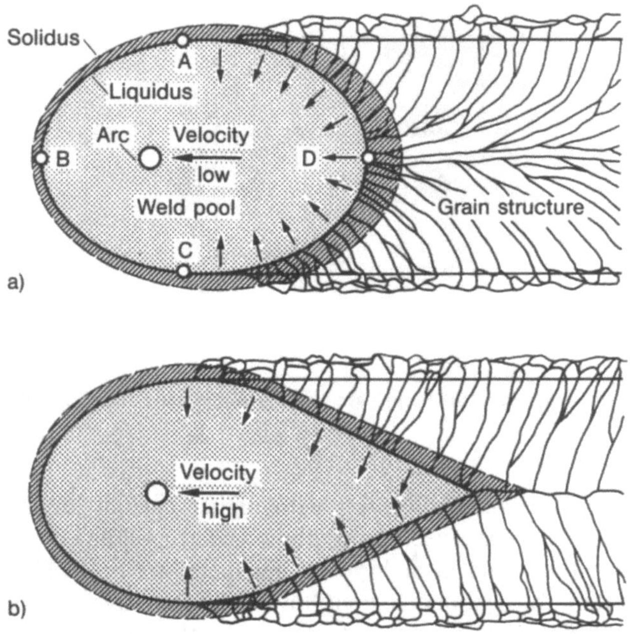

Fig. 59. Weld pool shape, partially molten boundary area and competitive grain growth in arc welding; low speed and current intensity (**a**), high speed and current intensity (**b**); after Savage [130]

direction in the fusion zone is controlled by the shape of the weld pool, and the grain shape follows the conditions of competitive growth (Fig. 59).

In the following sections, a survey is given on mathematical modelling of the processes, which control weld pool shape and size. Despite the remarkable progress achieved, these models give answers only on a very restricted class of questions from the point of view of practical application. To know the physical properties and material characteristic values of the fluids at the elevated temperatures is an additional problem in this connection. Concerning the physical background of the models, the book by Lancaster [72] should be consulted. The survey below does not comprise models referring to electrode melting, drop formation and transport of the liquid metal.

2.3.1.4.2 Welding arc modelling

The specific feature of the welding arc is the electrical discharge occurring between a rod tip and a plane or deformed plate surface. The arc may spread laterally, thus generating a plasma jet. The jet is augmented by added gas. Besides the overall electrical behaviour in terms of the current-voltage characteristic and the arc efficiency, more detailed field information is required in respect of weld pool modelling: temperature profiles, stream lines, current density, heat flux and arc pressure.

Arc modelling [72, 81, 83] is based on Maxwell's equations describing the distribution of magnetic and electric, field and flux intensities including current density in a plasma, on continuity of mass equations, on Navier-Stokes equations for viscous (here laminar) flow extended by an electromagnetic force term, on the equation expressing the conservation of energy (in therms of enthalpy) and on the equations of state. The partly non-linear and temperature-dependent system of differential equations is solved for the relevant boundary conditions.

The results of an analysis performed by Szekely [81] comprising temperature profiles, anode current density and anode heat flux are presented for a plane weld pool surface in Fig. 60 and for a depressed weld pool surface in Fig. 61. The important finding is that even a relatively slight pool depression may markedly change the heat flux incident on the weld pool. The off-centre peak of the flux curve at the edge of the surface depression seems to be a characteristic feature.

In the above investigation, the surface depression was arbitrarily prescribed. In reality the arc itself deforms the surface, mainly through the arc pressure. Arc pressure curves measured on a plane pool surface are presented in Fig. 62. The Gaussian normal distribution is confirmed.

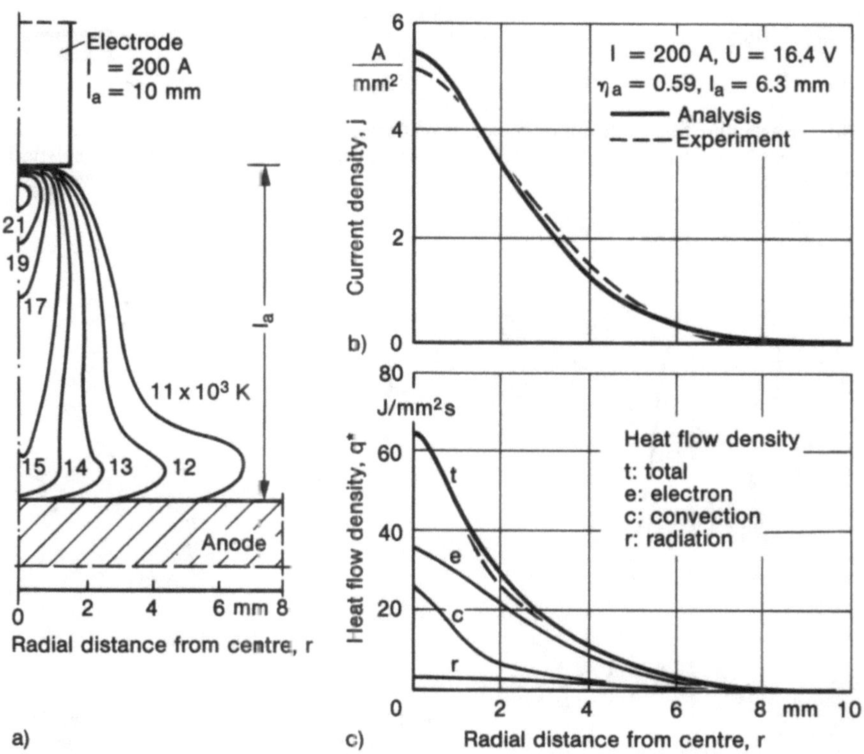

Fig. 60. Temperature field isotherms (**a**), anode current density (**b**) and anode heat flow density (**c**) in stationary gas tungsten arc ($I = 200$ A, plane pool surface, arc length 10 mm), analytical and experimental results; after Szekely [81]

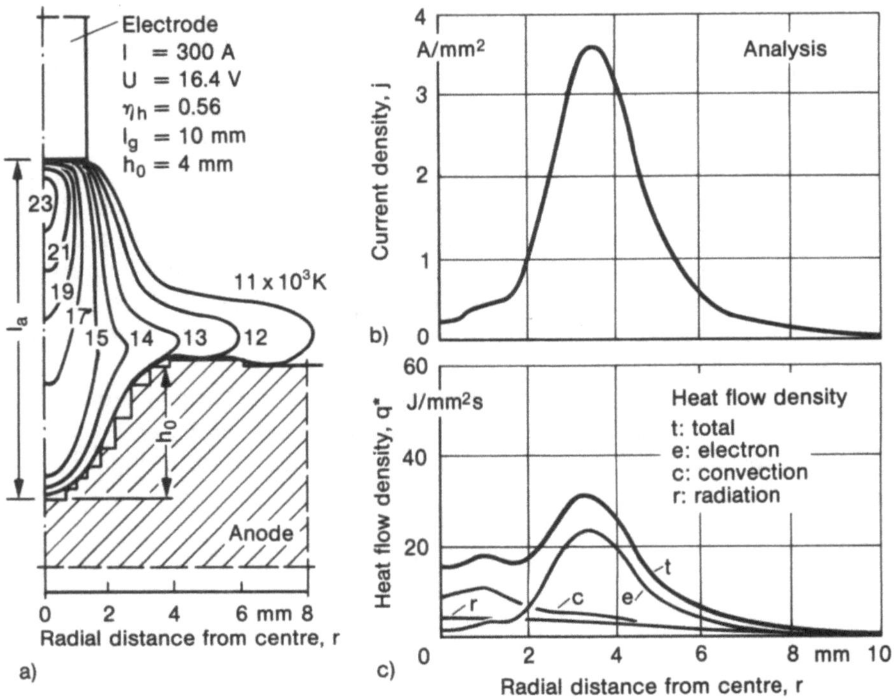

Fig. 61. Temperature field isotherms (**a**), anode current density (**b**) and anode heat flow density (**c**) in stationary gas tungsten arc (I = 300 A, pool surface 4 mm depressed, arc length 10 mm), analytical results; after Szekely [81]

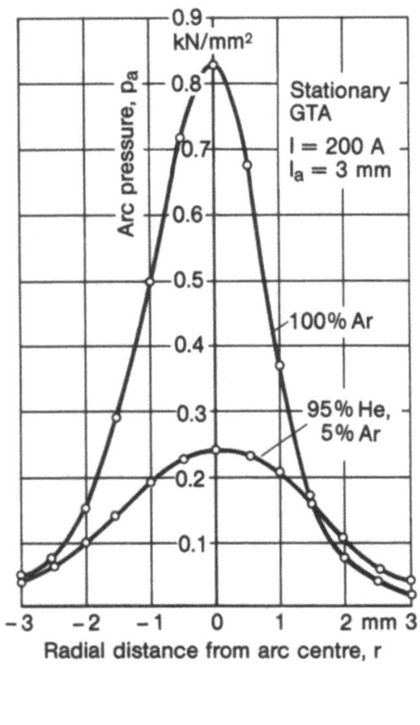

Fig. 62. Arc pressure on plane pool surface in stationary gas tungsten arc (I = 200 A, electrode diameter 3.2 mm, arc length 3 mm, vertex angle 30°), argon and helium shielding gases compared, experimental results; after Lancaster [72]

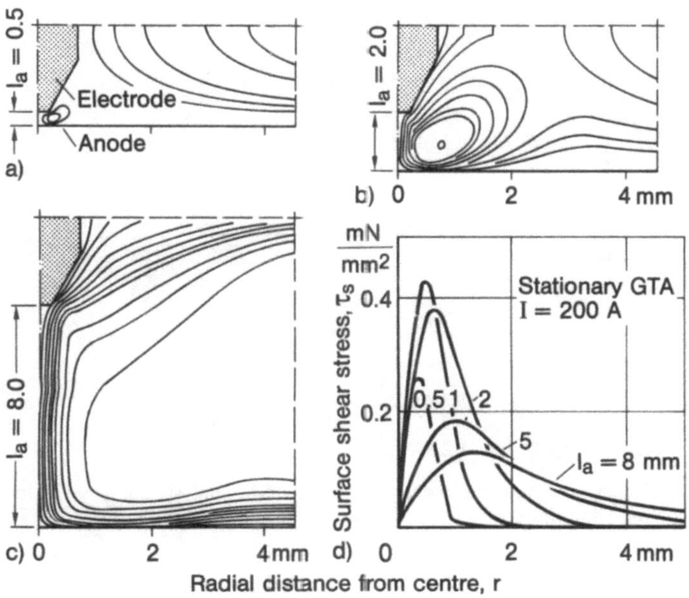

Fig. 63. Stream lines (**a**, **b**, **c**) and surface shear stress (**d**) dependent on radial distance from centre for different arc lengths of stationary gas tungsten arc (I = 200 A, arc spot diameter 2, 4 and 10 mm); after Matsunawa and Yokoya [83]

The results of an analysis by Matsunawa and Yokoya [83] assuming a plane pool surface show stream lines and surface shear stresses dependent on arc length (Fig. 63).

2.3.1.4.3 Hydrostatic surface tension modelling

As already stated, the surface tension σ^* and its temperature dependency plays a major role in weld pool flow. The chemical elements related to welding materials have a very high surface tension in the vicinity of their melting points (i.e. much higher than mercury at ambient temperature). Obvious examples of the effect of surface tension in welding are the stable suspension of the molten pool without any backing system in full penetration welding of thin plates or the smooth metal transfer from the filler wire to the weld pool in overhead position welding.

Analysis results have been achieved on the basis of a plane hydrostatic surface tension model by Nishiguchi and Ohji (presented by Matsunawa [87]). The fundamental relation, on which the model is based, is the force balance between gravitational and surface tension forces resulting in a curved surface. Only plane cross-sectional models have been investigated. The plane model may be acceptable in cases of low current welding, when directional flow in the pool is not strong enough to distort the cross-section profile.

The relation between bead reinforcement height h_r and bead width w_b for two slope angles θ is shown in Fig. 64. At first, height and width increase together but then the height asymptotically reaches a constant maximum value. The

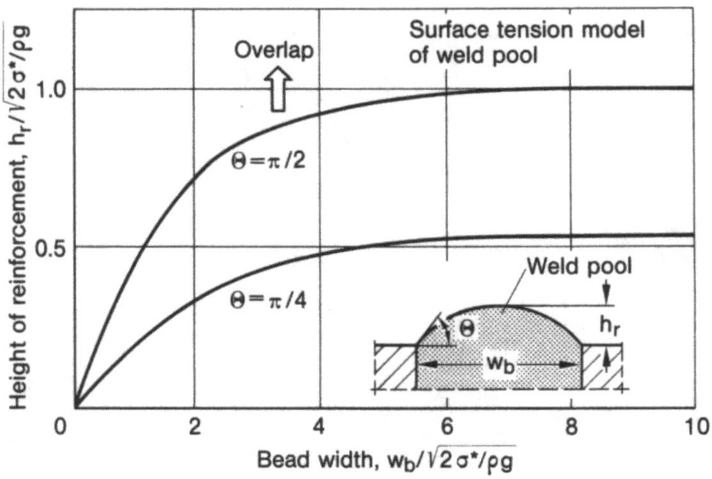

Fig. 64. Molten pool reinforcement height dependent on bead width, analysis based on plane hydrostatic surface tension model, overlap with $\theta > \pi/2$; after Nishiguchi and Ohji in Ref. [72, 87]

shape and size of the bead reinforcement may now be calculated on the basis of the heat input rate, which determines the bead width, and of the volumetric burn-off rate divided by welding speed, which determines the cross section area. The two-dimensional approximation seems to give too low values compared with the three-dimensional welding reality.

The effect of the surface profile of the pool on the allowable root bead width for self-supporting molten pools is shown in Fig. 65 based on the same model. The allowable width decreases inverse proportionally with the increase in plate thickness. If the top-surface is depressed, higher width values are allowable.

The bead profiles of horizontal fillet welds, with equal leg lengths, have been calculated based on the surface tension model (Fig. 66). The parameter area of favourable profiles avoiding both undercut and overlap has been defined. It will be evident from this diagram that the deposition rate may only vary within narrow limits if an unfavourable profile is to be avoided.

The same authors [98] have calculated the shape of the fusion zone for butt welding of thin plates. First the temperature distribution is determined, assuming a linearly moving circular heat source with a uniform heat input rate. The weld pool profiles at the top and root side are calculated proceeding from the force balance between surface tension, gravitational force and stagnation pressure. Welding in vertical down and vertical up position is compared.

2.3.1.4.4 Hydrodynamic weld pool modelling

Hydrodynamic modelling of the weld pool is necessary in cases of high current and deep penetration welding [82, 88, 93–97]. As stated formerly, electromagnetic forces, surface tension, drag forces, stagnation pressure, drop impingement, buoyancy and gravitational forces should be taken into account.

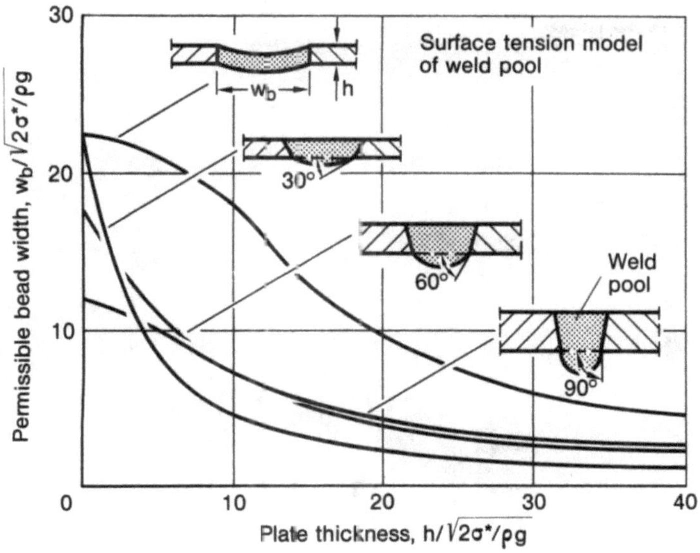

Fig. 65. Effect of plate thickness and pool surface profile on permissible root bead width, analysis based on plane hydrostatic surface tension model; after Nishiguchi and Ohji in Ref. [87]

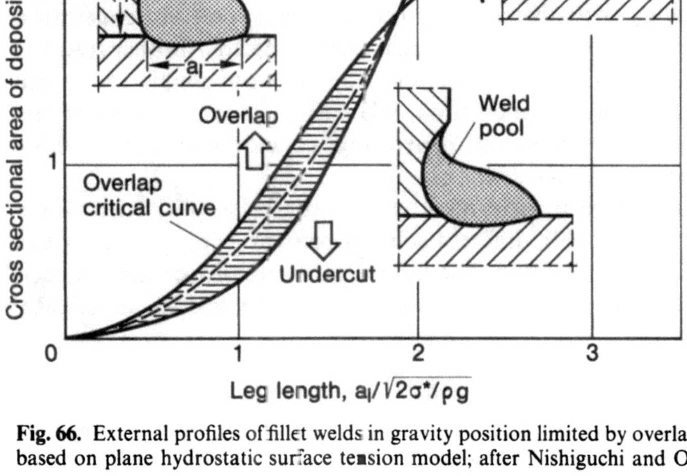

Fig. 66. External profiles of fillet welds in gravity position limited by overlap and undercut, analysis based on plane hydrostatic surface tension model; after Nishiguchi and Ohji in Ref. [72, 87]

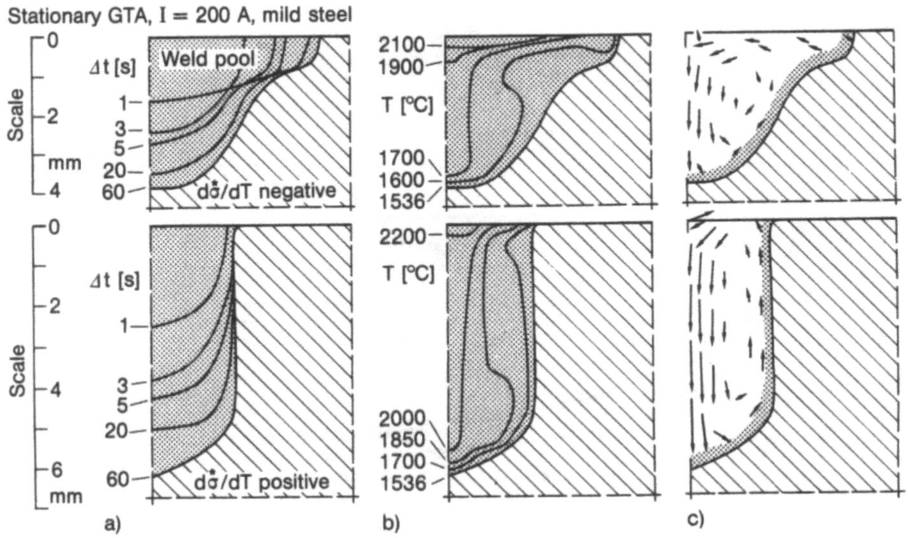

Fig. 67. Combined effect of electromagnetic, buoyancy, surface tension and aerodynamic forces on weld penetration (**a**), isothermal curves (**b**) and velocity field (**c**), negative (upper row) and positive (lower row) surface tension versus temperature gradient, stationary gas tungsten arc ($I = 200$ A, $q = 1.57$ J/s), mild steel; analytical results; after Matsunawa and Yokoya [83]

The analysis by Matsunawa and Yokoya [83] is based on the Navier-Stokes equation extended by electromagnetic and buoyancy forces. The pool surface is assumed as plane. Aerodynamic drag forces and shear forces by surface tension gradients are acting in the surface. What has been analysed is the axisymmetric molten pool produced by a stationary gas tungsten arc directed on the plane surface of mild steel using different arc lengths (surface deformation neglected). The underlying arc behaviour has been shown in Fig. 63. The combined effect of electromagnetic, buoyancy, surface tension and aerodynamic forces is presented in Fig. 67. Introducing a negative and a positive surface tension temperature gradient $d\sigma^*/dT$, the different pool shapes are clearly revealed. Another result of the investigation is that in short arc welding the electromagnetic forces are predominant, generating deep penetration, whereas in long arc welding the aerodynamic forces are predominant, generating a weld pool with shallow centre and deeper peripheral penetration.

A similar analysis has been performed by Zacharia et al. [82] on stainless steel of two melting charges with different sulphur content using stationary gas tungsten arc and laser beam heating. The influence of sulphur content on pool fluid flow, i.e. deeper penetration by higher sulphur content, is numerically confirmed.

2.3.1.4.5 Hydrostatic weld shape modelling

This type of hydrostatic weld pool and resulting weld shape modelling neglects the fluid flow but goes into detail in respect of other essential effects such as pool surface deformation in three dimensions proceeding from the quasi-stationary conditions around the linearly moving arc.

The numerical analysis by Sudnik [86] seems to be the most advanced (and application related) contribution in this field. These are the additional modelling features of the approach: Conservation of energy in terms of enthalpy, consideration of the two-phase zone around the weld pool, surface heat sources (Gaussian distribution), evaporation, convection and radiation on surface, force balance on top-side and root-side comprising surface tension, hydrostatic pressure and arc pressure (Gaussian distribution), mass balance of the liquid phase based on groove shape, groove dimensions, gap width and transverse shrinkage, welding speed and deposition rate of filler metal. The system of partly non-linear and temperature-dependent differential equations is coupled insofar as surface tension depends on temperature, temperature on surface shape and surface shape on surface tension. Iterative finite difference methods are applied for solving the set of equations, obviously within a moving reference frame securing quasi-stationary conditions.

As an application example gas tungsten arc welding of sheet metal has been considered. The realistic input of the heat source distribution is complicated by the reciprocal dependency of voltage $U = U(I)$, efficiency $\eta_h = \eta_h(U, I)$ and standard deviation $\sigma_h = \sigma_h(U, I)$ of the Gaussian distribution of heat sources. Simple formulae which avoid the above reciprocal dependency are derived based on regression analysis of experimental findings, $\eta_h = \eta_h(I, l_a, v)$ and $\sigma_h = \sigma_h(I, l_a)$ with arc length l_a

The calculated weld pool and finished weld geometry of a butt weld between 2.2 mm thick CrNi-steel plates produced by gas tungsten arc welding ($I = 430$ A, $v = 30$ mm/s, $l_a = 1$ mm) is shown in Fig. 68. The result is confirmed by the experimentally determined weld profile. Weld imperfections such as lack of penetration, excessive undercut and burn through have been predicted on the basis of the above model by way of "numerical experiments" (Fig. 69). Undercut occurs for current $I > 300$ A caused by increasing arc pressure, but the undercuts with a depth less than 10% of the plate thickness occurring for $v < 30$ mm/s seem to be acceptable.

Narrow gap tandem welding of 10 mm thick CMn-alloyed steel plates has also been numerically investigated by Sudnik [86]. The heading gas tungsten arc secures complete penetration, and the trailing gas metal arc produces the weld profile (Fig. 70). The lay-out of the welding process was performed on the basis of numerical analysis.

A numerical analysis similar in some aspects to the above has been conducted by Pardo and Weckmann [99]. The modelling is incomparably less sophisticated so that questions on weld imperfections cannot be answered. The basis is the heat conduction equation for quasi-stationary conditions (equation (14)). Convective heat diffusion is taken into account by artificially increasing the conductivity within the weld pool. The weld reinforcement is determined proceeding from the filler metal deposition rate assuming a parabolic reinforcement shape. The heat efficiency of the arc is assumed as a fixed known value.

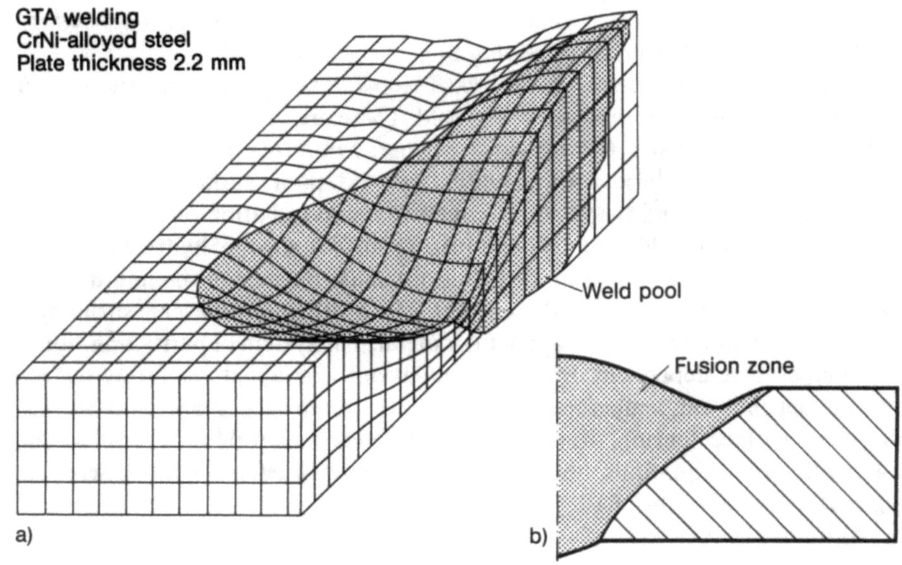

Fig. 68. Weld pool (**a**) and finished weld shape (**b**), analytical results based on hydrostatic modelling, gas tungsten arc welding (I = 430 A, v = 30 mm/s, l_a = 1 mm) of CrNi-alloyed steel (plate thickness 2.2 mm); after Sudnik [86]

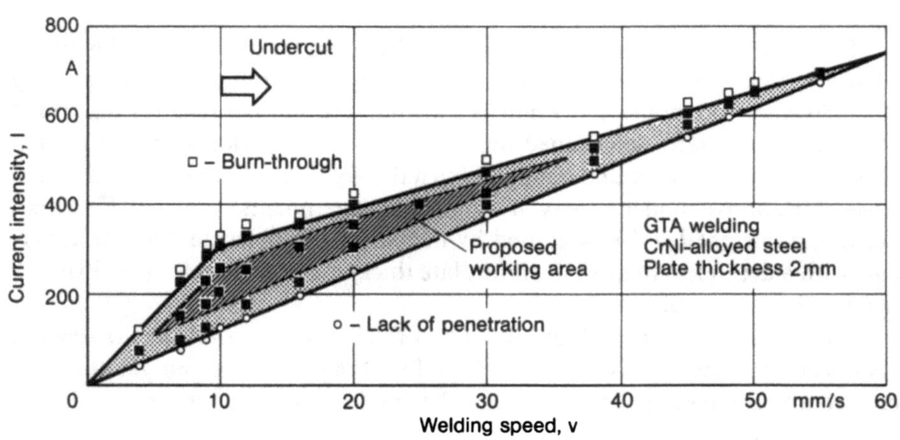

Fig. 69. Weld imperfections in gas tungsten arc welding dependent on welding speed and current intensity, CrNi-alloyed steel, plate thickness 2.0 mm, analytical results based on hydrostatic modelling; after Sudnik [86]

The temperature distribution and the weld reinforcement in plasma arc welding of the aluminium die casting alloy GD-AlSi8Cu3 has been numerically analysed by Na and Ruge [100]. The model takes surface stagnation pressure of the plasma jet (Gaussian distribution) into account. The pores are assumed as

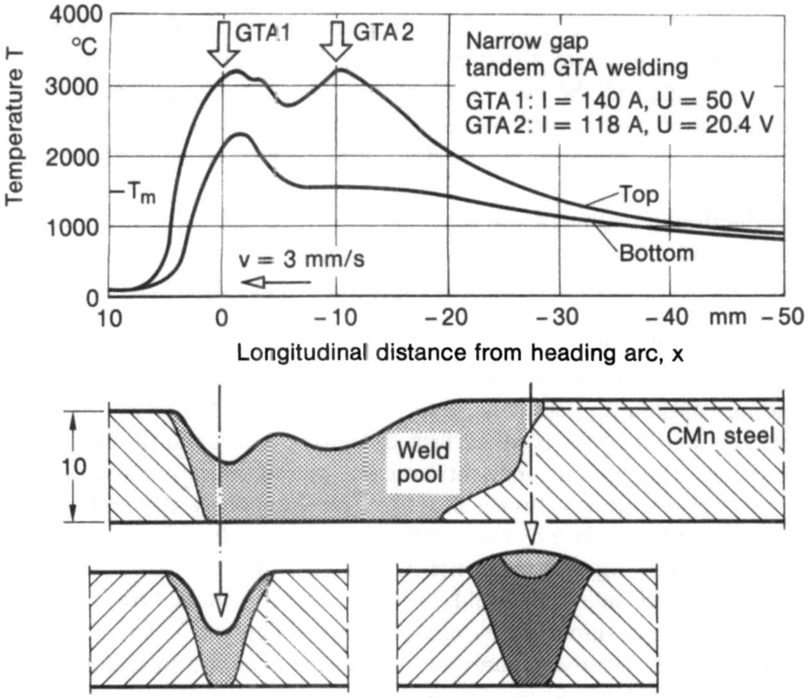

Fig. 70. Molten pool and finished weld shape in narrow gap tandem welding of 10 mm thick CMn-alloyed steel plates, analytical results based on hydrostatic modelling; after Sudnik [86]

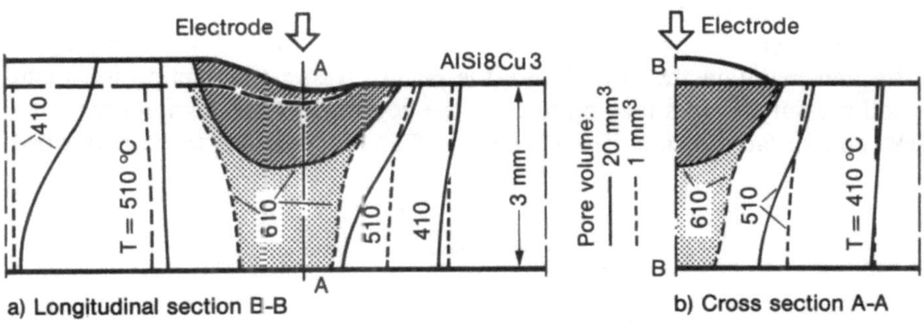

Fig. 71. Temperature field isotherms and weld reinforcement in plasma arc welding of aluminium die casting alloy GD-AlSi8Cu3 (liquidus temperature 610 °C, solidus temperature 510 °C) for low and high pore volumes (1 and 20 mm^3 per 23.2 mm weld length), longitudinal section (**a**) and cross-section (**b**), analytical results based on hydrostatic modelling; after Na and Ruge [100]

equally large and distributed at random. Heat conduction is completely suppressed in points of pore formation. The volume increases by pore formation so that the reinforcement is generated. The finite difference analysis is performed in three dimensions for the quasi-stationary state referenced to the moving heat source. As can be seen from Fig. 71, the temperature field is completely changed by pore formation. The predominantly one-dimensional (in the figure) temperature variation is changed into a two-dimensional field (in the figure) showing a fusion zone restricted to the near-surface area. This is accomplished by the heat conduction barrier generated by the pores and by increased heat dissipation by convection and radiation at the weld reinforcement generated also by the pores.

Heat conduction impediment and mass movement during void formation in encapsuled material solidifying to a denser state has been numerically investigated by Alexiades [85] in a more fundamental than application related sense.

2.3.1.4.6 Keyhole modelling

The weld pool behaviour is completely changed in the case of high power density welding (plasma arc, electron beam, laser beam). If the power density absorbed locally in the weld pool surface exceeds a particular level, which depends on the welding process, on continuous or pulsed operation and on the size of the heat spot, vapourization will occur. Simultaneously a depression is formed in the weld pool surface by the beam pressure, extending first to a crater and then to a thin hole right through the molten pool. Part of the beam may then appear on the back side of the weld. If the workpiece is moved transversely to the beam, the hole assumes an elongated shape ("keyhole"). The metal is melted at the front side, it moves round the sides of the hole and solidifies at the rear side thus forming the weld bead (Fig. 72).

The diameter of the keyhole is approximately 0.5 mm in electron beam welding, 1.0 mm in laser beam welding and markedly larger in plasma arc welding [72].

The stability of the keyhole may be studied considering the involved forces (see Lancaster [72] and Hashimoto and Matsuda [105]). Forces tending to form and maintain the keyhole are the beam pressure, vapour pressure and recoil pressure. Forces tending to close the keyhole are the gravitational pressure and surface tension. The balancing of these forces has to be performed

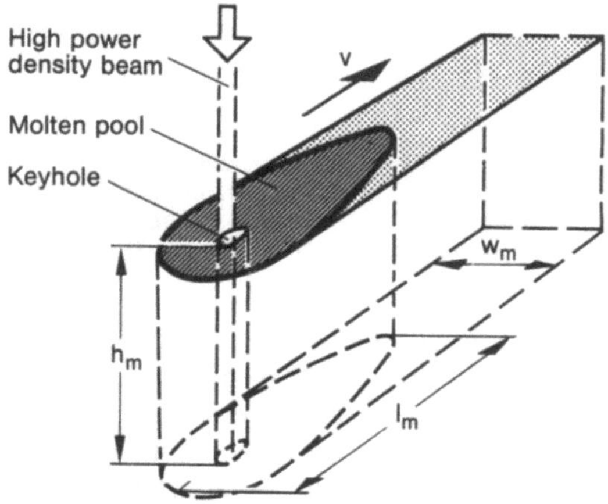

Fig. 72. Molten pool shape in high power density beam welding with keyhole formation (molten zone width w_m, molten zone length l_m, depth of weld penetration h_m, velocity of workpiece v); after Quigley in Ref. [72]

separately at the bottom of the closed keyhole and at the inside of the open or closed keyhole. A more refined numerical analysis seems still to be missing.

It is to be noted that the heat conduction considerations of Section 2.3.1.3.2 resulting in a limit thermal efficiency of 48.4% are valid also for keyhole welding.

2.3.2 Flame as a welding heat source

2.3.2.1 Physical-technical fundamentals

In the flame used for welding, acetylene C_2H_2 is combusted with oxygen O_2. The stated gas combination produces the highest flame temperatures. Acetylene and oxygen are combined in the torch nozzle. The ignitable gas mixture flows out through the nozzle at a high velocity and combusts in the flame under the inflow of oxygen from the surrounding air. The heat produced is transferred to the workpiece primarily by convection, and to a lesser extent also by radiation.

Three zones are distinguished in the acetylene-oxygen flame (Fig. 73). In the core zone, immediately adjoining the nozzle, the gas mixture is prepared for ignition and combustion. A thin layer of glowing carbon particles, which are produced as a result of pyrogenic decomposition, surround the core zone. The decomposition of the acetylene into carbon and hydrogen is accelerated in the intermediate zone adjoining to the outside (also called reduction zone), whereby the carbon combusts (incompletely) with the added oxygen producing carbon monoxide CO (primary flame):

$$C_2H_2 + O_2 = 2CO + H_2 + 0.021\ \mathrm{J/mm^3}. \tag{85}$$

Combustion is finally completed with atmospheric oxygen in the flame column adjoining to the outside (secondary flame):

$$2CO + H_2 + 1.5O_2 = 2CO_2 + H_2O + 0.027\ \mathrm{J/mm^3}. \tag{86}$$

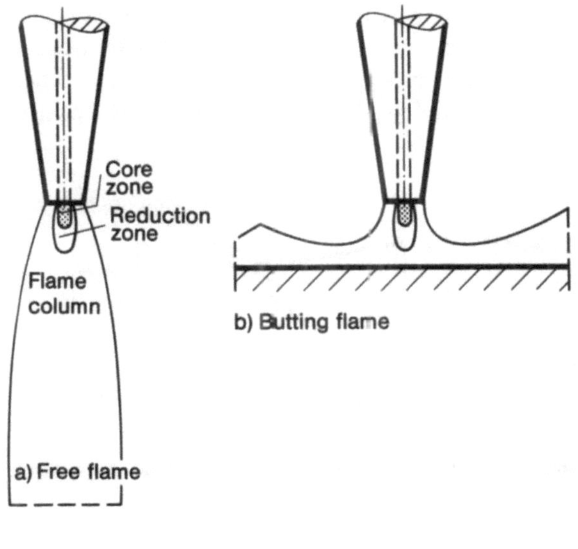

Fig. 73. Internal structure of acetylene-oxygen flame, free-burning (a) and butting on plate (b); after Rykalin [1]

Consequently, 44% of the heat corresponding to the (lower) calorific value of the acetylene is released in the intermediate zone and 56% in the flame column. The maximum temperature of the flame (approx. 3100 °C) is reached immediately ahead of the tip of the intermediate zone. The temperature in the flame column is lower (approx. 2500 °C in the centre of the free-burning flame). The (volumetric) mixing ratio of $O_2/C_2H_2 = 1.0$, required in accordance with equation (85), is increased in practice to approximately 1.2.

The size of the flame depends on the amount of mixture input. The mixture input, in turn, is determined by the nozzle diameter and the gas velocity at the nozzle outlet. The diameter determines the flame width, the outlet velocity the flame length.

The hottest part of the flame is used for heating the workpiece and the welding groove. The tip of the flame core should nearly touch the surface. The distance between nozzle outlet and surface of workpiece should be $1.2–1.5 \times l_c$, where l_c designates the length of the flame core.

2.3.2.2 Heat balance and heat flow density

The flame heat is transferred to the workpiece primarily by convection (approx. 85%) and only to a lesser extent by radiation (approx. 15%). The high convection portion is explained by the high outlet velocity of the gas.

As a consequence of the convection effect, the amount of heat transferred from the flame to the workpiece depends additionally on the degree of local heating of the workpiece at the location of the flame. The heat output transferred decreases as the workpiece heats up. With a stationary flame, the heat output q decreases sharply with the heating time t, whereas the heat content Q of the workpiece approaches a limit value Q_{li} (Fig. 74a). With a moving flame, there is only a slight reduction in the heat output q because the flame steadily touches new cold material whereas the heat content Q of the workpiece increases approximately linearly (Fig. 74b).

The effective heat output q of the flame increases with the size of the flame, in other words with the mixture input per unit of time and thus with nozzle diameter and gas velocity at the nozzle outlet (Fig. 75a). At the same time, though, the efficiency η_h of heat generation and heat transfer decreases heavily

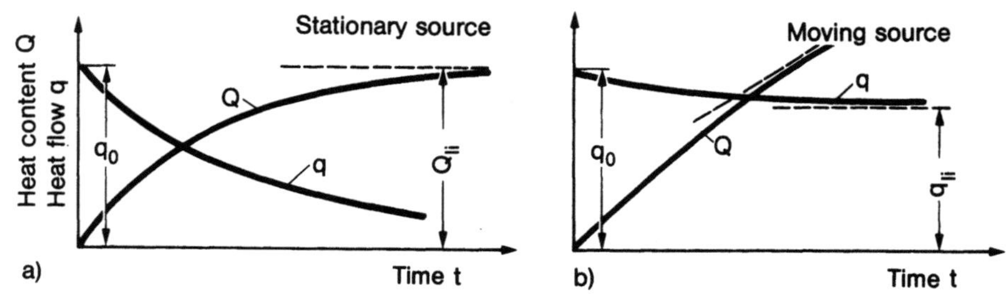

Fig. 74. Effective heat flow q of stationary source (**a**) and moving source (**b**) as well as heat content Q of workpiece, dependent on heating time t; after Rykalin [1]

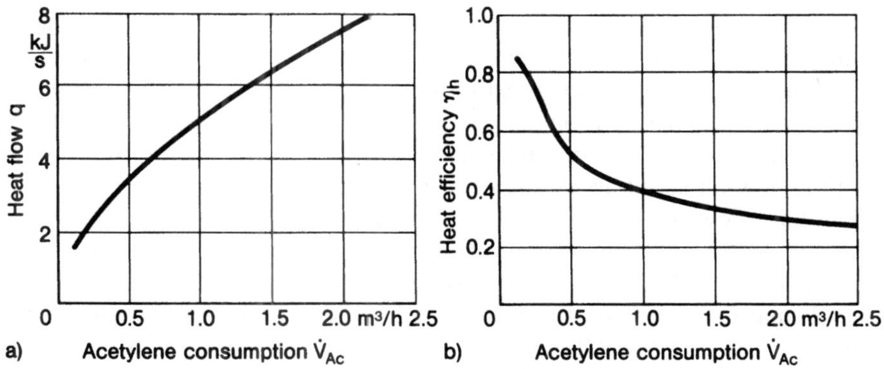

Fig. 75. Effective heat flow q (**a**) and heat efficiency η_h (**b**) of acetylene-oxygen flame moving on steel plate, dependent on acetylene consumption (controlled through nozzle size of torch); after Rykalin [1]

(Fig. 75b). The deterioration in efficiency is the result of a less favourable inflow of atmospheric oxygen as well as of reduced convection. Torch nozzles with a small outlet diameter are particularly efficient.

The heat efficiency η_h of the flame is defined as the ratio of effective heat output q at the workpiece to the (lower) calorific value q_c of the amount of acetylene consumed per unit of time:

$$\eta_h = \frac{q}{q_c}. \tag{87}$$

The effective heat output of the flame is influenced not only by the acetylene consumption but also by the following additional factors (according to Ref. [1]):

- Mixing ratio O_2/C_2H_2: maximum q where $O_2/C_2H_2 = 2.0$–2.4 compared to the usual setting of 1.15–1.2 (because of improved combustion).
- Outlet velocity of gas: high q as a result of high outlet velocity or small nozzle diameter (because of improved combustion and more intensive convection).
- Distance between nozzle and workpiece: decrease in q with enlarged distance (because of the temperature drop in the flame).
- Angle of inclination of the torch: maximum q at an angle of inclination of 60° (because of intensified convection), with moving flame, the torch inclined over the finished weld, so that the flame is directed to the colder part of the surface.
- Speed of movement of flame: high q as a result of high velocity (because the flame touches the colder part of the surface).
- Wall thickness of workpiece: high q as a result of high wall thickness (because of more rapid heat diffusion).
- Thermal conductivity or thermal diffusivity of the material: only slight rise of q with higher conductivity or diffusivity (experimental finding).

The (effective) heat flow or heat source density q^* is very unevenly distributed over a circular area (torch in perpendicular position) of the surface of the workpiece (heat spot), maximum value in the centre and sharp drop to the edges, due in particular to the rapid decrease in flow velocity and convection over distance from centre. Mathematically, the heat flow density is described as a Gaussian normal distribution (equations (16) to (19)). The heat flow density is greater if the flame temperature is higher, if the heat spot temperature is lower and if the flow-off speed above the heat spot is increased.

The heat flow density distribution can be determined experimentally by evaluating the temperature distribution in a thin steel plate over which the flame is moved rapidly. What is evaluated is the temperature in the cross-section perpendicular to the moving direction of the flame at the momentary position of the flame centre (Fig. 76). Allowance should be made for the slight deviation of the temperatures on the back surface from those on the front surface, the flame side, when evaluating the temperatures measured on the back surface of the plate. Measurement results are recorded in Ref. [1, 70, 71].

The heating covers a significantly larger area and is less intensive at the same effective heat output q, when the workpiece is heated by a welding flame compared to a welding arc. This has already been presented in Fig. 16 in which the flame reaches only approximately one eighth of the maximum value of the heat flow density with a heat spot diameter which is approximately three times as large. As one of the prime concerns in welding is to achieve the maximum possible heat concentration, this indicates that gas welding is inferior to arc welding. Hence, for modelling the heat sources, it follows that the usual point or line source should sooner be replaced by an appropriate normal distribution source in flame welding than in arc welding.

The heating of a larger area by the flame (which is additionally controllable) is, on the other hand, a decisive advantage for non-welding applications, for

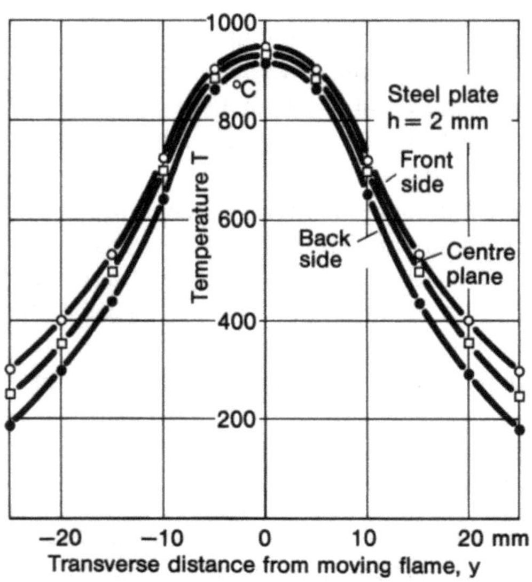

Fig. 76. Temperature profiles in thin steel plate with rapidly moving source, cross-section in centre of flame; after Rykalin [1]

instance for surface hardening, soldering, flame straightening, flame spraying and flame cleaning, as well as for pre- and post-heating. The nozzle with circular outlet may be replaced by a nozzle with circular or straight slots. Several nozzles may be arranged in line or in block. The arrangement may be plane or curved. Instead of acetylene, methane, propane or hydrogen may be used [70, 71]. Reference [1] should be consulted for modelling these interesting non-welding applications.

2.3.3 Resistance heating of weld spots

The heat generation Q in the case of resistance spot welding is given in accordance with equation (4) as a result of amperage I, voltage U (between the electrodes), welding time Δt and heat efficiency η_h. The heat losses to the surrounding air are negligibly small, $\eta_h \approx 1.0$. A small amount of heat is, of course, diffused into the metal plates and into the electrodes, thus being lost for the welding process. In accordance with Ohm's law, it follows from equation (4) for direct current (alternating current cycles should be converted to effective values of amperage and voltage, taking into account the phase relation) that:

$$Q = \eta_h I^2 R t_c. \tag{88}$$

The resistance R is composed of the Ohmic material resistance of the metal plates and electrode ends as well as the contact resistance between the metal plates and between metal plates and electrodes. The contact resistance is large at the start of the welding process, although it rapidly reduces with the temperature rise after the current is switched on. The contact resistance is dependent on the surface condition (films, oxides, roughness) and on the electrode force. Increased electrode force reduces the contact resistance so that amperage or current flow time must be increased in order to generate the same heat quantity Q.

A shorter current flow time may be selected with increased amperage or a longer with reduced amperage. Short-time welding has the advantage of a high heat concentration with correspondingly small losses by heat conduction and, consequently, long electrode life. A disadvantage is the increased hardening tendency in the heat-affected zone as a result of higher cooling rates.

In technological terms, the welding process proceeds in the following manner. The plane or curved electrodes press the metal plates locally one onto the other. The (alternating) current is switched on and maintained for a number of cycles. Heat is generated initially primarily by contact resistance; with a flat electrode more strongly at the outer edge of the electrode face, less strongly in the centre of the plate contact zone, the reverse being true for a curved electrode. Whereas the heat at the electrodes is rapidly carried off, it is retained in the contact zone of the plates and results in fusing the metal plates in this area. The contact resistance collapses as a consequence, and heat generation is now controlled by the electric current density, which is particularly high at the outer edge of the contact zone (i.e. theoretically infinitely high). The weld nugget is thus formed, proceeding from the edge zone, and only then covers the centre zone. The typical ovaloid shape of the weld nugget is finally determined by the enveloping curve of the melting temperature isotherms (resulting from the liquidus and solidus line). In contrast to the open weld bath in the case of arc or flame welding, the fusion zone in the case of resistance spot welding is enclosed all-around elastically and produced under transverse pressure. The heat generation is finally terminated by switching off the current, with the electrode pressure being maintained for a short cooling period (holding time) to avoid material separation as a result of cooling contraction. The undesired spraying during resistance spot welding is caused by excessive contact resistance in the electrode face and the surface melting caused by this. If the electrode pressure is too high, the electrode is pressed into the surface of the plate, with the consequence that the weld pool is extruded between the metal plates, which separate slightly and form a gap.

The result of an older combined current and heat conduction analysis [50] (finite difference method based on an electrical analogy) relating to the heat source and temperature distribution in the metal plate and electrode restricted to a symmetry-quarter, is presented in Fig. 77. The heat concentration at the edge of the contact zones and the onset of the ovaloid fusion zone ($T_m = 660$ °C) are recognizable. Further analysis models and results are referenced in Sections 2.2.7 and 3.2.3.

An improvement in the quality of the spot-welded joint is possible by pre- and post-treatment of the weld spot by means of a current and electrode force programme (Fig. 78). Such pre- and post-treatments are used for materials

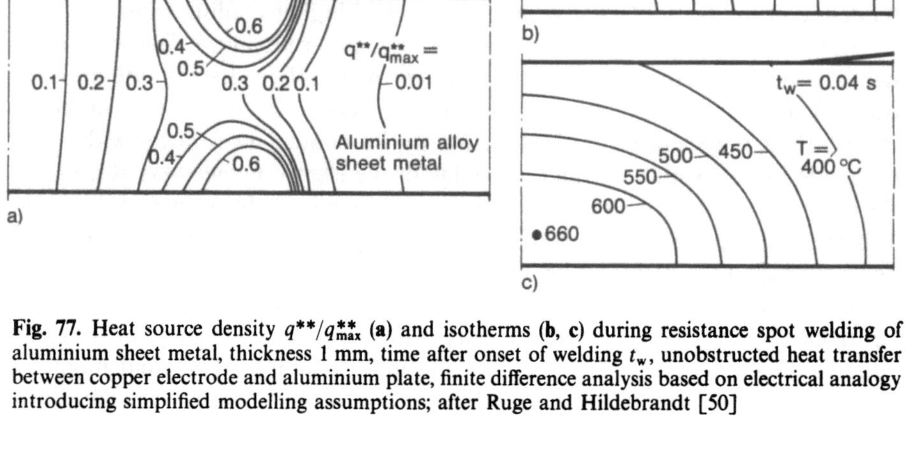

Fig. 77. Heat source density q^{**}/q^{**}_{max} (**a**) and isotherms (**b, c**) during resistance spot welding of aluminium sheet metal, thickness 1 mm, time after onset of welding t_w, unobstructed heat transfer between copper electrode and aluminium plate, finite difference analysis based on electrical analogy introducing simplified modelling assumptions; after Ruge and Hildebrandt [50]

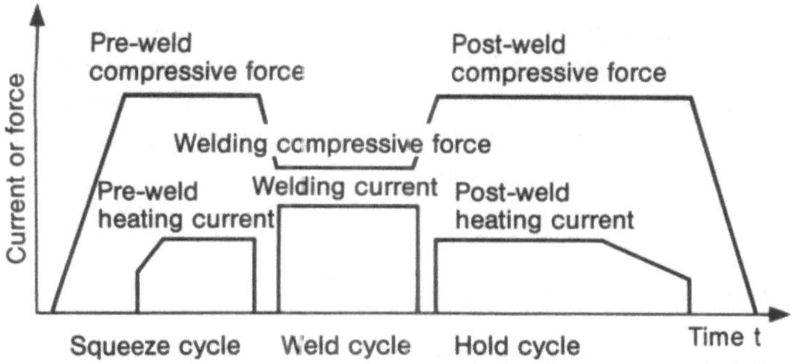

Fig. 78. Programme of current and electrode force for spot welding with pre- and post-treatment (schematically); after Pfeifer [76]

which are less suitable for welding because of their hardening tendency or for thick plate welding.

The electric and thermal fields during resistance projection welding are particularly unclear because the contact area depends in a complicated manner on the thermoplastic softening process.

2.3.4 Heat generation in friction welding

Heat generation in friction welding has been analysed by Potente et al. [68, 69] in respect of partly crystalline thermoplasts. Both the rotational friction welding process of polyethylene pipes and the vibration friction welding process of non-rotary polyethylene components are considered. Friction is produced by rotary or vibratory movement of the jointing faces against one another while a normal pressure force is applied. The friction heat causes melting of the contacting surfaces followed by jointing and cooling. The temperature cycle consists of four phases (Fig. 79). Instationary heating by solid surface friction occurs in phase I and instationary heating by molten layer shearing in phase II.

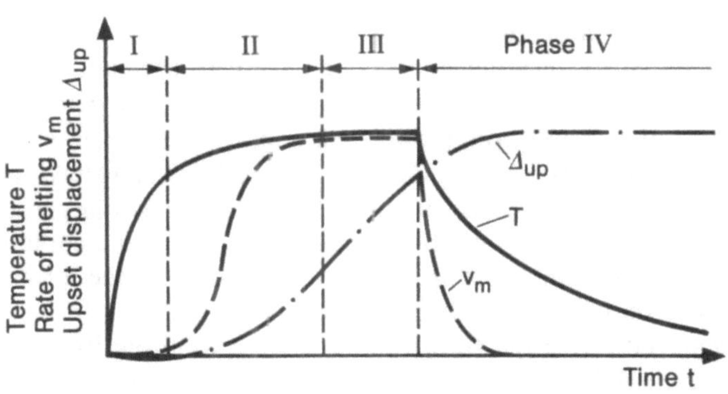

Fig. 79. Heating and cooling phases, temperature, rate of melting and upset displacement versus time in friction welding (schematic plot); after Potente and Kaiser [69]

In phase III, the temperature profile and the molten layer thickness remains stationary, and in phase IV cooling takes place. The upset displacement increases steadily in phases II and III and remains constant thereafter, while the normal pressure is taken away in phase IV. Concerning heat generation in friction welding of metals, see Ref. [448–450].

2.4 Local heat effect on the base metal

2.4.1 Microstructural transformation in the heat-affected zone

2.4.1.1 Thermal cycle and microstructure

In fusion welding, the base metal is fused on the (groove) surface and, together with the filler metal, constitutes the fusion zone of the weld seam. The base metal adjoining the fusion zone is subjected to welding-typical, short-time temperature cycles with a steep rise over time up to almost melting temperature followed by a flatter drop. Peak temperature and steepness decrease with distance from the weld.

As a consequence, different microstructural states are produced in layers in the base metal adjoining the melted and partially melted zones, i.e. in the heat-affected zone (Figs. 80 and 81). This zone is characterized by peak temperatures between melting temperature and A_{c1} temperature (approx. 750 °C) in the case of steels. Depending on the alloy, a wide variety of thermally activated changes in microstructure and properties may occur. Included are normal grain growth, recrystallization, solution and precipitation phenomena, as well as microstructural changes associated with allotropic transformations.

The most critical part of these zones is the partially melted and the grain coarsened layer. Strength and ductility may be reduced to the extent that cracks (e.g. hot or cold cracks, hardening cracks, corrosion cracks) may initiate during or after welding. Formation of cracks means that the welding suitability of the material or the weldability of the structure is lost. Attempt is therefore made to avoid the unfavourable microstructural states in the fusion and heat-affected zone by material production measures (composition, melting, heat treatment) and welding process measures (heating and cooling control, pre- and post-heating). Limitation of grain size, hardness and cooling rate are common engineering measures in this connection. However, a cooling rate which is too low may also be damaging. Particularly favourable mechanical behaviour results from a high portion of bainitic structure. Too high a cooling rate promotes the formation of martensite and should be avoided. Too low a cooling rate increases the austenitizing time, and in consequence promotes the formation of coarse grain and should likewise be avoided.

The heat treatment and microstructural transformation resulting from the welding temperature cycle differs significantly from common heat treatment of the material in respect of peak temperature and high temperature austenitizing time (Fig. 82). It is therefore necessary to record and present it separately. The

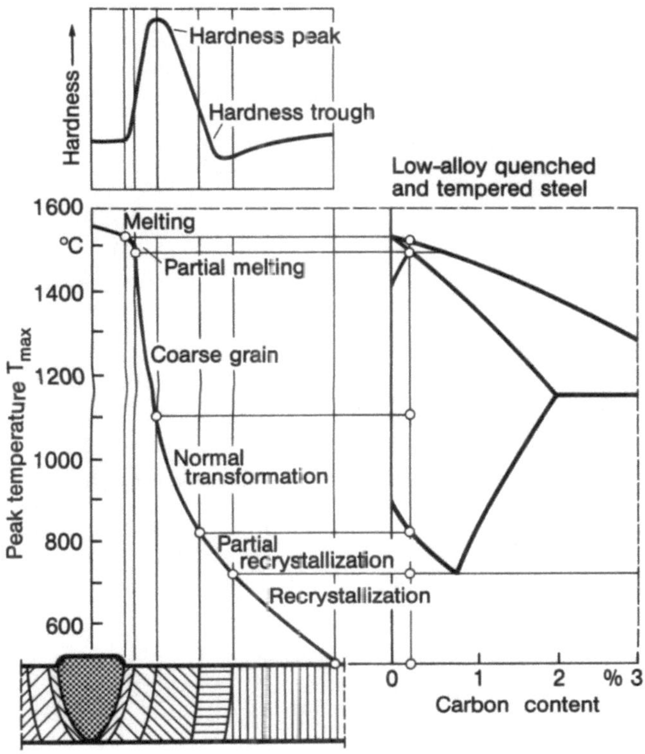

Fig. 80. Microstructural zones and hardness profile in heat-affected zone after welding of low-alloy quenched and tempered steel, peak temperature profile and iron-carbon diagram; after Ruge [119]

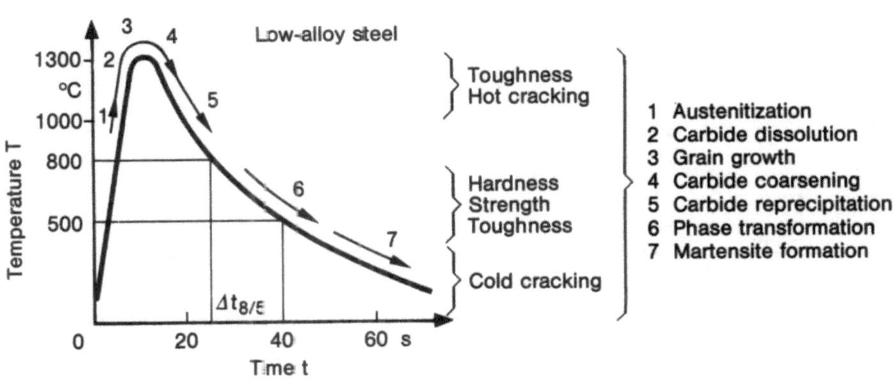

Fig. 81. Microstructural change in heat-affected zone of low alloy steel and its effect on strength and toughness, overview restricted to the most important processes; after Buchmayr and Cerjak [125]

related microstructural transformation diagrams are explained below, taking the example of mild steels. They disclose those temperature field parameters which primarily determine transformation behaviour and contain figures relating to the mechanical effects of transformation. Figures and approximation

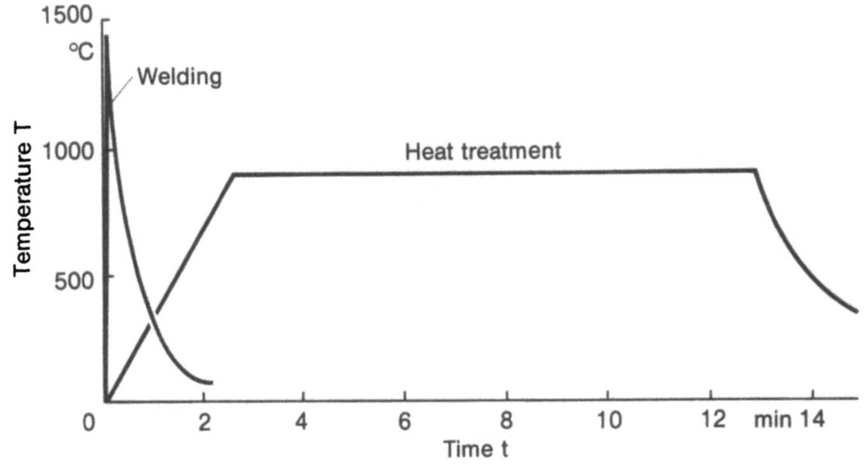

Fig. 82. Austenitizing conditions during welding compared to those during common heat treatment

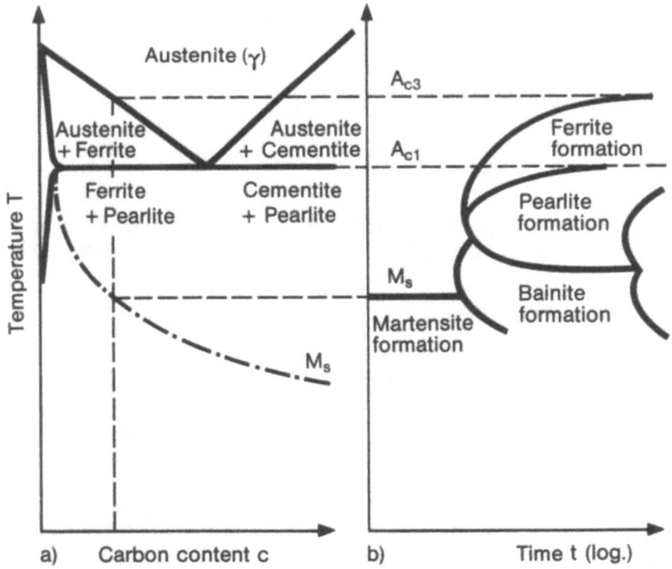

Fig. 83. Correlation between iron-carbon diagram (**a**) and TTT diagram (**b**) for isothermal test conditions (schematically); after Rose [106]

formulae for these temperature field parameters follow in Sections 2.4.3 and 2.4.4.

Depending on temperature and carbon content, mild steel exists in different microstructural states (or phases), which are presented in the well-known iron-carbon diagram (Fe-C diagram), (Fig. 83a). The diagram indicates states of equilibrium, which are approximately achieved after a prolonged annealing period (Fig. 83b). If the annealing period is shorter and cooling down more rapid, the transformation points shift and the microstructural states tend to be undercooled. The transformations proceed differently and, in some cases, result in other transformation products. An overview on the possible decomposition processes from the supercooled (i.e. non-equilibrium) states of the austenite are indicated by time-temperature transformation diagrams (TTT diagrams) in which the microstructural states are plotted for different cooling profiles in coordinates of temperature and cooling time [106–111; 115–117]. The austenitization peak temperature versus cooling time diagrams (PTCT diagrams) take into account the fact that the microstructural states depend additionally on peak temperature and on austenitizing time [112]. Finally, the principal mechanical characteristic values for different cooling profiles over time are also determined [110, 113, 114].

2.4.1.2 Time-temperature transformation diagrams

Depending on the temperature profile of the heat treatment, a distinction is made between isothermal, continuous and welding time-temperature transformation diagrams. The three types of diagrams thus contain the result of different ways of conducting the test and are, therefore, not easily transformable into one another.

In the case of the isothermal TTT diagram, the steel specimens are cooled from approximately 50 °C above A_{c3} (hardening temperature, approx. 900 °C) as rapidly as possible to different temperatures below A_{c1} (approx. 730 °C). This temperature is then held constant until microstructural transformation is completed. The connection of the dilatometrically measured transformation points in the temperature-time diagram results in transformation curves which approximate asymptotically the equilibrium temperatures of the iron-carbon diagram (Fig. 83b). The curve bulging furthest to the left designates the temperature of lowest austenite stability Such TTT diagrams, originally developed for the heat treatment of steels, are preferred in fundamental investigations of materials science because of the precise definition of the final states. They are not suitable, though, for welding engineering purposes.

In the case of the continuous cooling time-temperature transformation diagram shown in Fig. 84, the steel specimens are cooled starting from the hardening temperature (approx. 900 °C) at different rates controlled by convection (cooling by gas flow). The cooling temperatures over the logarithmically divided time axis display an initially flat, later on steep drop and finally approximate asymptotically the ambient temperature. The common starting temperature at $t = 0$ (infinitely far away on the left of the diagram), i.e. the hardening temperature, is not visible. In this case, also, the transformation curves separate the areas of austenite (A), ferrite (F), pearlite (P), bainite (B) and martensite (M). Transformation of the supercooled austenite starts at the intersection point of the cooling curve with the first transformation curve. The respective transformation is completed at the intersection point of the cooling

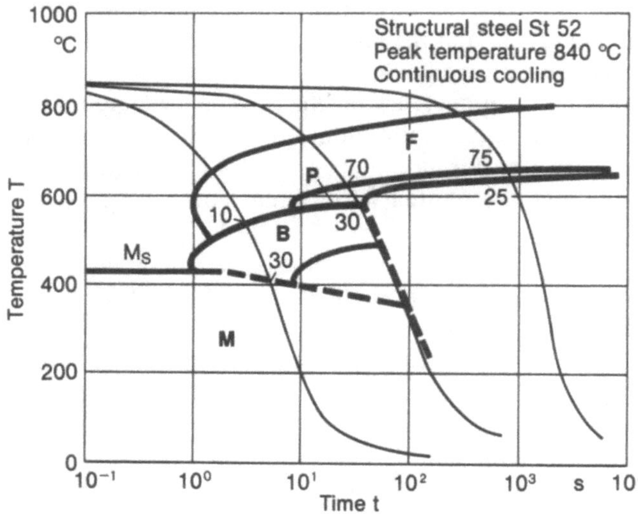

Fig. 84. TTT diagram of structural steel St52 (19Mn5) for continuous cooling, austenitizing temperature 850 °C; after Rose [106]

curve with the next transformation curve (the percentage of microstructural state portions is noted at the intersection point), and the new type of transformation starts. Where the cooling curve drops below the martensite line (M_s), the martensite portion results as the difference to 100% of the already transformed microstructural portion, provided residual austenite does not occur. The cooling curve which is tangent to the transformation curves at the furthest left, at the point of lowest austenite stability (450–600 °C), is designated as critical because it limits the range of still permissible hardening. The hardness attained [HV] is often noted at the ends of the cooling curves.

The transformation into ferrite and pearlite proceeds as diffusion-controlled. Martensite is formed as a result of an abrupt lattice folding process. Bainitic microstructure comprises both processes. The transformation processes are predictable on the basis of the theory of microstructural reaction-kinetics, see Section 2.4.2.

The continuous cooling TTT diagram covers the common heat treatment methods [109]. In respect of welding, though, it does not permit any sufficiently accurate statement regarding the particularly critical part of the heat-affected zone which is rapidly overheated within a short austenitizing time with temperatures up to nearly melting temperature (approx. 1350 °C) (Fig. 82). The increased austenitizing temperature makes the steel less transformable, the transformation curves being shifted to the right. Martensite is already formed at a lower cooling rate; the austenite grain size increases. Rapid heating and extremely short austenitizing time cause additional deviations. Where higher accuracy requirements prevail, the needs of welding engineering are satisfied only by the welding TTT diagrams.

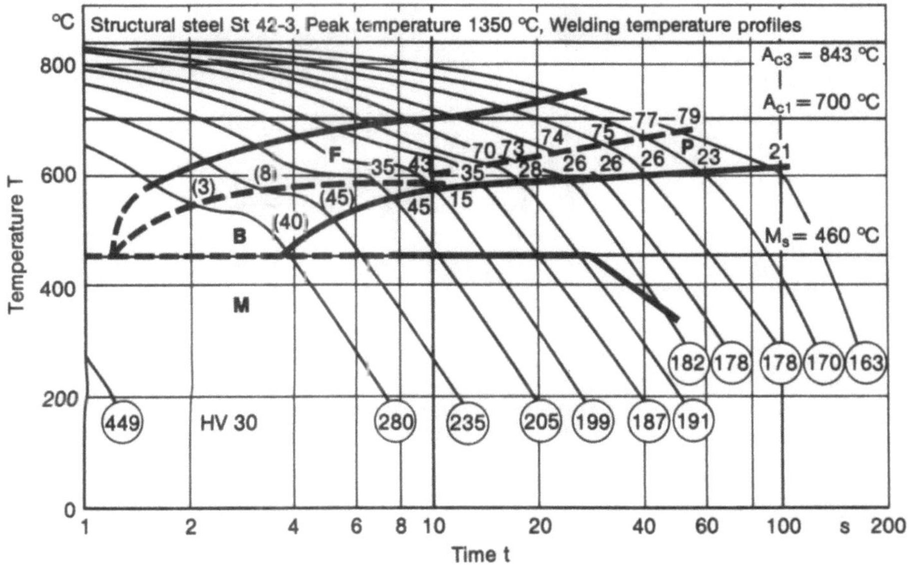

Fig. 85. Welding TTT diagram of steel St42-3 (0.18% C, 0.21% Si, 0.49% Mn), austenitizing temperature 1350 °C; after Seyffarth [116]

In the case of the welding TTT diagram shown in Fig. 85, the steel specimens are subjected to welding temperature profiles with a peak temperature of 1350 °C and different cooling rates in the transformation range between 850 °C (or 800 °C) and 500 °C for determining the transformation behaviour. The point at which the cooling curve passes through 850 °C is selected as the zero point of the time axis. The small swings of the cooling curves in the transformation range can be explained by the transformation heat which is released in this range. Approximation formulae are stated in Ref. [116] for determining the TTT diagram from the chemical composition of the steel. Welding TTT diagrams of steels suitable for welding and of filler metals are compiled in Ref. [115, 116]. A continuous TTT diagram with a peak temperature of 1300 °C, as used for the finite element analysis of a welding process, is shown in Fig. 86. The comparison of this diagram for steel St37 with that for steel St52 in Fig. 84 shows that St37 with a higher austenitizing temperature tends more to hardening than steel St52 with a lower austenitizing temperature. The transformation curves of steel St37, compared to those of steel St52, are shifted toward higher time values. The critical cooling times t_{cM} for 100% martensite formation and t_{cF} for 0% ferrite formation are presented in Fig. 86.

2.4.1.3 Evaluation of time-temperature transformation diagrams

As information from several TTT diagrams of the same steel with different austenitizing peak temperatures is required for assessing the various layers of the heat-affected zone, the most important information is summarized in a peak

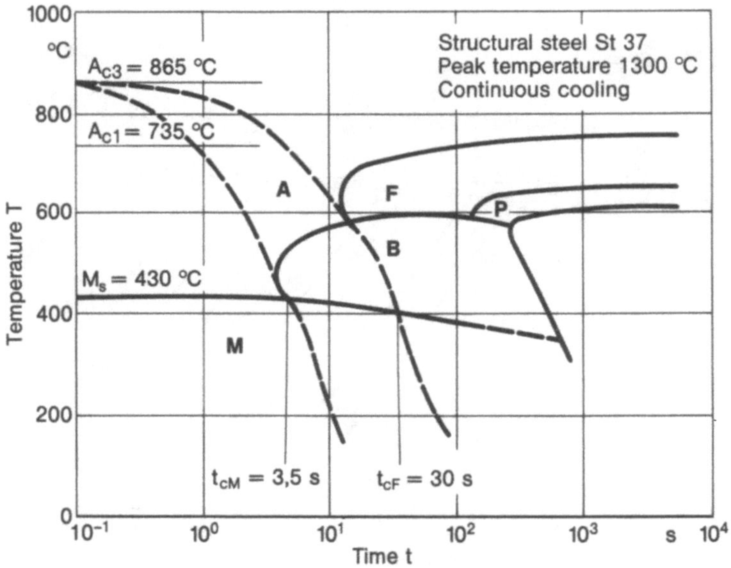

Fig. 86. TTT diagram of structural steel St37 for continuous cooling, austenitizing temperature 1300 °C; continuous cooling curves at end of 100% martensite range (t_{cM}) and at beginning of ferrite range (t_{cF}); after Argyris, Szimmat and Willam [248]

temperature versus cooling time diagram (PTCT diagram) (Fig. 87). The coordinates in this case are the austenitizing peak temperature T_{amax} and the cooling time $\Delta t_{8/5}$. The cooling time $\Delta t_{8/5}$ is the time required for cooling down from 800 °C (occasionally also from 850 °C) to 500 °C. It replaces the less descriptive indication of the cooling rate in the temperature range of transformation. The PTCT diagram shows the types of microstructures which occur with different combinations of T_{amax} and $\Delta t_{8/5}$. The hardness relating to the particular type of microstructure is stated. Assuming the approximation of equal $\Delta t_{8/5}$ values with different T_{amax} values in the heat-affected zone of the particular weld seam, the relevant microstructural states lie on a vertical line.

The TTT and PTCT diagram can be supplemented by a diagram in which the most important local strength characteristic values of the material such as hardness, tensile strength, yield limit, contraction in area at rupture, elongation at rupture and notch impact toughness are plotted over parameters which are decisive for the microstructural state, e.g. the cooling time. The correlation between microstructural state, hardness and further strength characteristic values is given on the basis of empirically developed approximation formulae supported by theoretical considerations [143–148]. Strength tests with specimens subjected to the critical welding temperature cycles, i.e. simulating the most critical microstructural states homogeneously (Gleeble test), may additionally be performed [113, 114, 131].

The austenitizing time Δt_a, particularly that above (A_{c3} + 100 °C), is decisive for the austenite grain growth. The latter impairs post-weld ductility and

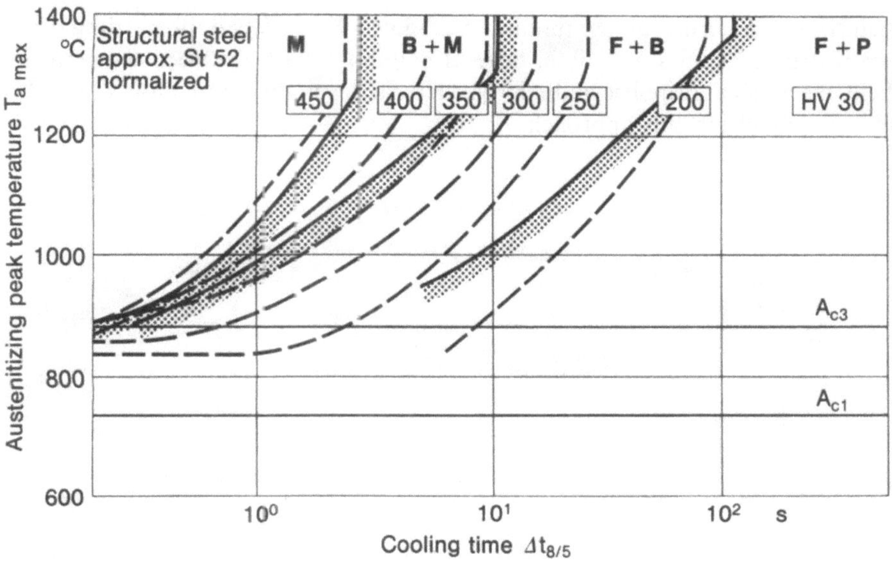

Fig. 87. PTCT diagram of normalized structural steel (approx. St52, 0.16% C, 1.50% Mn, 0.40% Si) with $\sigma_Y = 360$ N/mm^2 and $\sigma_U = 540$ N/mm^2; after Berkhout and Leut [112]

increases the susceptibility to cracking. Excessively long austenitizing times should therefore be avoided.

In respect of the evaluation of the temperature fields of weld seams, the following parameters are of interest concerning steels:

- the cooling rate at the temperature of lowest austenite stability (450–600 °C),
- the cooling time $\Delta t_{8/5}$ in the temperature range of austenite transformation (800–500 °C),
- the austenitizing peak temperature $T_{a\max}$,
- the austenitizing time Δt_a,
- the cooling rate at 400–150 °C in respect of hydrogen diffusion, which promotes cold cracking.

These quantities are dependent on the parameters of the welding method and component geometry. Approximation formulae relating to this dependence are stated in Sections 2.4.3 and 2.4.4.

2.4.2 Modelling of microstructural transformation

A more recent methods development [118, 120, 121, 128, 129, 130, 132, 142, 149–153], with emphasis on low-alloy steels cooling from the austenitized state describes the growth and transformation of microstructure on the basis of microstructural thermokinetics. The sets of poorly-known kinetic constants which appear in such a treatment are determined by fitting the equations to data from real or simulated welds at certain fixed points. Thus, the macrostructural behaviour in terms of continuum mechanics is modelled proceeding from microstructural laws and data. Several coupling effects between temperature field, mechanical response and microstructural development (see Fig. 4) are also considered within this approach.

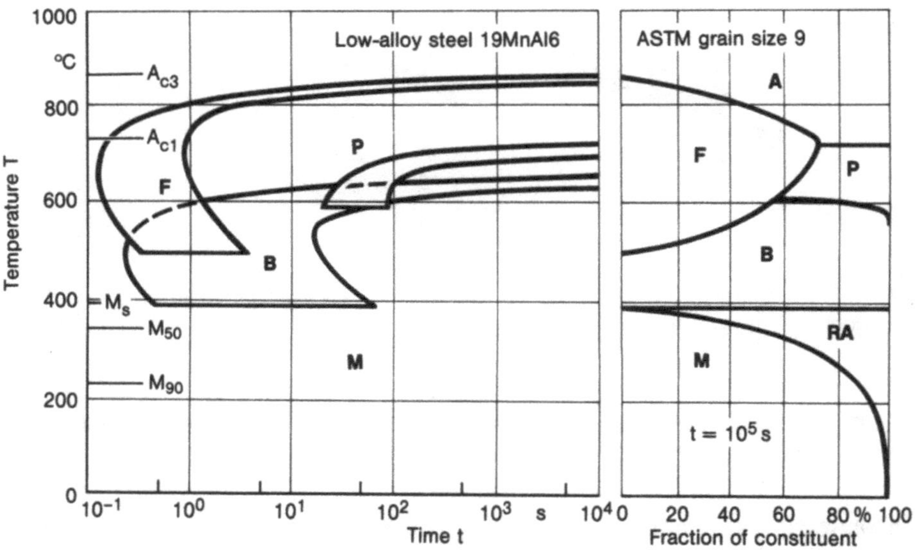

Fig. 88. Calculated TTT diagram of low-alloy steel 19MnAl6 supplemented by tractions of microstructural constituents after 10^5 s; after Buchmayr and Cerjak [125]

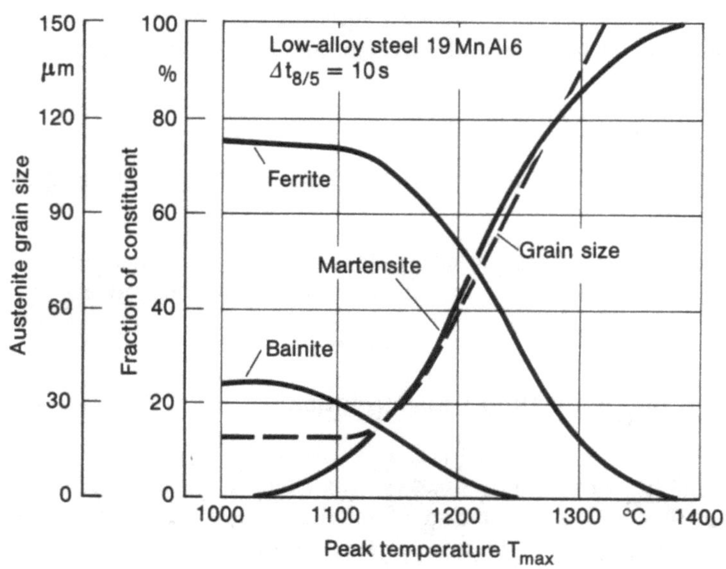

Fig. 89. Calculated microstructural constituents of low-alloy steel 19MnA16 as a function of peak temperature for cooling time $\Delta t_{8/5} = 10$ s; after Buchmayr and Cerjak [125]

The transformation behaviour is described starting from chemical composition, austenitizing conditions and initial austenite grain size, proceeding with the thermodynamical conditions of equilibrium of chemical potentials and the laws of transformation kinetics in respect of incubation time, nucleation rate, grain, precipitate and phase growth rate (mainly diffusion controlled with the exception of martensite and bainite formation) and ending with the integration over time of differential equations to obtain both the iron-carbon phase diagram and

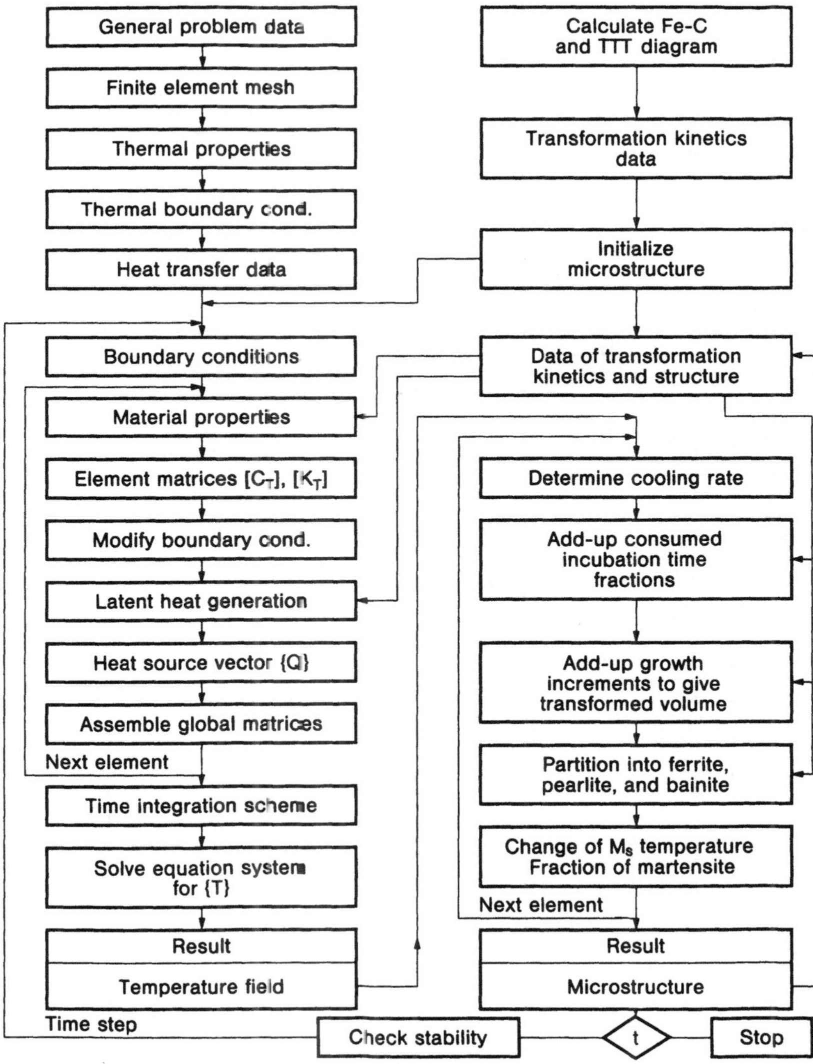

Fig. 90. Flow chart of finite element analysis of coupled transient temperature and microstructural transformation field during cooling from austenitization temperature; after Buchmayr and Kirkaldy [138]

the isothermal TTT diagram. As an example, the calculated isothermal TTT diagram for low-alloy steel 19 MnAl6, assuming ASTM grain size 9, is shown in Fig. 88. The curves in the diagram are 0.1% start and end curves of phase transformation.

The transfer from isothermal to continuous cooling or welding cycle conditions is performed using a time-temperature integration of the Avrami equation [124, 125, 127, 138]. Other authors [122, 123, 126] offer continuous and welding cycle TTT diagrams on the computer on a more empirical basis. Besides generating the TTT diagram, the dependency of grain size, martensite content

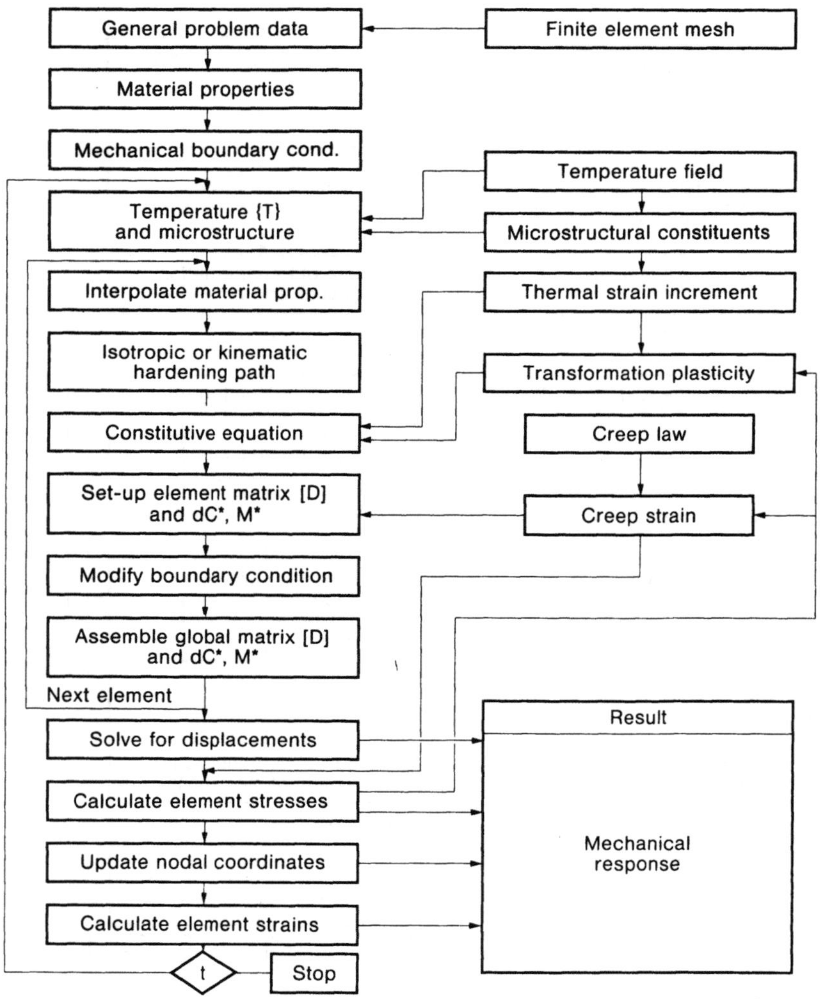

Fig. 91. Flow chart of finite element analysis of coupled thermal, mechanical and microstructural effects during heat treatment processes including welding; after Buchmayr and Kirkaldy [138]

and hardness on peak temperature can be correctly calculated (Fig. 89). The hardness is determined on the basis of approximative empirical relations. Calculations of the considered type should take the scatter range of the chemical composition into account if practical requirements are to be met.

It may be necessary for obtaining accurate results to take coupling effects between microstructural transformation and temperature field into account when performing thermodynamical analysis. The material characteristic values which control the temperature field may depend on the type and extent of microstructural transformation. On the other hand, microstructural transformation depends on the temperature field. A similar coupling effect holds true for the latent heat involved in microstructural transformation. A flow chart for the finite element analysis of temperature fields coupled with data from transformation processes suitable for the heat effects of welding is shown in Fig. 90.

It may also be necessary to take coupling effects between microstructural transformation and deformation field into account when performing mechanical analysis. Martensitic transformation plasticity should especially be modelled in such a way. The appertaining flow chart for finite element analysis is shown in Fig. 91.

The above modelling methods are available in the form of software packages with interactive operation partly on personal computers [122, 124, 126, 127]. These packages comprise capabilities for finding optimal welding parameters such as heat input per unit length of weld, preheating temperature and post-weld annealing temperature.

2.4.3 Cooling rate, cooling time and austenitizing time in single-pass welding

2.4.3.1 Cooling rate in solids and thin plates

The temperature cycles in single-pass welding as a function of the process and design variables are considered first proceeding from the temperature field equations for the rapidly moving high power source [1]. Two limit cases are investigated first in this connection:

- the point source on the semi-infinite solid according to equation (45) as a model for the surfacing weld on solid bodies,
- the line source in the infinite plate according to equation (48) with $b = 0$ as a model for the butt weld in thin metal sheets.

The question is posed, what the cooling rate dT/dt is for various momentary temperatures T. To simplify the equations, only points on the weld centre line are considered. This is justified because the cooling rate at a definite temperature in the temperature range under consideration of 850–500 °C decreases only slightly with increasing distance from the weld. The workpiece mean temperature T_0 is assumed to be constant (without or with preheating).

For the semi-infinite solid with a point source, this results, in accordance with Ref. [1] (ibid. equation (231)), in:

$$\frac{\mathrm{d}T}{\mathrm{d}t} = 2\pi\lambda\frac{(T - T_0)^2}{q_w}. \tag{89}$$

For the infinite plate with a line source, this results, in accordance with Ref. [1] (ibid. equation (232)), in:

$$\frac{\mathrm{d}T}{\mathrm{d}t} = 2\pi\lambda c\varrho\frac{(T - T_0)^3}{(q_w/h)^2}. \tag{90}$$

The cooling rate thus depends, with a given material on the type of heat propagation, on the preheating temperature T_0 and on the (effective) heat input per unit length of weld, q_w. It rises with $T - T_0$ and it drops with q_w (Figs. 152 and 154 in Ref. [1]). The effect is more pronounced with the infinite plate than with the semi-infinite solid.

The above formulae derived for the rapidly moving high power source are identical with corresponding equations for instantaneously deposited weld seams.

2.4.3.2 Cooling rate in thick plates

The point source on the plane layer (see Section 2.2.6) serves as model for the surfacing weld on a thick plate. The last passes of multi-pass welds and the fillet welds on plates are also included in this.

A simple-to-use approximation method, which approximates the (thick) plate with surface point source, is stated in Ref. [1] (ibid. equations (233) and (234)). It interpolates between the semi-infinite solid with point source and the infinite (thin) plate with line source on the basis of the dimensionless parameters ω and θ characterizing $\mathrm{d}T/\mathrm{d}t$ (dependent on T) and h.

$$\omega = \frac{\frac{\mathrm{d}T}{\mathrm{d}t}q_w}{2\pi\lambda(T - T_0)^2}, \tag{91}$$

$$\theta = \frac{\pi h^2 c\varrho(T - T_0)}{2q_w}. \tag{92}$$

The correlation between ω and $1/\theta$ is presented in Fig. 92 for points of the weld centre line on the semi-infinite solid, the infinite thick plate and the infinite thin plate as well as for points next to the weld centre line (the latter characterized by $y/h = 0$) for the thick plate. On the weld centre line, the solution for the semi-infinite solid is adequate for $1/\theta < 0.4$ according to Ref. [1] and the solution for the thin plate adequate for $1/\theta > 2.5$. The values of the thick plate apply in the range $0.4 < 1/\theta < 2.5$ for $y/h = 0$ and also beyond the stated $1/\theta$ range for $y/h \neq 0$. Obviously, though, the values for $1/\theta \lesssim 0.6$ of the semi-infinite solid and the values for $1/\theta \gtrsim 0.6$ of the thin plate also result in useful approxima-

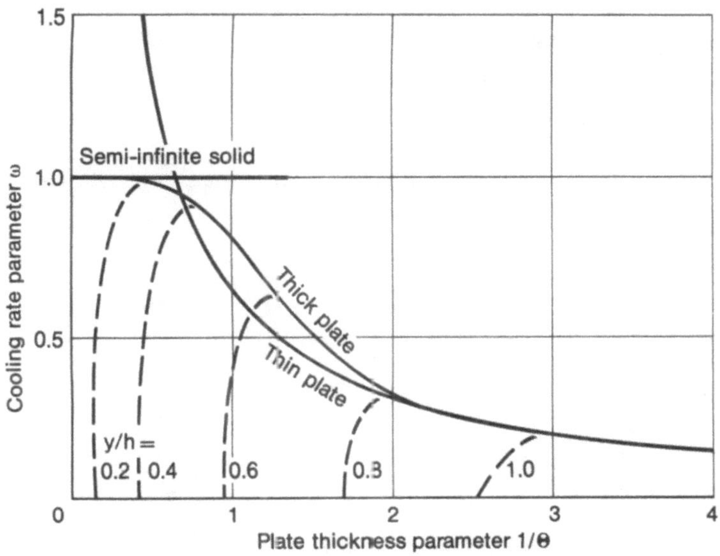

Fig. 92. Momentary cooling rate dependent on plate-thickness during deposition of surfacing welds; after Rykalin [1]

tion statements (for $y/h \approx 0$). For determining the momentary cooling rate, $1/\theta$ is calculated first in accordance with equation (92) the relevant value of ω is taken from Fig. 92, and equation (91) is then solved for dT/dt by putting in this value. The influence of plate thickness h is covered additionally to the influencing variables already mentioned, $T - T_0$ and q_w. The cooling rate (for $y/h = 0$) increases with h at a given q_w (Fig. 151 in Ref. [1]).

Similar derivations in respect of the influence of plate thickness on cooling rate have been performed in Ref. [40], including considerations on heat dissipation by convection and radiation. The cooling rate in the (point source) weld centre line on the top surface of the plate is once more expressed dependent on the momentary temperature. The cooling rate on the bottom surface underneath the heat source is additionally considered. The results given in Fig. 93 are more or less identical with those in Fig. 92 but the scale on the abscissa is reversed.

For any given (point) heat source and material, the centre line cooling rate is a maximum when the plate is very thick. This maximum is assigned the arbitrary value of 1.0 and the cooling rates which occur in plates of finite thickness are expressed as fractions of this value.

Starting from the left-hand side of Fig. 93, as plate thickness increases, heat flow downward into the plate begins to have an effect which is additive to the heat flow parallel to the plane of the plate. The resulting cooling rate is higher than that predicted for the thin plate with line source, equation (90). Starting from the right-hand side of Fig. 93, as plate thickness progressively decreases, there is a gradual inhibition of heat flow in the downward direction and cooling rates become lower that those predicted by equation (89).

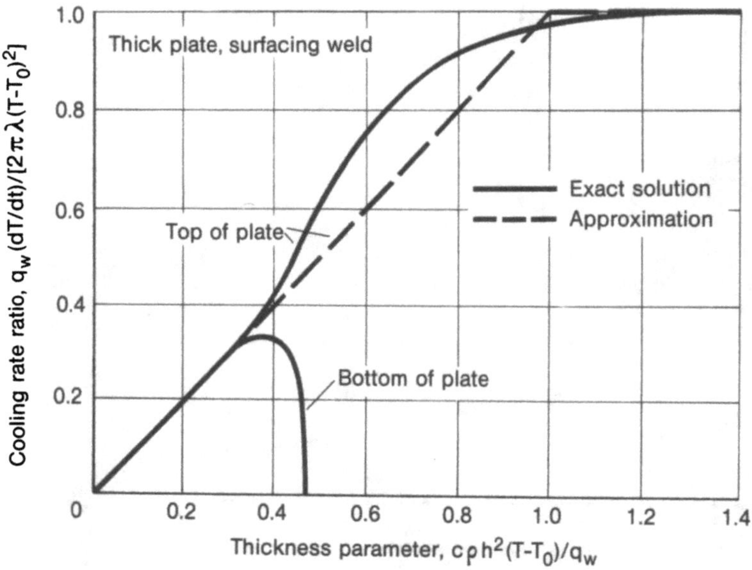

Fig. 93. Momentary cooling rate dependent on plate thickness during deposition of surfacing welds; after Jhaveri et al. [40]

Earlier work describing the use of equations (89) and (90) stated the general rule that whichever of the two gave the lower cooling rate was most nearly correct in the case of thick plates with surface point source. From the intersection point of the curves for semi-infinite solid with point source and infinite thin plate with line source, $(2/\pi) \times \theta = 1.0$, the "transition thickness" h_t is derived:

$$h_t = \sqrt{\frac{q_w}{c\varrho(T - T_0)}}. \tag{93}$$

It is concluded from Fig. 93 that the error from such a procedure might become as great as 20% in the middle region of the thickness parameter, $0.3 \leq (2/\pi) \times \theta \leq 1.0$.

The bottom surface of a very thin plate undergoes the same thermal cycles as the top surface. As plate thickness increases, the bottom surface temperatures are lower. The appertaining cooling rate initially increases with thickness, goes through a maximum and decreases to zero. The value on the abscissa where the cooling rate is zero gives the maximum temperature T_{bmax} on the bottom surface underneath the heat source [40].

$$T_{bmax} - T_0 = \frac{0.47\, q_w}{c\,\varrho h^2} \tag{94}$$

2.4.3.3 Cooling time in solids and plates

For practical application the cooling time $\Delta t_{8/5}$ between 800 and 500 °C governing the low temperature transformation behaviour of steels is better suited than the cooling rate $\mathrm{d}T/\mathrm{d}t$ at some temperature in the stated range. What results in accordance with Ref. [1] (ibid. equation (230b)) for the cooling time $\Delta t_{8/5}$ between 800 and 500 °C for the semi-infinite solid and for the infinite plate (with net heat input per unit length of weld, q_w) is:

$$\Delta t_{8/5} = \frac{q_w}{2\pi\lambda}\left(\frac{1}{500 - T_0} - \frac{1}{800 - T_0}\right), \tag{95}$$

$$\Delta t_{8/5} = \frac{q_w^2}{4\pi\lambda\varrho c}\frac{1}{h^2}\left[\left(\frac{1}{500 - T_0}\right)^2 - \left(\frac{1}{800 - T_0}\right)^2\right]. \tag{96}$$

The "transition thickness" h_t, obtained according to Ref. [161–165] by equating $\Delta t_{8/5}$ from equations (95) and (96) and solving for h with identical q_w, delimits the recommended application ranges of equations (95) and (96) derived for three- and two-dimensional heat diffusion. It indicates the reversal point for the ratio of the cooling times $\Delta t_{8/5}$ in the semi-infinite solid with point source and in the infinite plate with line source. For $h < h_t$, $\Delta t_{8/5}$ is smaller in the semi-infinite solid than in the infinite plate. For $h > h_t$, $\Delta t_{8/5}$ is larger in the semi-infinite solid than in the infinite plate.

Expressing the general rule mentioned above to always use the lower cooling rate from equations (89) and (90), in terms of cooling time, i.e. always to use the higher cooling time from equations (95) and (96), another relation for the transition thickness h_t is derived. Equating $t_{8/5}$ from equations (95) and (96) with identical q_w results in

$$h_t = \sqrt{\frac{q_w}{2c\varrho}\left(\frac{1}{500 - T_0} + \frac{1}{800 - T_0}\right)}. \tag{97}$$

This formula is identical with that of equation (93) if $500 - T_0$ and $800 - T_0$ are substituted by $T - T_0$. In respect of the $\gamma\alpha$-transformation, on which $t_{8/5}$ refers, the medium value $T = 650$ °C might be used for evaluating $\mathrm{d}T/\mathrm{d}t$ so that nearly identical values of h_t are determined from equations (93) and (97).

Slightly modified formulae for cooling time and transition thickness for submerged-arc welding of low-alloy high-tensile structural steels are given in Ref. [161–165] on the basis of experimental findings, stated below with gross heat input per unit length of weld, $\bar{q}_w$, and ignoring the weld geometry factor (Figs. 94 and 95):

$$\Delta t_{8/5} = (0.67 - 5\times 10^{-4}T_0)\bar{q}_w\left(\frac{1}{500 - T_0} - \frac{1}{800 - T_0}\right), \tag{98}$$

$$\Delta t_{8/5} = (0.043 - 4.3\times 10^{-5}T_0)\frac{\bar{q}_w^2}{h^2}\left[\left(\frac{1}{500 - T_0}\right)^2 - \left(\frac{1}{800 - T_0}\right)^2\right]. \tag{99}$$

It is now possible to resolve the critical remarks of Ref. [170] (included in the German edition of this book) concerning transition thickness. The transition thickness delimits the validity ranges of simplified formulae for cooling rate or

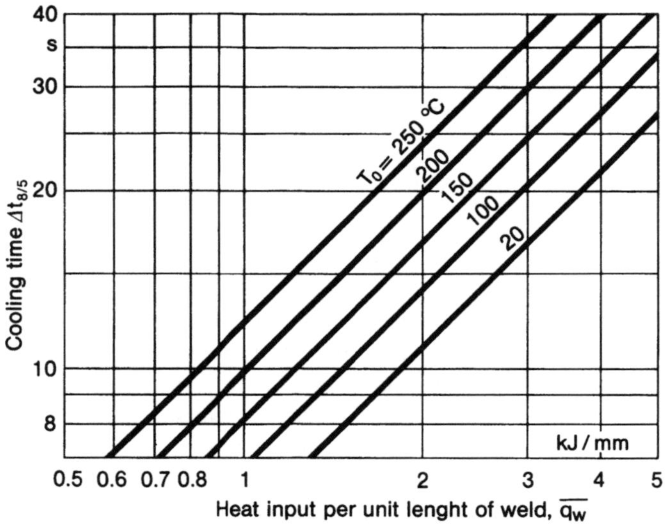

Fig. 94. Cooling time $\Delta t_{8/5}$ for three-dimensional heat diffusion dependent on gross heat input per unit length of weld $\bar{q}_w$ and preheating temperature T_0 for submerged arc welding of low-alloy high-tensile structural steels; after Uwer and Degenkolbe [161–165]

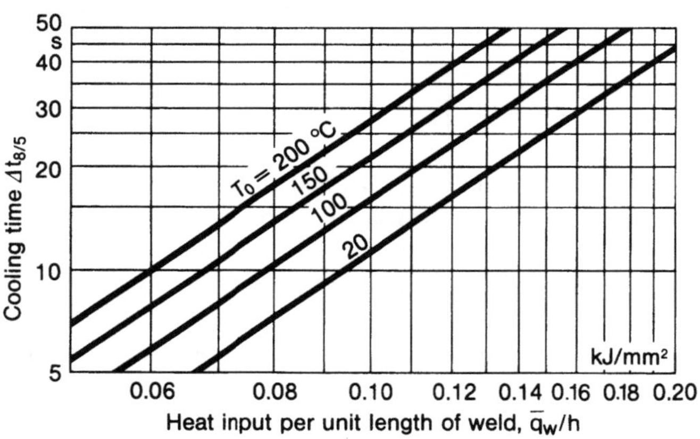

Fig. 95. Cooling time $\Delta t_{8/5}$ for two-dimensional heat diffusion dependent on gross heat input per unit length of weld $\bar{q}_w/h$ and preheating temperature T_0 for submerged arc welding of low-alloy high-tensile structural steels; after Uwer and Degenkolbe [161–165]

cooling time of plates with a one-sided moving point source, i.e primarily not of one- and two-dimensional heat diffusion. Obviously the transition thickness depends on heat input and temperature irrespective of the actual dimensionality of the heat flow.

On the other hand, the stated rule always to consider the lower cooling rate (or higher cooling time) on which the transition temperature concept is based, must be restricted to cases for which the surface point source model is appropriate from the heat flow point of view. If, contrary to this assumption, a deep penetrating weld needs a line source simulation, the resulting higher cooling rate (or lower cooling time) must be taken into account.

Further relations between cooling time, heat input, plate thickness and preheating temperature, which can be found in the literature, are now reviewed.

The monograms in Figs. 96 to 98 have been taken from Japanese investigations (Ref. [166–168]) relating to covered electrode welding, gas metal arc welding and submerged arc welding of different types of welds. The cooling time $\Delta t_{8/5}$ can be simply measured off, dependent on (gross) heat input per unit length of weld (gross), plate thickness and preheating temperature. Further diagrams for $\Delta t_{8/5}$ are compiled in Ref. [116]. The numerical evaluation and graphical presentation of welding temperature field equations (including spot welding) in dimensionless parameters is dealt with in Ref. [160] (not very descriptive). Dimensionless parameters, with which test results can be favourably interpolated and extrapolated, as a result of which the number of necessary tests is reduced, are obtained in Ref. [159] by reducing heat diffusion in welding to a most simple rod or cylinder model.

In the case of the different welds, joints and plate thicknesses which occur in practice, it is a problematical matter to respectively assign them to semi-infinite solid or infinite plate or also to test results with a specific surfacing, butt or fillet weld. To get round this problem, use is made of estimated geometry factors. For example, it was possible to standardize the measurement results relating to $\Delta t_{8/5}$

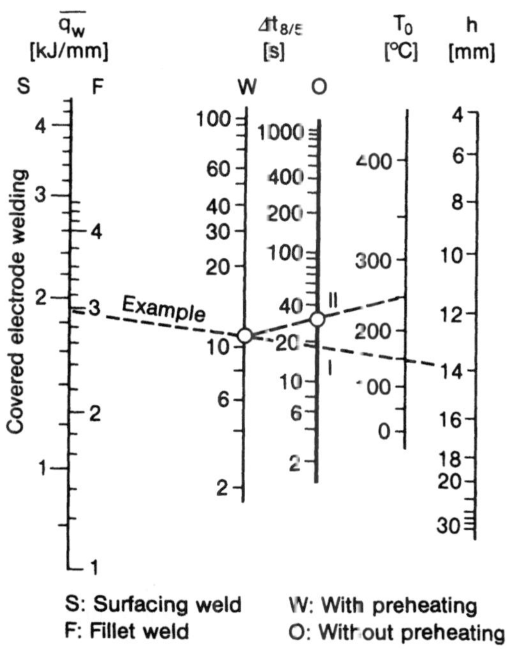

Fig. 96. Cooling time $\Delta t_{8/5}$ for covered electrode welding (surfacing or fillet weld) dependent on gross heat input per unit length of weld $\bar{q}_w$, plate thickness h and preheating temperature T_0; after Inagaki, Nakamura and Okada [166–168]

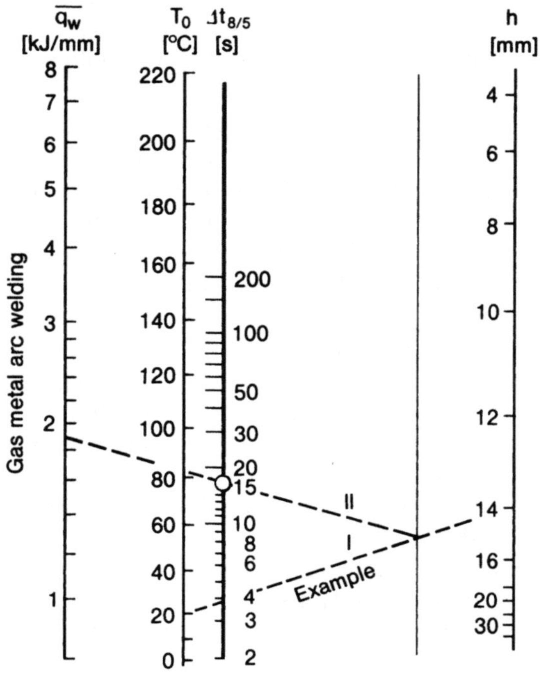

Fig. 97. Cooling time $\Delta t_{8/5}$ for gas metal arc welding dependent on gross heat input $\bar{q}_w$, plate thickness h and preheating temperature T_0; after Inagaki, Nakamura and Okada [166–168]

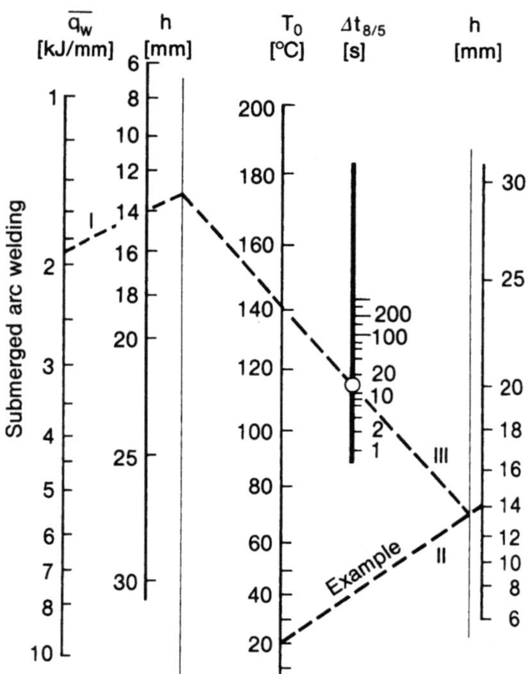

Fig. 98. Cooling time $\Delta t_{8/5}$ for submerged arc welding dependent on gross heat input $\bar{q}_w$, plate thickness h and preheating temperature T_0; after Inagaki, Nakamura and Okada [166–168]

with the geometry factor $1/\sqrt{n^* h}$ ahead of q_w [106], in which $n^* = 1$ for the root pass in the V groove, $n^* = 2$ for the surfacing weld and $n^* = 3$ for the fillet weld at the T joint (rising with the angular range of heat diffusion). Geometry factors relating to equations (98) and (99) are stated in Ref. [161–165]. The reduction of cooling time when welding fillet welds compared to welding butt welds is also taken into account by increasing the numerical plate thickness [116].

Geometry correction factors k^* ahead of q_w are also used in order to employ the semi-infinite solid solution in accordance with equations (89) and (95) or the plane layer solution according to equations (91) and (92) to the fillet weld or to the root pass of a V groove butt weld [1]. The factor k^* is the result from the heat flow estimation relative to the semi-infinite solid. From the one and a half way heat diffusion at the edge fillet weld and T joint fillet weld, it follows that $k^* = 2/3$ for equal plate thickness. From the two way heat diffusion at the cruciform joint fillet weld last welded, it follows that $k^* = 1/2$ for equal plate thickness. If the plate thicknesses in the joint are not equal, the k^* values above should be modified according to the heat diffusion possibilities. For the root pass in the V groove with groove-included angle α, $k^* = \pi/(\pi - \alpha)$ is introduced, i.e. $k^* = 3/2$ for the common value $\alpha = \pi/3$.

2.4.3.4 Austenitizing time in solids and plates

The austenitizing peak temperature T_{amax} can, in the case of a rapidly moving high power source, be approximately equated with $T_{\max}$ according to equation (46) for the semi-infinite solid and according to equation (49) for the infinite plate. In the fusion zone, $T_{\max}$ is not precisely limited by the melting temperature T_m, for overheating of the weld pool is possible.

A simple and clear presentation is found in Ref. [1] for the austenitizing time Δt_a, this being the local dwell time above a certain temperature T_a in the temperature range of the austenite. The limit cases of the point source on the semi-infinite solid and the line source in the infinite plate are considered. The

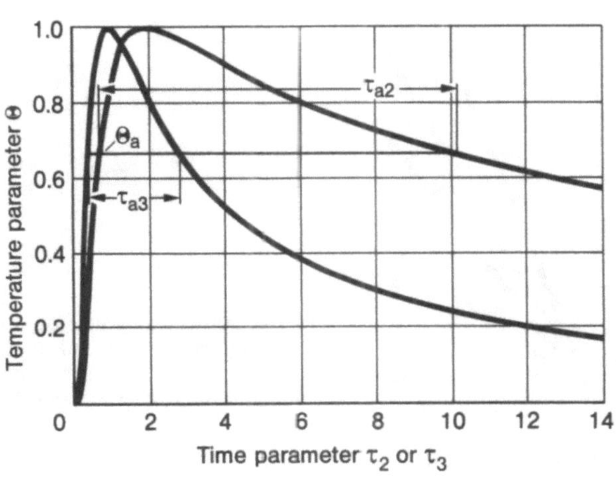

Fig. 99. Temperature profile in points alongside the weld for rapidly moving high-power source in dimensionless parameters θ and τ, point source on semi-infinite solid (τ_3, τ_{a3}) and line source in plate (τ_2, τ_{a2}); after Rykalin [1]

basis is the solution for the rapidly moving high power source, which is more generally justified for the considered points close to the weld axis (i.e. also for more slowly moving sources). It is possible with the aid of the dimensionless temperature and time parameters θ and τ, which are largely identical with the parameters used in Ref. [159],

$$\theta = \frac{T - T_0}{T_{\max} - T_0}, \tag{100}$$

$$\tau_3 = \frac{4at}{r^2}, \quad \tau_2 = \frac{4at}{y^2}, \tag{101}$$

to reduce the variety of temperature cycles in arbitrary points of the semi-infinite solid or infinite plate, which are dependent on material and process parameters, to a single line in each case, from which the dimensionless austenitizing time $\Delta\tau_{a3}$ and $\Delta\tau_{a2}$, respectively, can be measured off (Fig. 99).

The austenitizing time Δt_a in the semi-infinite solid and in the infinite plate is given by (e = 2.7183: base of the natural logarithms):

$$\Delta t_{a3} = \frac{\Delta\tau_{a3}}{2\pi e} \frac{q_w}{\lambda(T_{\max} - T_0)}, \tag{102}$$

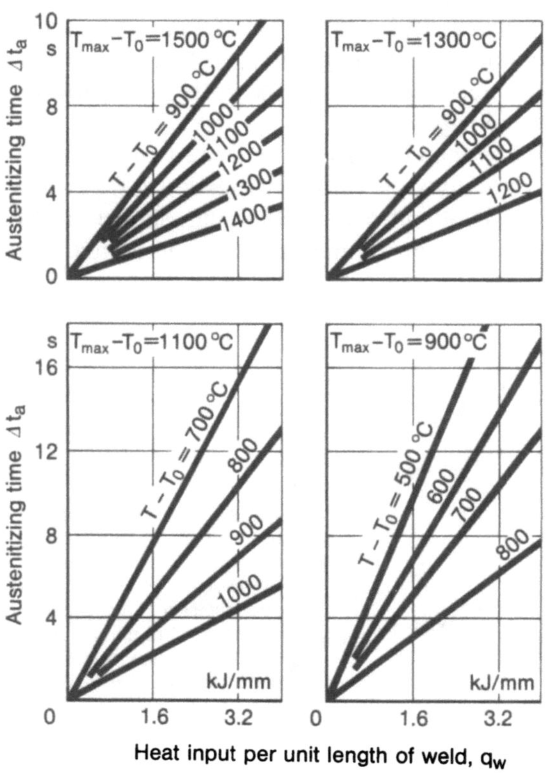

Fig. 100. Austenitizing time Δt_a above temperature T, dependent on q_w, T_0 and $T_{\max}$ for point source on semi-infinite solid; after Rykalin [1]

$$\Delta t_{a2} = \frac{\Delta \tau_{a2}}{8\pi e} \frac{1}{\lambda c \varrho} \left[\frac{q_w}{h(T_{max} - T_0)} \right]^2 . \tag{103}$$

The evaluation for steel (with $\lambda = 0.042$ J/mmsK) is summarized in Figs. 100 and 101. The austenitizing time Δt_a above the temperature T rises with the heat input per unit length of weld, q_w, with the preheating temperature, T_0, and with the peak temperature T_{max}.

From the equations stated for cooling rate, cooling time and austenitizing time, it can be seen that heat input per unit length of weld and preheating temperature are available for controlling the local temperature profiles. The margin in the case of single-pass welding is narrow, however, because cooling time and austenitizing time are changed in the same direction with the result that the microstructural improvement as the result of a longer cooling time on the one hand is offset by a microstructural impairment as a result of a longer austenitizing time on the other. It is, consequently, also not worthwhile to enlarge the groove-included angle in the case of the butt weld in order to

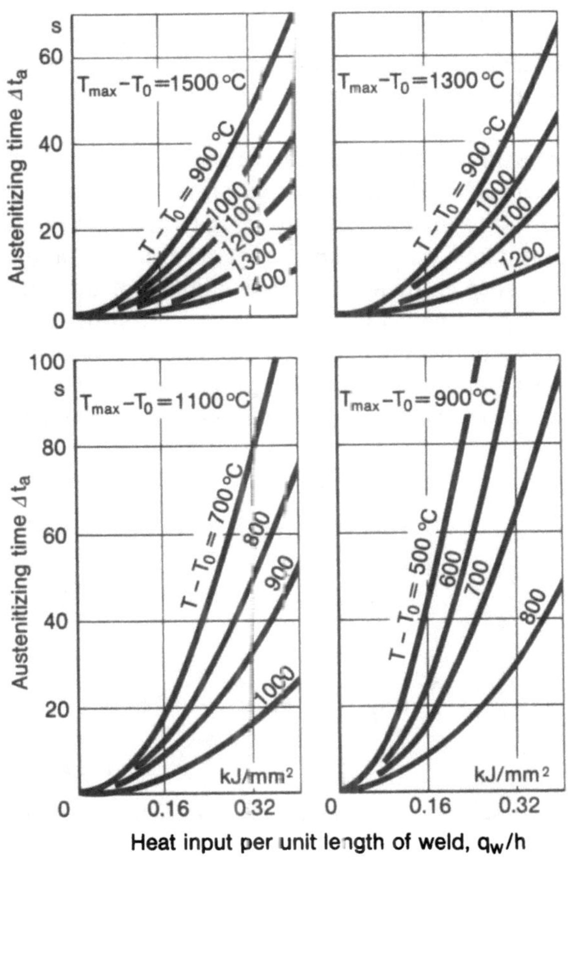

Fig. 101. Austenitizing time Δt_a above temperature T, dependent on q_w/h, T_0 and T_{max} for line source in infinite plate; after Rykalin [1]

increase the heat input per unit length of weld. Multi-pass welding offers a way out of this situation.

2.4.4 Temperature cycles in multi-pass welding

In multi-pass welding, the weld groove is filled by several weld beads laid one on top of the other with the effect that the weld beads are heated several times. Even in the double-sided single-pass fillet weld at tee joint, cruciform joint or lap joint, some kind of multiple heating exists.

Two limit cases require to be distinguished in respect of the phase position of local superimposition of multiple heating [1]. If the individual passes are very long (e.g. longer than 1 m, as is the case in submerged-arc welding), the subsequent pass encounters material which has largely cooled down. If the individual passes are very short, e.g. 100 mm, the subsequent pass encounters highly heated material, the "preheating temperature" of which depends in particular on the length of the weld portion.

The temperature cycle obtained for the limit case of long weld passes is the sawtooth pattern presented in Fig. 102. The austenitizing temperature A_{c3} is exceeded only once in each case in the base metal close to the weld. If martensitic structure has been produced, it is annealed by the subsequent pass. The annealed martensite is less hard so that the strength behaviour is more favourable. However, cold cracks may be produced even before welding the subsequent pass.

The local distribution and repetitive sequence in time of peak temperature in the cross-section area of the heat-affected zone of a multilayer weld joint is schematically drawn in Fig. 103. Up to three thermal cycles may occur, depending on the position of the considered point. The peak temperatures of the individual cycles differ. In consequence, there is a large number of different microstructural states with correspondingly varyingly mechanical properties.

The temperature cycle in the weld area may be analysed in accordance with the model of the rapidly moving point source on the plane layer, allowance being made for the geometry factors k^* relating to the heat input per unit length

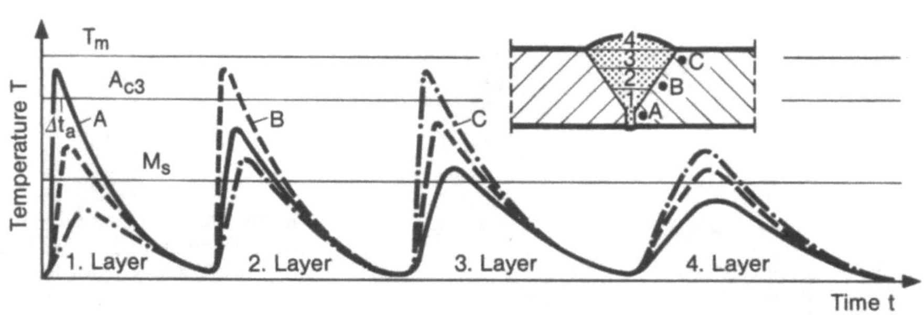

Fig. 102. Temperature cycles in multilayer welding of long welds in three points of the groove edge close to the individual layers; after Rykalin [1]

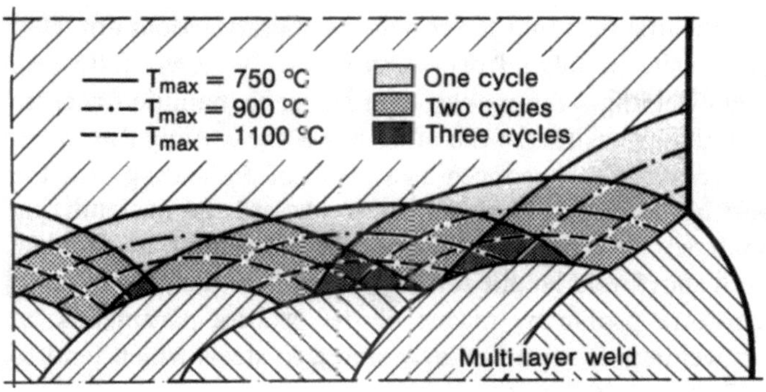

Fig. 103. Distribution and reiteration of peak temperatures in heat affected zone of multi-layer weld; after Pätzold, Hou and Ruge [131]

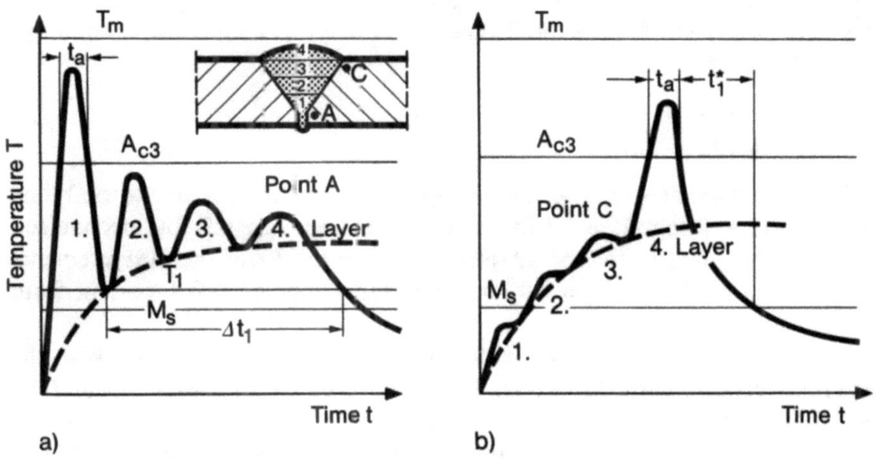

Fig. 104. Temperature cycles in multilayer welding of short welds in two points of the groove edge close to the first layer (**a**) and close to the final layer (**b**); after Rykalin [1]

of weld, q_w. The heat input per unit length of weld, q_w, is approximately proportional to the cross-section area of the deposited weld pass when welding with melting electrode. Its reciprocal value determines the cooling rate. The cross-section of the individual pass should, therefore, not be selected too small in order to avoid impermissible hardening. As to the influence of $T - T_0$, q_w and h on dT/dt, the statements made in respect of the single-pass weld (Fig. 162 in [1]) apply.

The temperature cycle obtained for the limit case of short weld passes is presented in Fig. 104 for the first and last pass. The first pass does not cool down below a certain increased temperature while the subsequent passes are being applied. The last pass is welded in a state preheated as a result of the preceding passes. If the welding parameters and the weld length are selected accordingly, it is possible to ensure for the first pass that the cooling temperature does not drop initially below the martensite point and decreases relatively slowly following completion of multi-pass welding. This enables the favourable bainitic structure to be generated instead of martensite. On the other hand, the cooling rate when the last pass is applied is favourably reduced as a result of preheating. For the first and last pass (and likewise for the intermediate passes), the austenitizing time is relatively small so that unfavourable grain coarsening is avoided. Multi-pass welding with short weld length consequently offers the possibility of controlling the temperature profile by means of the process parameters, in contrast to single-pass welding. It is particularly suited to steels which tend to harden or tend toward grain coarsening.

Proceeding from the temperature field equation of the rapidly moving line source in the plate and from further simplifications relating to the heating and cooling process, approximation formulae are derived in Ref. [1] for weld length l with cooling of the first pass to temperature T_1 (desired slightly above the martensite point) and for the dwell time Δt_1 above this temperature (equation (247) in Ref. [1]):

$$l = 0.7\frac{k_t^2 k_m q_w^2 v}{h^2(T_1 - T_0)^2}. \tag{104}$$

The factor k_t designates the effective arc operation time relative to the total time of multi-pass welding ($k_t \approx 0.7$ for manual welding, $k_t \approx 0.9$ for mechanized welding) and k_m is a factor for adaptation to the results of the measurements ($k_m = 1.5$ for butt joints, $k_m = 0.9$ for T and lap joints, $k_m = 0.8$ for cruciform joints).

Cooling temperature T_1 rises with q and q_w, respectively (i.e. also with the pass cross-section), with T_0, k_t, $1/v$ (if v is covered by q_w), with $1/h$ and $1/l$ (Fig. 167 in Ref. [1]). Dwell time Δt_1 rises with h, $1/T_1$ and groove-included angle α, while the behaviour of q or q_w, l and k_t is not uniform (Fig. 173 in Ref. [1]).

2.5 Hydrogen diffusion

A further important phenomenon in zones influenced by the heat effects of welding is hydrogen diffusion. Hydrogen is easily dissolved in the weld pool because the solubility is high when the metal is in the liquid state. The dissolved hydrogen has its origin in hydrogen-containing compounds in the filler metal or in the welding auxiliary materials, in moisture of the electrode coating, in impurities on the surface of the welded part and in moisture of the ambient air in hand welding. The molten zone of a welded joint may thus contain large amounts of dissolved (i.e. atomic) hydrogen. When cooling down, the solubility decreases (Fig. 105). It drops rapidly when changing from the liquid to the solid state. Pores containing effused (i.e. molecular) hydrogen may be found especially under conditions of extremely rapid cooling (i.e. more often in aluminium alloys

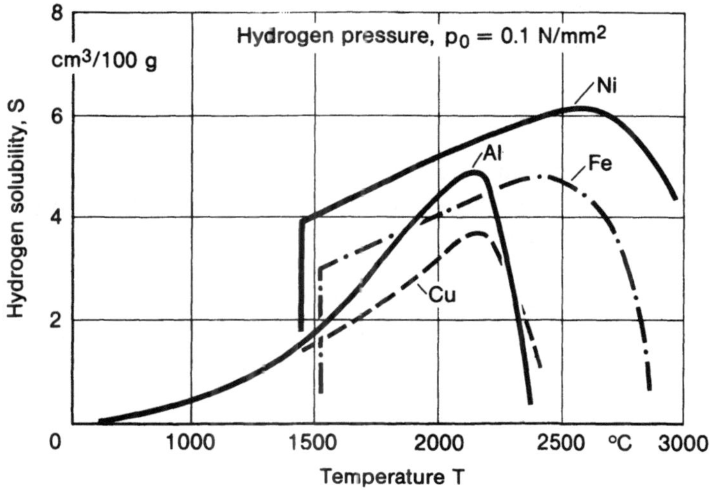

Fig. 105. Hydrogen solubility dependent on temperature for different metals; after Howden and Milner in Ref. [72]

than in steel). With clean pore surfaces, hydrogen may once more be dissociated; it then flows into the metal lattice. Plastic deformation accelerates the hydrogen input because the surface is partly torn open by the slip bands and hydrogen is transported into the interior by means of moving dislocations. The further process of hydrogen diffusion is controlled by the changing temperature field. After a sufficient long period of time, nearly all the diffused hydrogen has left the welded part.

The engineer's interest in hydrogen diffusion results from the fact that cold cracking in the weld and heat affected zone (see Section 5.2) is highly promoted by the local content of diffusible hydrogen. Besides hydrogen, the microstructural state and the level of stress (or strain) is essential for crack initiation and propagation. The initiated crack is first of all stably enlarged before occasionally resulting into an unstable instantaneous fracture. Very often the cold cracks are arrested before final fracture takes place (dormant cracks). This is especially the case with cracks initiated by residual stresses during cooling in connection with temporarily high hydrogen contents. Therefore the structure should not be loaded immediately after welding.

In specimen testing this process is termed "static fatigue" because the formal aspects are similar to cyclic fatigue. An endurable stress versus elapsed time curve is observed in cases of cold crack fracturing which resembles the Wöhler fatigue curve. The endurable stresses drop with the elapsed time to fracture. There is a low stress level below which no cracks are initiated even after a very long elapsed time (endurance limit for cold cracking).

The modelling of hydrogen diffusion and comparisons with test results are considered in Ref. [171, 172, 132].

Hydrogen diffusion is numerically analysed proceeding from the well known diffusion equation

$$\frac{\partial C}{\partial t} = D\left(\frac{\partial^2 C}{\partial x^2} + \frac{\partial^2 C}{\partial y^2} + \frac{\partial^2 C}{\partial z^2}\right) + \frac{\partial H_v}{\partial t} \tag{105}$$

In this equation, which applies for a constant solubility, C is the hydrogen concentration, D the diffusion coefficient and H_v the volumetric hydrogen source (or sink) density. The equation without the last term corresponds to Fick's second law.

In the case of inhomogeneous material or varying temperatures, the inhomogeneous distribution of solubility S must be taken into account [132] (besides introducing D as material and temperature dependent):

$$\frac{\partial(C/S)}{\partial t} = D\left(\frac{\partial^2(C/S)}{\partial x^2} + \frac{\partial^2(C/S)}{\partial y^2} + \frac{\partial^2(C/S)}{\partial z^2}\right) + \frac{\partial(H_v/S)}{\partial t} \tag{106}$$

Another formulation [173–175, 178] proceeds from the relation given by Sieverts between hydrogen concentration C and hydrogen partial pressure ratio p/p_0 ($p_0 = 0.1\ \mathrm{N/mm^2}$, ambient air pressure) and introduces it into equation (105):

$$C = \kappa\sqrt{\frac{p}{p_0}} = \kappa\Pi \tag{107}$$

The factor κ is the equilibrium coefficient and Π is a harmonic (or potential) function valid under the condition of equation (105).

Hydrogen concentration is generally measured in [$\mathrm{cm^3/100\ g}$]. Standardized procedures (DIN 8572, ISO 3690) are applied for determining the total amount of hydrogen in a weld seam immediately after the end of welding. A specimen with surfacing weld bead is water quenched, supercooled in liquid nitrogen, cut to blocks for separating the weld ends and put into a measuring chamber for hydrogen effusion at ambient temperature.

If molecular hydrogen is accumulated in pores and other "traps" at low temperature, enabling the hydrogen to rediffuse through the trap surface, this process can be taken into account by fictitiously increasing the solubility at low temperatures accordingly. This presumes that the volume of the traps is much smaller than the volume of the welded joint.

The diffusion equation (106) (or an equivalent formulation based on equation (107)) is solved introducing the initial distribution of hydrogen and its solubility, and the boundary condition at the free surfaces stating that the hydrogen partial pressure here is the same in the metal and in the ambient air [178]. Comparing equations (105) or (106) for hydrogen diffusion with equation (10) for heat conduction, the similarity in terms of mathematics is obvious. Both are of the potential type if temperature independent material characteristic values are introduced. Both are solved by finite difference or finite element methods mainly for the plane cross-section model. Temperature-dependent values of D may be found in Ref. [175, 176, 178].

Numerical analysis of hydrogen diffusion is demonstrated by way of example considering the specimen with surfacing weld standardized for measuring the total hydrogen content after welding. The base metal is fine-grained

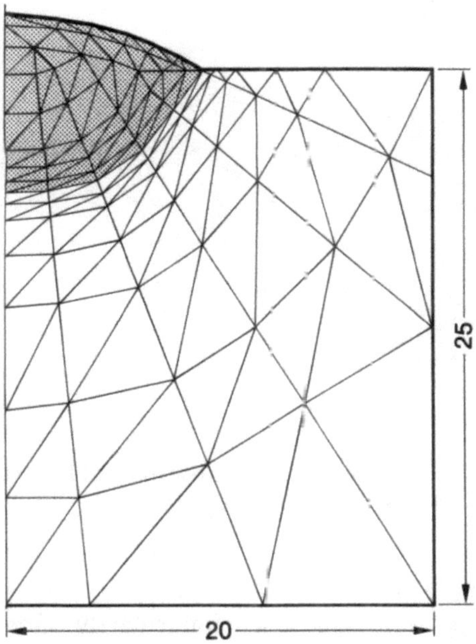

Fig. 106. Finite element cross section model simulating hydrogen diffusion in bead specimen; after Michailov et al. [178]

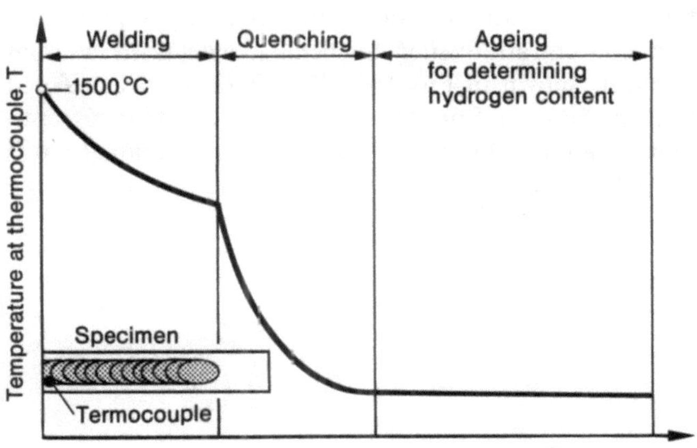

Fig. 107. Temperature profile over time (schematic representation) for weld surface point in centre section of bead specimen standardized for measuring hydrogen content after welding; after Dickehut and Ruge [176]

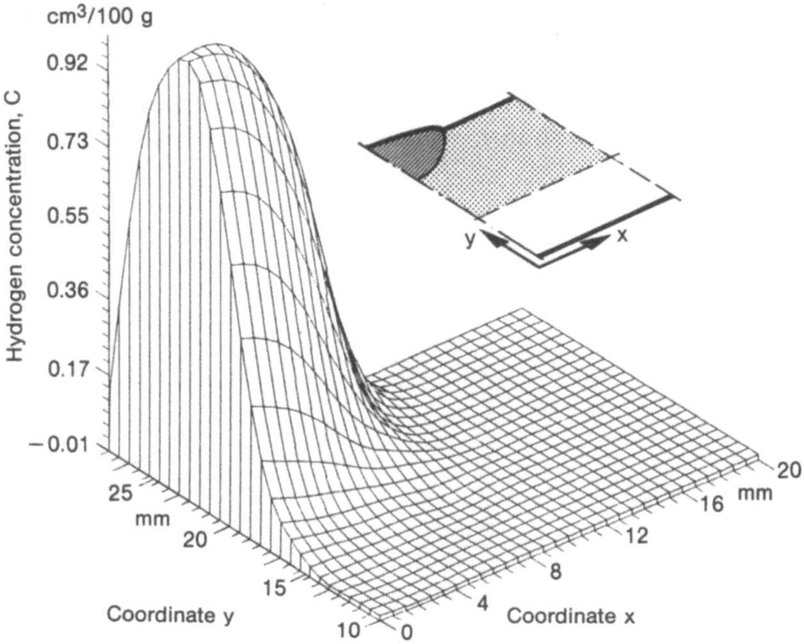

Fig. 108. Hydrogen concentration in standardized bead specimen made of structural steel StE460 after quenching, results of finite element analysis based on the model in Fig. 106; after Michailov et al. [178]

structural steel, StE460. The specimen is water quenched immediately after welding with a final temperature of 100 °C. The finite element cross-section model used in the analysis is shown in Fig. 106. The temperature profile over time for the weld surface in the centre section of the specimen is plotted in Fig. 107. The hydrogen concentration after quenching, calculated without martensitic transformation, is shown in Fig. 108. The distribution calculated with transformation is nearly the same. For assessing the cold cracking susceptibility, the hydrogen concentration field must be related to the residual stress field including data on limit stresses.

3 Welding residual stress and distortion

3.1 Fundamentals

3.1.1 Temperature field as the basis

A characteristic of application-oriented analysis of welding residual stress and distortion is the mainly decoupled representation of the thermal, mechanical and microstructural processes (illustrated by the line arrows in Fig. 3). The basis of stress and distortion analysis is the temperature field during welding, determined numerically or by measurement, as well as the microstructural changes caused by the temperature field (presented in Chapter 2). Coupling effects are only occasionally taken into account.

The temperature change in solid bodies is connected with thermal strain. The relation which applies for adequately small temperature changes ΔT with the linear thermal expansion coefficient α to the linear, in total volumetric thermal strain ε_T in the three coordinate directions, is:

$$\varepsilon_T = \alpha \Delta T. \tag{108}$$

With major temperature changes ΔT, α should be inserted as temperature-dependent and equation (108) should be applied on a temperature-incremental basis, or mean values α_m should be used in a specified temperature range.

The linear thermal expansion coefficient α or α_m is determined using a dilatometer. Dilatometer curves for an austenitic steel without transformation and a pearlitic steel with transformation are shown in Fig. 109. When heating the pearlitic steel, the A_{c1} transformation temperature close to 800 °C is marked as a discontinuity. Depending on the austenitizing parameters, i.e. the austenitizing peak temperature $T_{a\max}$, the austenitizing time Δt_a, and the cooling time $\Delta t_{8/5}$, the transformation temperature during cooling differs in its level (see also Fig. 123). The transformation strain ε_{tr} in the opposite direction of the thermal strain can both be evaluated directly from the diagram and also taken into account in the temperature dependence of α.

The mean thermal expansion coefficient α_m results from the mean gradient $\tan\theta$. The momentary or differential thermal expansion coefficient α follows from the local gradient (see also Fig. 122).

The time- and position-dependent thermal and transformation strains cause elastic or elastic-plastic stress fields as well as related local and global deformations. The essential fundamentals and applications of elastic and inelastic

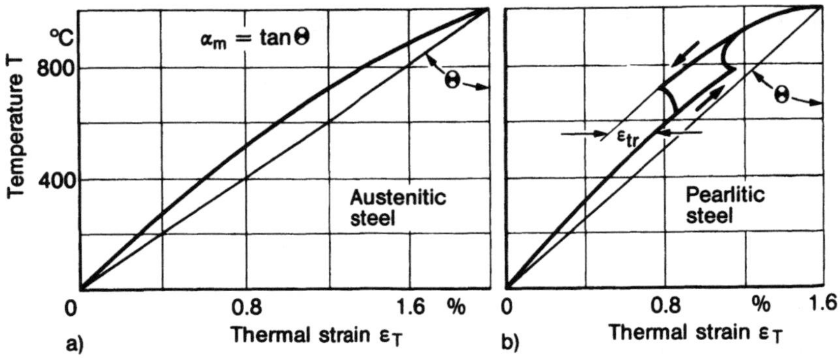

Fig. 109. Dilatometer curves for austenitic steel without transformation (**a**) and pearlitic steel with transformation (**b**), transformation strain ε_{tr}; after Vinokurov [8]

thermomechanics have been summarized by Melan and Parkus [180], Parkus [181], Boley and Weiner [182], Nowacki [183] and Ziegler [184].

3.1.2 Elastic thermal stress field

The thermal and transformation stress fields during and after welding are to a large extent non-linearly inelastic. In continuum mechanics, however, the solution for a non-linearly inelastic field problem is generally obtained by proceeding from a linearized elastic field problem. Fundamental questions relating to the stress field can be clarified with the linear-elastic solution. Moreover, this solution may be the starting point for further solution steps in the non-linear range. The linear-elastic solution is limited in the case in question to the thermal stresses without the necessarily non-linearly acting transformation influence.

Linear-elastic thermal stress fields can be presented based on functional analysis if the model is sufficiently simplified. In the case of more complicated models, the elastic finite element solution is appropriate. The elastic solution may be the basic step of elastic-plastic analysis (see Section 3.1.3).

The principle of linearized thermal stress representation based on functional analysis consists in allowing the volumetric thermal stress sources, which are proportional to the position- and time-dependent temperature changes, i.e. the elastic stress from triaxial suppression of thermal strain ε_T,

$$\sigma_T = \frac{-\alpha \Delta T E}{1 - 2\nu} \tag{109}$$

to act on the elastic structure while releasing the initially introduced complete thermal strain suppression.

First of all, the thermal stress field is considered in the infinite plate subjected to a momentary line heat source (line perpendicular to plate plane) with temperature field according to equation (25) with $b = 0$ and $T_0 = 0$. The result after Ref. [8] is:

$$\sigma_r = -\frac{\alpha EQ}{8\pi\lambda ht}\frac{4at}{r^2}(1 - e^{-r^2/4at}), \tag{110}$$

$$\sigma_t = \frac{\alpha EQ}{8\pi\lambda ht}\left[\frac{4at}{r^2}(1 - e^{-r^2/4at}) - 2e^{-r^2/4at}\right]. \tag{111}$$

The contents of equations (110) and (111) are plotted dimensionlessly in Fig. 110 ($A = 4\pi\lambda ht/\alpha EQ$). The stress state is biaxial and differs in part considerably from the temperature pattern. Radial and tangential compression prevails close to the line source. The tangential compression turns further to the outside into tension, whereas the radial compression decreases without any reversal of the sign. The process can be illustrated in such a way that the heated internal region presses radially on the colder external region where it is supported as tangential tension. By comparison, the (dimensionless) stress $-\alpha TEA$, which would result from uniaxial suppression of the local thermal strain αT, is presented. With biaxial suppression (for example in the plate plane), this stress is increased by the factor $1/(1 - \nu)$; with triaxial suppression by the factor $1/(1 - 2\nu)$. From Fig. 110, it is clear that the maximum values of σ_r and σ_t are only half as large as the maximum value of $-\alpha TE$. The thermal stresses are thus reduced effectively by the elasticity of the plate.

The quasi-stationary elastic thermal stress field of the line heat source moving uniformly and linearly (velocity v) in the infinite plate is gained by analogy to the procedure in the case of the temperature field by integrating the thermal stress effects of momentary sources in accordance with equations (110) and (111). After Ref. [8] this results (with Y_0 and Y_1 Bessel functions of the

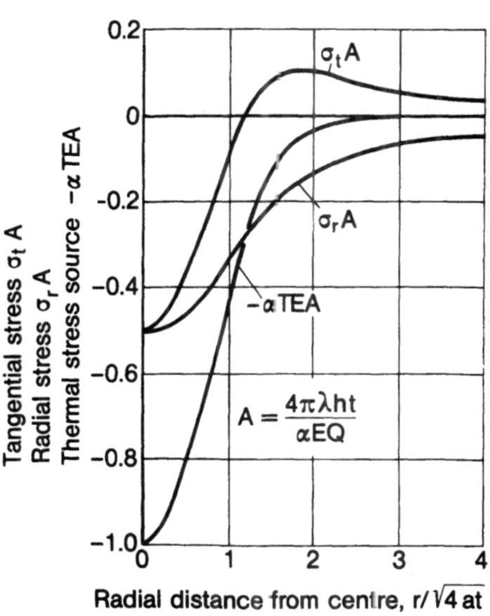

Fig. 110. Thermal stresses around momentary line source in infinite plate; after Vinokurov [8]

second kind and zero and first order) in:

$$\sigma_x = -\frac{\alpha E q}{4\pi\lambda h}\left\{e^{-vx/2a}\left[Y_0\left(\frac{vr}{2a}\right) - \frac{x}{r}Y_1\left(\frac{vr}{2a}\right)\right] + \frac{2a}{v}\frac{x}{r^2}\right\}, \tag{112}$$

$$\sigma_y = -\frac{\alpha E q}{4\pi\lambda h}\left\{e^{-vx/2a}\left[Y_0\left(\frac{vr}{2a}\right) + \frac{x}{r}Y_1\left(\frac{vr}{2a}\right)\right] - \frac{2a}{v}\frac{x}{r^2}\right\}, \tag{113}$$

$$\tau_{xy} = \frac{\alpha E q}{4\pi\lambda h}\left[e^{-vx/2a}\frac{y}{r}Y_1\left(\frac{vr}{2a}\right) - \frac{2a}{v}\frac{y}{r^2}\right]. \tag{114}$$

The longitudinal stresses σ_x and the transverse stresses σ_y in the motion line of the heat source as well as the longitudinal stresses σ_x in a cross-section immediately after the source are presented in Fig. 111 according to Ref. [8]. The fact that σ_x does not form an equilibrium system in the finite frame of Fig. 111b (this follows also from equation (112) for $x = 0$ with $T = 0$: σ_x is only negative) can be

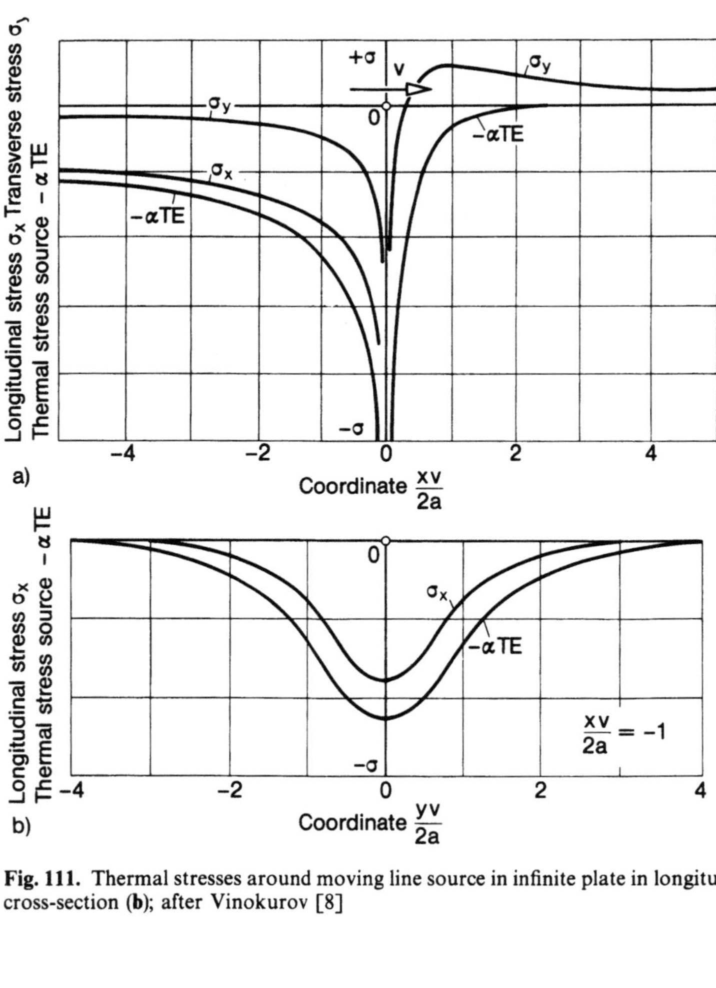

Fig. 111. Thermal stresses around moving line source in infinite plate in longitudinal section (**a**) and cross-section (**b**); after Vinokurov [8]

explained on the basis of the infinite extension of the plate. An incorrectness is imputed in Ref. [8] for the corresponding derivations of Parkus [181]. The stress reduction (in some cases to half the value), due to the elasticity of the plate, already observed in the case of the momentary source, follows from the comparison with the thermal stress source $-\alpha ET$.

By means of the functional analysis solution for the elastic stress field for two heat sources in the infinite plate moving against one another as a model for the weld end at the edge of the semi-infinite plate (method of mirror image source arrangement), it was verified that transverse tension occurs directly after the source where the two sources approach each other, which is equivalent to the approach of one source to the edge [232, 234].

Elastic thermal stress fields during and after welding are in many cases also determined by using the finite element method, both in the first step of an otherwise non-linear analysis as well as in the absence of further-going non-linear analysis capabilities [232, 234, 239].

3.1.3 Elastic-plastic thermal stress field

Essential for the formation of (permanent) welding residual stresses are the plastic, and in some cases also viscoplastic deformations, whereas the elastic thermal stresses considered hitherto disappear after removal of the temperature changes which caused them. The elastic field solutions may offer reference points where the yield stress is presumably exceeded, where consequently residual stress sources may exist, but they do not permit any quantitative statement regarding the elastic-plastic field. On the other hand, functional analysis solutions for the elastic-plastic thermal stress fields during and after welding are not known. Numerical solutions for simpler model cases, by contrast, are gained through finite elements and incremental loading (see Section 3.2).

To illustrate a number of basic facts of elastic-plastic thermal stress fields during and after welding, an axisymmetrical field with momentary heat source is presented quantitatively (in anticipation of Section 3.2.3) and a quasi-stationary field with uniformly moving heat source is presented qualitatively. The axisymmetrical field may simulate spot pressure welding, the quasi-stationary field seam fusion welding.

The circular plate with initial temperatures distributed over distance from centre according to a $\cos^2$-function in the inner plate area, maximum temperature $T_{max} = 700\,°C$ in the centre of the plate, without electrode pressure, is considered as an illustrative model for spot welding (Fig. 112). The structural steel used displays the marked yield limit, $\sigma_Y = 240\ N/mm^2$, without hardening. Following heating, the central compression yield zone is formed in the temperature range $\geq 200\,°C$. During cooling, characterized by decreasing T_{max}, three plastic zones occur temporarily, the $\sigma_r - \sigma_t$ tension zone in the centre, the σ_t compression zone on the outside and the σ_t compression against σ_r tension zone between them. Following complete cooling, only the central tension zone still remains. Yield is initiated as a result of the critical principal shear stress after

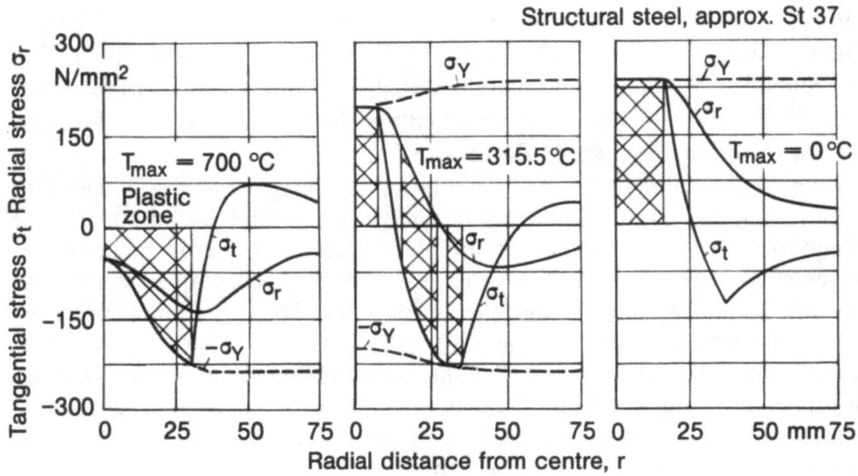

Fig. 112. Radial and tangential thermal stresses and residual stresses, respectively, around momentary heat source in circular plate (diameter 760 mm), plastic zone at the onset, at an intermediate state and at the end of the cooling process; after Gurney [212]

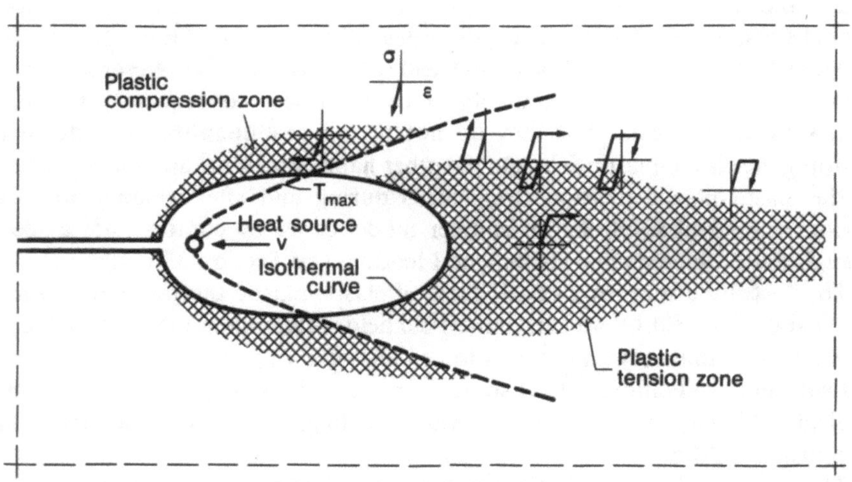

Fig. 113. Plastic zone and local stress-strain cycles in quasi-stationary temperature field of moving heat source

Tresca, so that yield is a matter of the difference of the stresses σ_r and σ_t to zero (inner and outer zone) or to each other (intermediate zone). The stress pattern described above is confirmed qualitatively by more precise models.

A plate model for seam fusion welding is proposed in Fig. 113. Within a certain isothermal curve around the moving line heat source (e.g. $T \geq 600\,°C$ with mild structural steel), the material is largely free of stress as a consequence of the reduced yield limit at elevated temperature. This area is therefore replaced by an oval opening in continuation of the groove gap. The parabola-like curve drawn as a broken line marks the local temperature maxima, local temperature rise ahead of the curve and local temperature drop behind it. As a consequence of the thermal expansion during heating, a plastic compression zone occurs ahead of the curve and, as a consequence of the thermal contraction during cooling, a plastic tensile zone occurs after the curve, the latter separated by a strip with elastic unloading from the compression state into the tension state. The respective local stress state is indicated for various points of the field by means of a schematized (cyclic) stress-strain curve, without registering its temperature dependence. For example, a point just above the boundary isotherm passes into compressive yield, unloads elastically into tensile stress, then experiences tensile yield and finally unloads elastically in opposite direction. A field solution for the quasi-stationary elastic-plastic stress field during seam welding is not available.

3.1.4 Basic equations of thermomechanics

According to the first fundamental law of thermodynamics, which postulates the equivalence of thermal and mechanical energy and the conservation of total energy, the thermal and the mechanical field (i.e. the temperature field and the thermal stress field) are coupled. The law states for the volume element that the heat stored per unit of time $c\varrho\dot{T}$, and the heat supplied or carried off, $q^{*}_{i,i}$ (related to surface area), are equal to the heat released or consumed per unit of time, $\dot{Q}_v$ (related to volume), and to the energy from elastic and viscoplastic deformation rate, $\dot{\varepsilon}_{e\,ij}$ and $\dot{\varepsilon}_{vp\,ij}$. Hence, the fundamental equation of thermomechanics of elastic-viscoplastic continua follows according to Ref. [182, 184] ($\sigma_{d\,ij}$ deviatoric stress tensor):

$$c\varrho\dot{T} + \dot{q}_{i,i} = \dot{Q}_v - \frac{E\alpha T}{1-2\nu}\dot{\varepsilon}_{e\,ii} + \xi\sigma_{d\,ij}\dot{\varepsilon}_{vp\,ij}. \tag{115}$$

The loss factor, $\xi \leq 1.0$, allows for the fact that not all inelastic deformation energy is dissipated in heat but that a part may appear in the microstructural change. The terms with the deformation quantities in equation (115) are designated as "mechanical coupling terms". As already stated in the preceding text, however, the coupling between thermal and mechanical field which is expressed in equation (115) can be ignored in welding problems which are dominated by the heat input from the outside (comparison calculation including coupling in Ref. [193]). The decoupled solution of the temperature field problem is presented in Chapter 2 (equation (12) corresponds to equation (115) without coupling terms). The mechanical fields are analysed below, proceeding from the as-decoupled determined, i.e. already known, temperature fields.

In this section the basic equations for the isotropic thermoelastic-plastic continuum are stated in tensor form (in accordance with Ref. [194]) and the resulting material equation then in the matrix form of the finite element method.

Space does not permit either completeness nor adequate detailing. The equations serve only to illustrate the principal relations. Following this, general information is given regarding the particularities of the finite element analysis methods in the case of elastic-plastic structural behaviour. The restriction to plastic behaviour without viscosity is adequate for the majority of welding residual stress problems (with the exception of hot stress relieving and multi-pass welding). For a more detailed theory of the elastic-plastic continuum, Ref. [185–188] should be consulted, for creep and relaxation theory Section 4.4.3.2.5, for the basic equations of welding residual stress analysis Ref. [193–197], for the elastic-plastic finite element method Ref. [189–192].

The total strain $d\varepsilon_{ij}$ is composed of elastic strain $d\varepsilon_{e\,ij}$, conventional plastic strain $d\varepsilon_{p\,ij}$, plastic strain from transformation plasticity $d\varepsilon_{tp\,ij}$, thermal strain $d\varepsilon_{T\,ij}$ and transformation strain $d\varepsilon_{tr\,ij}$ (the differential form being selected in respect of the finite element analysis which is to be applied incrementally):

$$d\varepsilon_{ij} = d\varepsilon_{e\,ij} + d\varepsilon_{p\,ij} + d\varepsilon_{tp\,ij} + d\varepsilon_{T\,ij} + d\varepsilon_{tr\,ij}. \qquad (i, j = 1, 2, 3) \tag{116}$$

Hooke's law, which is written subdivided according to deviatoric and dilatoric or volumetric portion, applies to elastic strain:

$$\varepsilon_{d\,ij} = \frac{1}{2G}\sigma_{d\,ij}, \tag{117}$$

$$\varepsilon_{v\,ii} = \frac{1}{3K}\sigma_{v\,ii}. \tag{118}$$

Shear modulus G and compression modulus K can be expressed by the elastic modulus E and Poisson's ratio ν:

$$G = \frac{E}{2(1+\nu)}, \tag{119}$$

$$K = \frac{E}{3(1-2\nu)}. \tag{120}$$

The combination of yield condition, yield law and hardening law applies to the plastic strain. The yield condition designates the initiation of yield in the multi-axial stress state. The yield law correlates the plastic strain increments with the momentary stress state and with the stress increments. The hardening law states how the yield limit is changed by the yield deformation.

The yield condition mostly used is the von Mises distortion energy hypothesis:

$$\frac{1}{2}\sigma_{d\,ij}\sigma_{d\,ij} - \frac{1}{3}\sigma_Y^2 = 0. \tag{121}$$

The yield law states the plastic strain increment as coaxial and proportional to the deviatoric stress:

$$d\varepsilon_{p\,ij} = d\lambda\sigma_{d\,ij}. \tag{122}$$

The isotropic (or also kinematic) work-hardening (entering in $d\lambda$) is represented by the hardening modulus H and the effective plastic strain increment $d\varepsilon_p^*$:

$$H = \frac{d\sigma_Y}{d\varepsilon_p^*}, \tag{123}$$

$$d\varepsilon_p^* = \frac{\sigma_{dij} d\varepsilon_{pij}}{\sqrt{1.5\sigma_{dij}\sigma_{dij}}}. \tag{124}$$

The exponential law after Ramberg and Osgood is often used for the (isotropic) hardening (unless H is set constant at a specified temperature), according to which the relation between the equivalent strain ε_{eq} (after von Mises), yield stress $\sigma_Y = \sigma_Y(T)$ and yield stress maximum $\sigma_{Y\max} = \sigma_{Y\max}(T)$ (from tensile test, see Fig. 117) is established through the material constants $m \approx 2$ and $C = C(T)$:

$$\sigma_Y = \frac{C\varepsilon_{eq}}{\left[1 + \left(\dfrac{C\varepsilon_{eq}}{\sigma_{Y\max}}\right)^m\right]^{1/m}} \tag{125}$$

The thermal strain ε_T is given by the dilatatoric strain components:

$$d\varepsilon_{Tii} = 3d\varepsilon_T = 3\alpha dT. \tag{126}$$

The dilatoric transformation strain ε_{tr} depends on the defining quantities of the transformation process. In the case of Fig. 123 the martensitic transformation is connected with $\varepsilon_{tr} \approx 0.4\%$.

The transformation-plastic (deviatoric) strain ε_{tp} is set proportional to the stress σ (assuming uniaxial low stress):

$$\varepsilon_{tp} = K_t\sigma \tag{127}$$

The factor K_t is found proportional to the transformation related volume change $\delta = \Delta V/V$ and inversely proportional to the yield stress of the weaker phase, σ_{Ywp} (according to Greenwood and Johnson [155]):

$$K_t = \frac{5}{6}\frac{\delta}{\sigma_{Ywp}} \tag{128}$$

These formulae were extended introducing the relative volumetric content X of the newly formed phase according to Abrassart (equation (129)), Leblond (equation (130)) or to an empirically based relation (equation (131)), [140], taking into account the fact that transformation plasticity is largest at the start of transformation.

$$\varepsilon_{tp} = \frac{3}{4}\frac{\delta}{\sigma_{Ywp}}\sigma X\left(1 - \frac{2}{3}\sqrt{X}\right) \tag{129}$$

$$\varepsilon_{tp} = \frac{2}{3}\frac{\delta}{\sigma_{Ywp}}\sigma X(1 - \ln X) \tag{130}$$

$$\varepsilon_{tp} = K_t\sigma X(2 - X) \tag{131}$$

Multiaxial conditions are taken into account by considering strain increments $d\varepsilon_{tpij}$ dependent on volumetric transformation increments $d\delta$ or volumetric content increments dX and the deviatoric stresses σ_{dij} [138]:

$$d\varepsilon_{tpij} = \frac{3}{2} K_t(1 - X)dX\sigma_{dij} \tag{132}$$

The factor K_t has been determined experimentally and by microstructural calculations for the martensitic or pearlitic transformation of alloyed steels [139, 140] resulting in $K_t = 3\text{–}12 \times 10^{-5}$ mm²/N with a most probable value $K_t = 5 \times 10^{-5}$ mm²/N. The factor K_t is expected to be dependent on the hydrostatic prestress. Hydrostatic pressure reduces the transformation strain, hydrostatic tension increases it. The above formulations neglect the fact that transformation plasticity is partly a viscoplastic phenomenon.

Within a simplified algorithmic procedure, transformation plasticity is described by a severe reduction of the yield stress in the temperature range of transformation (see Figs. 125 and 127) setting strain hardening to zero in this range.

Following a series of merging analysis operations with equations (116) to (126), the constitutive equation of the thermoelastic-plastic continuum finally appears in matrix notation as follows:

$$\{d\sigma\} = [D]\{d\varepsilon\} + dC^*\{\sigma_d\} - M^*\{d\varepsilon_T\}. \tag{133}$$

The six components of the respective tensor are summarized in the column vectors $\{d\sigma\}$, $\{d\varepsilon\}$, $\{\sigma_d\}$ and $\{d\varepsilon_T\}$. The momentary thermoelastic-plastic stress-strain matrix $[D]$ and the coefficients dC^* and M^* depend in a complicated manner on the temperature-dependent material characteristic values G, K, H, σ_Y, α and also on the equivalent stress (see Ref. [194, 192]). Microstructural transformation is not explicitly included in equation (133).

With the known constitutive equation (equation (133)), finite elements and finite element systems are definable in the usual way. Excellent meshing, mesh grading and possibly remeshing capabilities should be available. It should be possible to add elements to simulate the addition of material during welding, and on the other hand to remove elements to simulate fusion. As the elastic strains during formation of welding residual stresses are small compared to the dilatatoric thermal strains (this is enabled as a result of large plastic strains), particular attention should be paid to compatibility between impressed dilatatoric thermal strain and dilatatoric total strain simulated by the shape function of the element. The shape function should be at least of the order of the temperature function. In addition, the mesh should have more degrees of freedom than are made dependent on each other by the thermal dilatation (see Ref. [194]). The added-up plastic strains are reset to zero during fusion and added up again following solidification.

The transient and temperature-dependent non-linear structural behaviour during welding, represented by the nodal point displacements, is simulated numerically by a time-stepped (i.e. incremental) loading of the finite element model. Such calculations are characterized by the fact that the stiffness matrix of the respective load increment depends not only on the momentary temperature, stress and strain state but also on the stress history. The displacement increments may be determined by a modified Newton iteration of the equilibrium condition with a (fictitiously) constant stiffness matrix. According to another procedure, the stiffness matrix is newly determined by stress integration within each integration step. In this case, Euler's (explicit or implicit) forward or backward integration [192] is used.

In addition to the above thermoelastic-plastic capabilities of non-linear finite element software, provisions are necessary for accurately modelling the transformation processes in welded joints if the transformation effects are severe (this is the case for martensitic transformation). From the flow chart in Fig. 91 both the common and the additional analysis steps can be seen. The latter comprise the generation of the data on the microstructural constituents, the data on transformation plasticity derived therefrom, the jumping back from the incrementally calculated stresses to the transformation plasticity module and a creep capability for cases of post-weld heat treatment or multilayer welding. These additional analysis steps are substituted by simply reducing the yield limit in the temperature range of transformation if simplified procedures are applied.

3.1.5 Thermomechanical material characteristic values

For theoretical-numerical welding residual stress investigations, the thermomechanical material characteristic values entering in the basic equations are required dependent on temperature (the thermal material characteristic values including the density have already been presented in Section 2.1.2.6, individual further values relating thereto are contained in the figures of this section). The stock of known values is not always adequate so that numerical solutions may encounter narrow boundaries alone for this reason.

For residual stress analysis, the following thermomechanical material characteristic values are required, in addition to the density ϱ already mentioned, depending on temperature:

- thermal expansion coefficient α [1/K],
- elastic modulus E [N/mm^2],
- Poisson's ratio ν [−],
- yield limit σ_Y [N/mm^2].

In the case of α, a distinction requires to be made between the momentary values α at the respective temperature and the mean values α_m between initial and final temperature. In addition, information relating to α and σ_Y is required at the transformation points. All the information is necessary initially only in the low temperature range of relatively high yield limit for the purpose of calculating residual stresses.

An excellent compilation by Richter [23] of values relating to iron and steel unfortunately comprises no data on the yield limit. Other relevant data, particularly also in text-books on welding residual stresses [1–10], are rather scanty

[198, 199]. A general deficiency consists in the fact that yield stress and tensile strength, yield strain and elongation at rupture are almost always related to the initial cross-section or initial length of the specimen, i.e. not the corresponding values of the "true" stress or "true" strain, which would be more appropriate to the problem in question as far as large strains occur. Consequently, it is possible to locally measure residual stresses which are above the nominal value of the ultimate tensile strength solely because the "true" tensile strength is higher.

The material characteristic values σ_Y, α, ϱ, λ, E, c (in one case also ν) are plotted dependent on temperature for three common structural materials suitable for welding, for a fine-grained structural steel (with transformation at 700 °C in Fig. 114 (after Ref. [234]), for an aluminium alloy in Fig. 115 (after Ref. [10]) and for a titanium alloy in Fig. 116 (after Ref. [215]). A similar representation for the nickel-chrome-iron alloy Inconel 600 can be found in Ref. [249].

The stress-strain curves as well as the initial and maximum yield stress gained therefrom, $\sigma_{Y0.1}$ and $\sigma_{Y\max}$, respectively, related to the initial cross-section of the tensile specimen, are presented dependent on temperature for the structural steel St 37 in Fig. 117 (after Ref. [193]). The "true" stresses are significantly higher only after the constriction has started (this stress region is not shown here). Particularly noteworthy are the especially high yield stresses in the region of blue tempering (with reduced elongation at rupture). Corresponding (simplified) curves for the high-alloy steel X4CrNiMo1913 are shown in

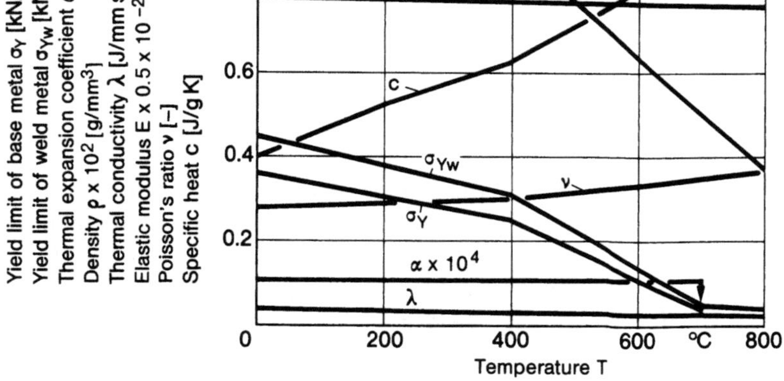

Fig. 114. Thermal and mechanical material characteristic values of fine-grained structural steel suitable for welding (micro-alloyed CMn steel) dependent on temperature (transformation heat at 700 °C covered by specific heat capacity c); after Karlsson [234]

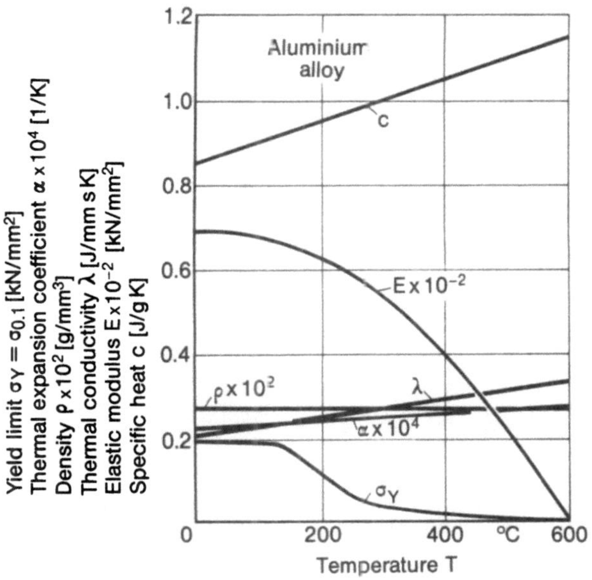

Fig. 115. Thermal and mechanical material characteristic values of aluminium alloy (5052-H32 according to US standard) dependent on temperature; after Masubuchi [10]

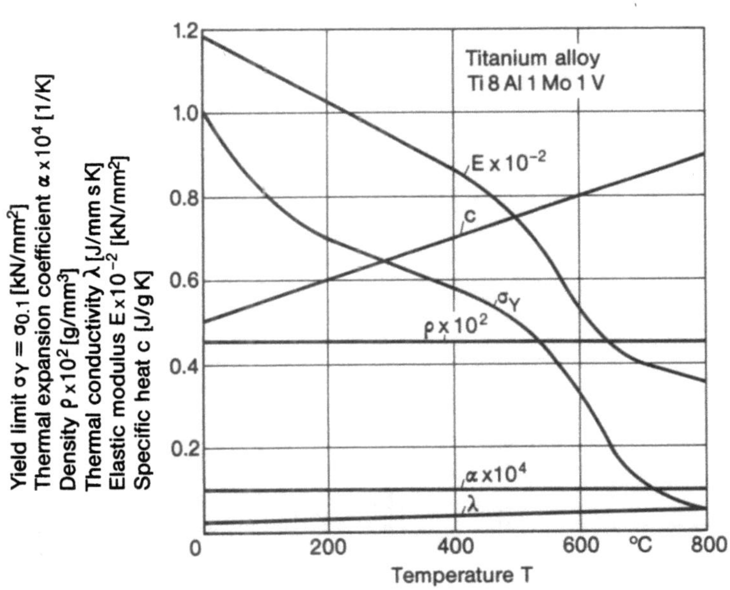

Fig. 116. Thermal and mechanical material characteristic values of titanium alloy (Ti8Al1Mo1V) dependent on temperature; after Lindh and Tocher [215]

Fig. 118. Obviously, the "true" stresses are presented related to the momentary cross-section. The curves shown may follow other courses during cooling as a result of transformation (for example dependent on the austenitizing conditions [251]).

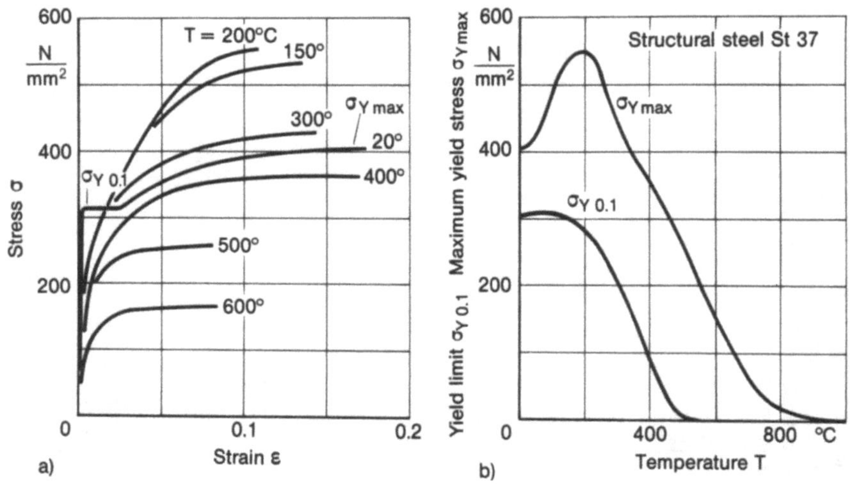

Fig. 117. Stress-strain curves of structural steel St37 (**a**), yield limit σ_Y (alternatively $\sigma_{0.1}$) and maximum yield stress $\sigma_{Y\max}$ (identical with ultimate strength σ_U) (**b**); after Argyris, Szimmat and Willam [193]

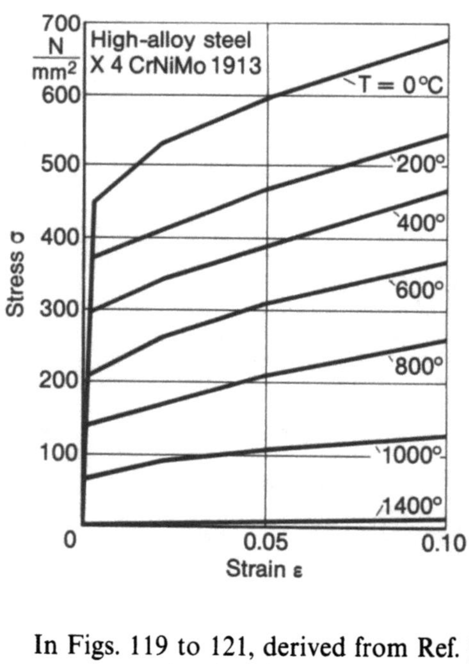

Fig. 118. Stress-strain curves of high-alloy steel X4CrNiMo1913, yield initiation and work-hardening dependent on temperature; after Rammerstorfer et al. [154]

In Figs. 119 to 121, derived from Ref. [23], the charateristic values α, E and ν are presented as banded areas for curve courses over temperature, on the one hand for mild (ferritic or eutectoid) and low-alloy steels, on the other hand for high-alloy (austenitic) steels, which localize the individual curves of Ref. [23]. The discontinuity of the curve for mild and low-alloy steel at 700 °C is due to the $\alpha\gamma$-transformation beginning here (A_{c1} temperature).

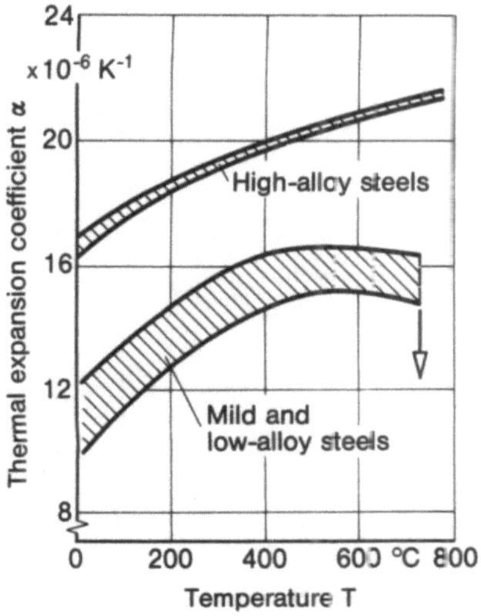

Fig. 119. Thermal expansion coefficient α of mild, low-alloy and high-alloy steels dependent on temperature, areas of individual curves after Richter [23]

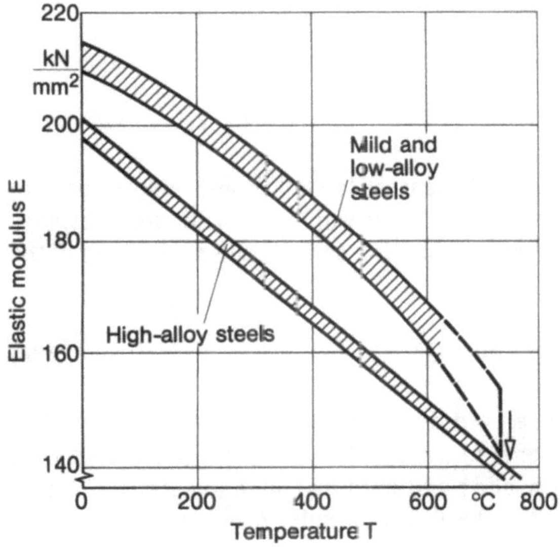

Fig. 120. Elastic modulus E of mild, low-alloy and high-alloy steels dependent on temperature; areas of individual curves after Richter [23]

The momentary ("differential") thermal expansion coefficient α, measured in the dilatometer (Fig. 119), rises with the temperature and is higher for high-alloy steels than for mild and low-alloy steels. At the transformation temperature, A_{c1}, an abrupt change to negative values occurs, caused by the transformation strain acting in opposite direction to the thermal strain.

At the melting temperature, a considerable increase in volume occurs and solidification has an opposite effect. This is generally of no consequence for the formation of welding residual stresses, however, because the yield stress drops towards zero at melting or solidification temperature, respectively.

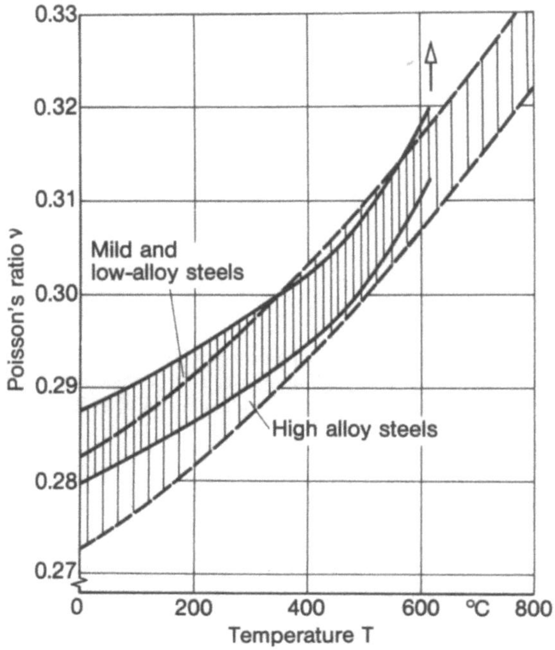

Fig. 121. Poisson's ratio ν of mild, low-alloy and high-alloy steels dependent on temperature, areas of individual curves after Richter [23]

The mean thermal expansion coefficient of low-alloy and mild structural steels in the temperature range up to 600 °C, in which the yield stress displays relatively high values, is stated for approximation calculations in the form of the combined quantity

$$\left(\frac{\alpha}{c\varrho}\right)_{\mathrm{m}} = 2.5 \times 10^3\text{–}3.0 \times 10^3\ [\mathrm{mm^3/J}] \tag{134}$$

The elastic constants E and ν, measured dynamically in the resonance test (the values measured statically differ only to a negligible extent) (Figs. 120 and 121), each show a narrow band for low-alloy and mild or high-alloy steels. Close to the transformation temperature A_{c1}, the elastic modulus drops rapidly (towards zero) while Poisson's ratio rises sharply (towards 0.5).

The dilatometer curves each for a steel without and with transformation have been presented and explained by way of example in Section 3.1.1. The thermal expansion coefficient α for a high-alloy nickel steel (11.6% Ni) with austenite-martensite transformation at around 350 °C is shown in Fig. 122 [154]. In the heat-affected zone of welded joints, greatly different maximum temperatures $T_{\mathrm{a\,max}}$ exist with only slightly differing cooling times $\Delta t_{8/5}$, so that diagrams of the type of Fig. 123 are required for analysis (further diagrams in Ref. [238]). The heat-affected zone should then be discretized in several layers using a fine mesh in the finite element model, which at the same time benefits the registration of the transformation-related temperature discontinuity.

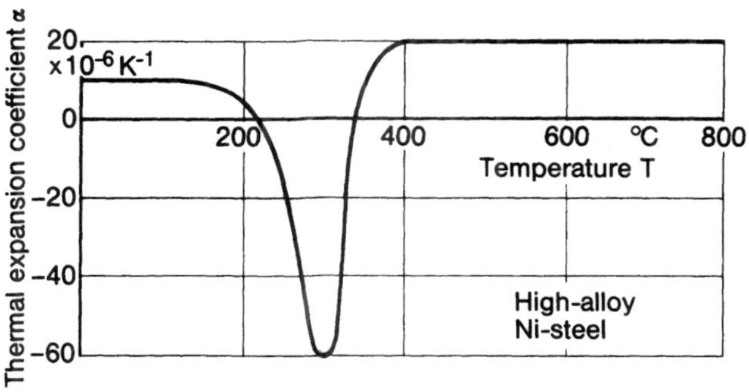

Fig. 122. Thermal expansion coefficient α dependent on temperature for high-alloy Ni-steel, change of sign in the temperature range of microstructural transformation; after Rammerstorfer et al. [154]

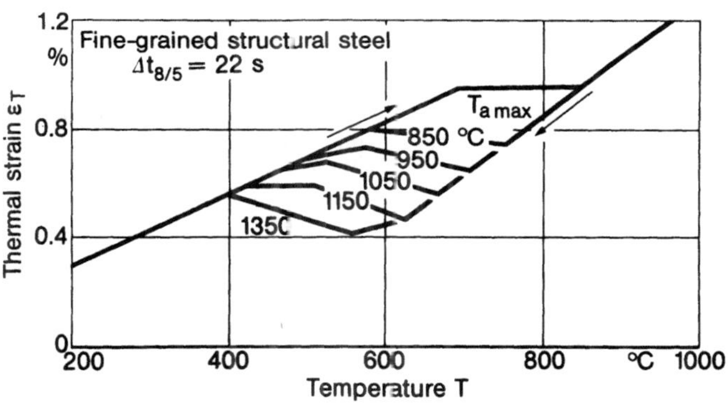

Fig. 123. Dilatometer curves of fine-grained structural steel suitable for welding showing microstructural transformation; after Jonsson, Karlsson and Lindgren [238]

The quantitative relation between microstructural change (identifiable by the dilatometer curve) and yield limit of the microstructural constituents austenite, pearlite and martensite during heating and cooling is depicted in Fig. 124 (including the elastic modulus). The considered high-temperature steel is CrMoV-alloyed. The differing yield limits of the various microstructural constituents are to be noted.

Less generally known is the also important fact for the formation of welding residual stresses that the yield stress may be greatly diminished in the temperature range of transformation compared to the lower yield stress of the two phases concerned (Fig. 125). The reduced yield limit during cooling of a transforming two-phase material compared to the same during heating (Fig. 126) is termed "yield strength hysteresis". In the case of microstructural transformation in the presence of stress, a plastic deviatoric transformation strain occurs in addition to the volumetric transformation strain.

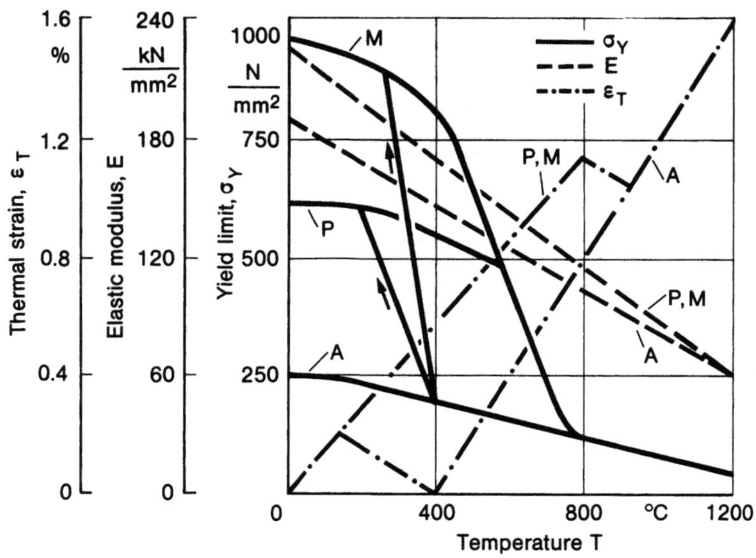

Fig. 124. Yield limit and elastic modulus of microstructural constituents austenite (A), ferrite and pearlite (P) and martensite (M) dependent on temperature, combined with dilatometer curve of thermal strain designating a typical microstructural transformation; after Michailov [179]

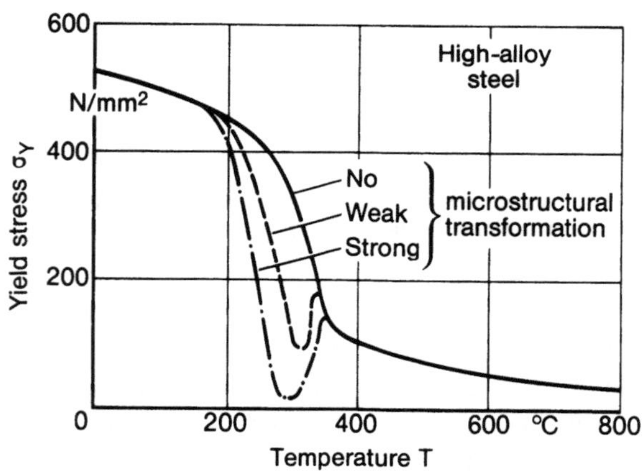

Fig. 125. Yield stress reduction in temperature range of microstructural transformation for high-alloy Ni-steel, after Rammerstorfer et al. [154]

This plastic anomaly, termed "transformation plasticity", is caused by second order residual stresses resulting from the discontinuous distribution of transformation progress in neighbouring crystallites [139]. Explanations on the basis of atomistic or dislocation theory concepts have also been given. Transformation plasticity has the characteristics of superplasticity but the latter occurs here only under the non-constant test conditions of transformation.

The martensitic transformation in the fusion and heat-affected zone of mild steel is modelled in Ref. [193] with material characteristic values according to

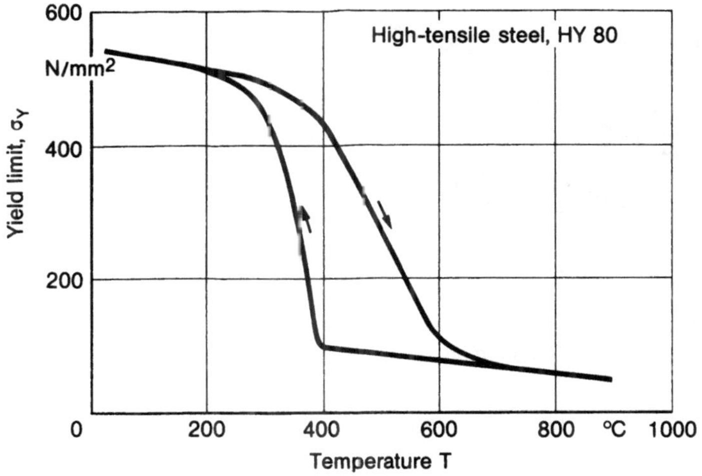

Fig. 126. Yield limit versus temperature for high-tensile steel HY80, showing yield strength hysteresis; after Oddy, Goldak and McDill [257]

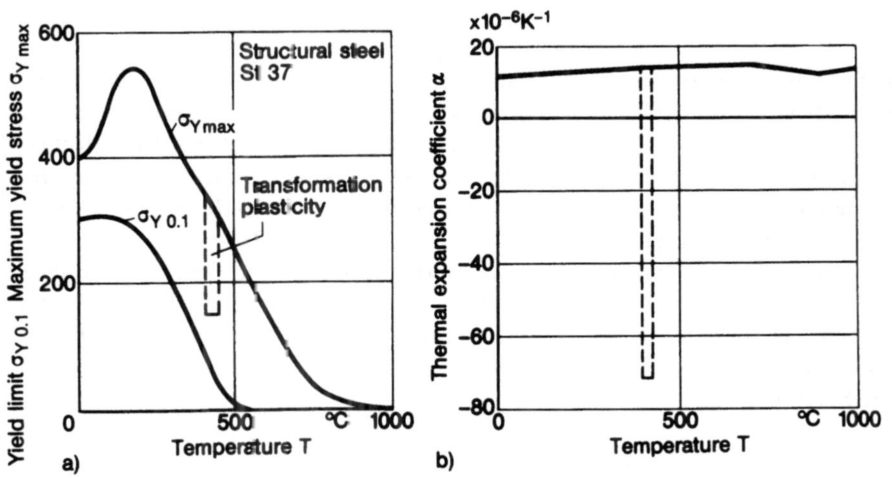

Fig. 127. Yield limit $\sigma_{Y0.1}$ and maximum yield stress $\sigma_{Y\max} = \sigma_U$ (a), just as thermal expansion coefficient α (b) of structural steel St37 dependent on temperature covering microstructural transformation to martensite; after Argyris, Szimmat and Willam [212]

Fig. 127. It is questionable whether allowance for the transformation plasticity solely in $\sigma_{Y\max}$ as presented in Fig. 127 is adequate. The influence of stress on transformation strain in the dilatometer test is shown in Fig. 128. Obviously, the transformation strain is reduced by compressive stresses or even reversed.

Time-temperature transformation and peak-temperature cooling-time diagrams are used to determine whether transformation occurs in the heat-affected zone or in the weld metal, at which temperature and cooling rate this takes place

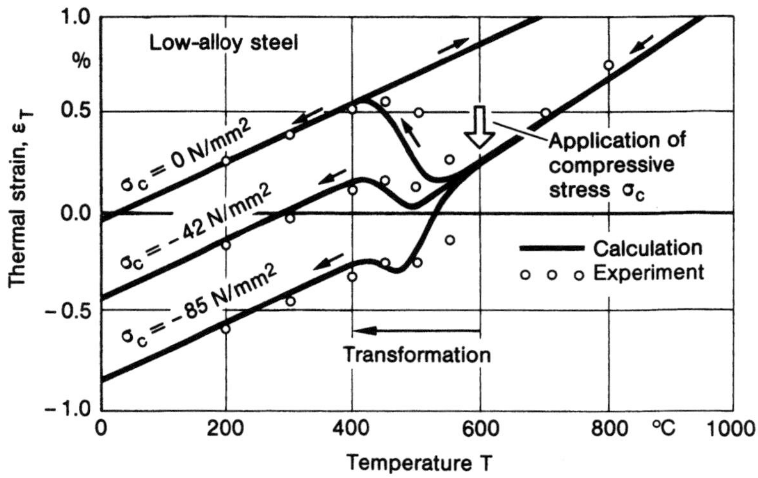

Fig. 128. Lower part of dilatometer curve of thermal strain indicating microstructural transformation, influence of transformation plasticity activated by compressive forces, low-alloy steel AFNOR15MND5 (French standard); after Leblond and Devaux [132]

depending on maximum temperature and temperature holding time, which microstructural constituents occur to what extent during this process (martensite formation being especially important) and what the resulting strength effects are. Alternatively, thermal and microstructural-kinetic calculations are performed to obtain these data including the coupling between thermal, mechanical and microstructural effects if necessary (see Section 2.4.2).

At elevated temperature, (time-dependent) viscoplastic deformation (creep or relaxation) occurs in addition to elastic and (time-independent) plastic deformation. The time-dependent processes may be ignored in single-pass welding. In multi-pass welding and hot stress relieving, however, they play a significant role (see Section 4.4.3.2).

3.2 Finite element models

3.2.1 Intelligent solution

Analysing welding residual stresses and welding deformations using finite elements in space and time (including finite difference methods) which simulate the thermal and mechanical (elastic-viscoplastic) material and component behaviour during welding, with the degree of detailing common from elastic component analyses, remains an insoluble task even in the age of supercomputers. The following features of the finite element task posed assist the expert in understanding the degree of difficulty and the immense effort involved in a general solution:

- The model should be designed three-dimensionally, at least in the weld area, in order to allow for the different cooling conditions in the interior and on the surface.
- The process to be modelled is transient to a high degree, with extremely different field gradients dependent on position and time because of the rapid heating-up and cooling-down events.
- The process to be modelled is non-linear and temperature-dependent to a high degree in respect of the thermomechanical material behaviour.
- The momentary, local material behaviour is dependent on the local thermal and mechanical stress and strain prehistory.
- Material is molten during welding, in some cases also added to the component, and, as it solidifies, changes the connectivity of the component.
- Changes in state and microstructure have to be simulated.
- Defects and cracks, which may occur at critical points, make the continuum concept doubtful.

The numerical solution of this extremely complex problem in a general form would require exceedingly powerful computers and solution algorithms as well as self-adaptive (three-dimensional) meshes and (time-step) procedures. Powerful supercomputers are, of course, available today although methods and software development has not kept pace with the hardware progress. And even if these means were available, the convergence proofs and error assessments would meet with difficulties which, at this time, appear insurmountable.

There is a further problem which may prohibit the use of finite element analysis of welding residual stresses in industrial product and process development. Analyses of this kind demand a multitude of material characteristic values and their temperature dependence, which, for the time being, exist only fragmentarily. Many a material characteristic value is dependent not only on the material but on the respective microstructural state. It may also be necessary to take account of local anisotropies or inhomogeneities in the material characteristic values.

The critical remarks presented above relate, however, only to the project of simulating the complex reality in the finite element model with the maximum possible detail ("unintelligent solution"). They do not apply if the problem is consequently reduced to its respective central issue and only the influencing parameters dominant in the respective case are presented in the finite element model ("intelligent solution"). The finite element method then permits statements of practical relevance. This is all the more important as the alternative to analysis, residual stress measurement, permits statements only to a very limited extent. If a non-destructive measurement procedure is adopted, only the stress-state in the surface of the component is registered and, even with a destructive procedure, it is not possible to determine with a sufficient high degree of accuracy the complete three-dimensional stress state in the interior of the component.

Simplifications of the type mentioned may, for example, be:

- Reduction of the desirably three-dimensional mechanical model to a two-dimensional or even to a one-dimensional mathematical model, e.g. by assuming axisymmetry, by considering only the plate plane or only the plane of section perpendicular to the weld seam, by reducing the problem to a rod element or shrinkage force model.
- Simplification of geometry, support and load conditions.
- Symmetrization or periodization of the model.
- Reduction of the non-linear thermoelastic-viscoplastic model to a linear thermoelastic model.
- Reduction of the transient process to a quasi-stationary process.
- Decoupling of the thermal and mechanical processes.
- Ignoring defect or crack formation.
- Omitting the fusion and solidification phase as well as the transformation processes which take place at elevated temperature and consequently at low yield stress.
- Registering the transformation at lower temperature only globally in the specific heat and thermal expansion coefficient.
- Ignoring creep and hardening as well as introducing simplifications in yield laws.
- Simplification of groove shape and layers structure.
- Replacement of heat source movement by momentary application of total heat quantity or by rapid heat source movement, ignoring heat propagation in the direction of movement.
- Replacement of the temperature-dependent material characteristic values by temperature-constant averaged values in the decisive temperature range.
- Modelling of the formation of residual stresses as a pure cooling process.

The simplifications which are acceptable in the individual case depend on the questions posed after being reduced to the respective central issue. In general, the possibilities of the most important measure, namely reduction in dimensionality, are limited because allowance has to be made in the application-oriented analyses for the different cooling conditions in the interior and on the surface, particularly in respect of the microstructural transformation. The situation in practice in respect of simplifications, on the other hand, is facilitated by the fact that analyses are primarily expected to provide relative statements to specific problems.

It should therefore be noted that even "intelligent" finite element solutions to welding residual stress problems permit generally valid relative statements only in exceptional cases. The number of possible parameter combinations and process variants is almost unlimited in practice. Finite element analysis can be performed only for a very limited number of single cases out of the total parameter field if the expenditure is limited. It is often only possible to deal with a single case of relevance in practice. This, admittedly, covers an essential part of industrial interest in development and improvement of products and manufacturing processes; from a scientific viewpoint, however, it is only a "numerical experiment" which does not permit any general statement. In addition, numerical solutions have to overcome the difficulty that the results of analysis and measurement do not correspond very well if the model has been greatly simplified. On the other hand, some authors (e.g. Ref. [258]) state better agreement between analysis and measuring results for the simplified model.

The following sections with the overview of finite element solutions to welding residual stress problems are arranged according to the mathematical dimensionality of the finite element models. The one-dimensional rod element model for longitudinal residual stresses can be used to preclarify questions relating to the influence of different parameters. The one- or two-dimensional axisymmetrical ring element model is a particularly successful procedure for determining welding residual stresses. Two-dimensional plate model solutions for welding residual stresses have a slightly higher degree of complexity and expenditure. They have been developed both for cross section models with single- and multipass weld (model plane transverse to weld) and also for seam welding of sheet metal plates with unrestrained edges (model plane identical to plate plane). Three-dimensional problem solutions for welding residual stresses have been achieved in recent years.

In addition to the survey which follows, reference is also made to the older status reports [200, 201] as well as to an older literature survey [202] with many contributions in Japanese.

3.2.2 Rod element model

The longitudinal residual stresses in and near welds (or also in flame cut edges) may, under certain preconditions, be regarded as being mainly uniaxial. This assumption is particularly justified if the components examined are rods or plates with longitudinal weld which are not too wide and have not too large a thickness. But even if these conditions are only insufficiently met, the one-dimensional model may correctly reflect the effect of influencing parameters according to tendency. Uniaxial modelling leads to discretization of the considered continuum into rod elements.

Uniaxial modelling of longitudinal residual stresses was first applied by American authors (literature references in Ref. [209]). It was further pursued and independently developed in particular in Russia (for example by Okerblom [2]). In Germany, contributions exist from Tall and Feder [207, 208] and Radaj [209, 210]. The model is reactivated in Ref. [211]. In the simplest form of the model, the following assumptions are made:

- The stress state is uniaxial. The component with weld seam can be considered to be composed of a large number of rod elements which carry only tension or compression (without bending) in the direction of the weld.

- The stress state is additionally plane, the component can be modelled as a membrane plate, the rod elements are arranged in a single plane.
- The stress state is formed under the deformation condition that the cross-sections of the component remain plane. The rod elements can be considered as being rigidly connected in their end points. With a large plate width and the weld positioned centrically, the condition is simplified to a rigid support.
- The deformation is elastic-plastic; creep is ignored.
- The temperature dependence of the yield limit is substituted by a horizontal line between ambient temperature and, for example, 500 °C in the case of mild steel or 250 °C in the case of aluminium, followed by a steeply linear, frequently vertical drop to zero. Hardening is ignored.
- All the remaining material characteristic values are set constant and identical to the value at ambient temperature in the considered temperature range.
- Microstructural transformation strains are ignored.
- A single heating and a single cooling step in the component cross-section (with temperature constancy over plate thickness) is taken as a basis.

The simplest form of the uniaxial membrane model, still without rod elements, is graphically depicted in Fig. 129 for a plate strip with centric longitudinal weld and in Fig. 130 for a plate strip with eccentric longitudinal weld. The material, mild structural steel, displays a constant yield stress for $0 < T < 500\,°C$ and a linear drop to zero for $500 < T < 600\,°C$. The temperature state in cross-section 1–1′ relatively close after the (moving) heat source in which the 600 °C isotherm reaches its greatest transverse extension is regarded as the decisive basis for residual stress formation. From this cross-section on, cooling stresses are built up; beforehand it was only heating stresses. We now imagine that heating and cooling are completed each in a single step with the cross-section remaining plane, so that the stresses follow the thermal strains up to the yield limit (which is set temperature-dependent in a simplified manner) and in so doing constitute an equilibrium system.

The heating strains αT below 600 °C for the centric longitudinal weld with temperature field according to Fig. 129b are shown in Fig. 129a on the left. Converted to stresses αTE, they must constitute an equilibrium system with horizontal median line m–m', whereby the yield limit σ_Y must not be exceeded. The cooling strains $-\alpha T$ in Fig. 129a on the right supply, after subtraction of the elastic heating strains, the thermal residual strains ε_R, shown crosshatched, which cause the residual stresses after cooling down. The latter are replotted over zero in Fig. 129c. They, too, after being converted to stresses by multiplication with E, must constitute an equilibrium system with horizontal median line m–m', without exceeding the yield limit σ_Y. The graphically determined results are the longitudinal residual stresses. The width of the plastic zone is designated w_{pl}. It corresponds approximately to the maximum width of the 200 °C isotherm. The median line m–m' identifies the shrinkage strain of the plate strip. In the plastic zone, the residual stresses reach the (tension) yield limit. A similar procedure is adopted for the eccentric longitudinal weld (Fig. 130). Only the

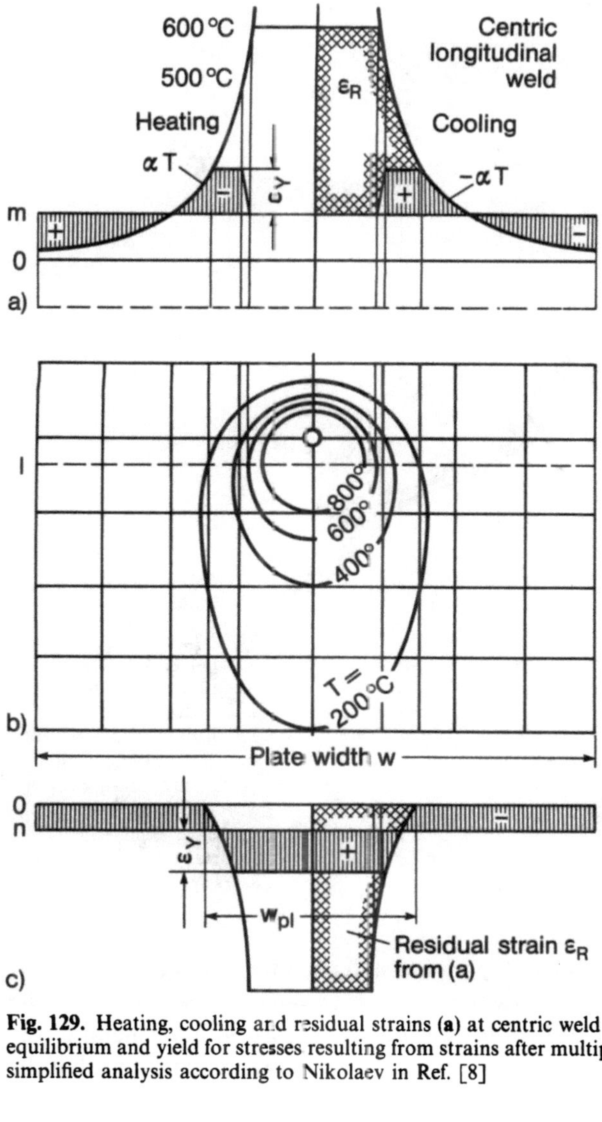

Fig. 129. Heating, cooling and residual strains (**a**) at centric weld in plate strip with isotherms (**b**), equilibrium and yield for stresses resulting from strains after multiplication with elastic modulus (**c**), simplified analysis according to Nikolaev in Ref. [8]

equilibrium systems have to be formed relative to the inclined median lines *m–m′* and *n–n′* (corresponding to freedom of resultant force and moment with cross-section remaining plane). The width of the plastic zone is reduced by virtue of the lower (bending) rigidity of the plate strip. It corresponds approximately to the width of the 450 °C isotherm. The above averaging can be performed simply by graphical means as estimations but requires computer support if the quantities are to be recorded more precisely.

The greatly simplified continuum model described above can be significantly improved by conversion into a finite rod element model and through use of

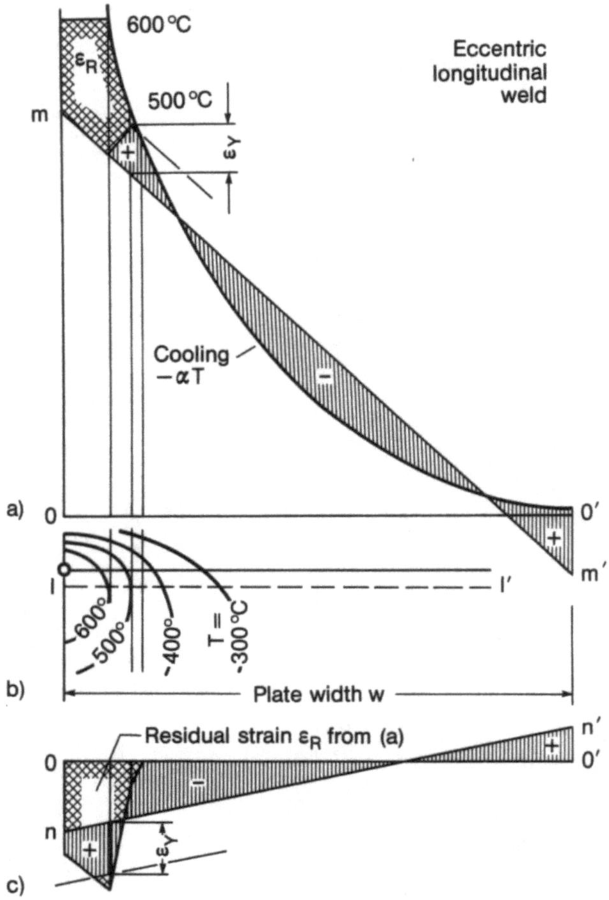

Fig. 130. Cooling and residual strains (**a**) at eccentric weld in plate strip with isotherms (**b**) equilibrium and yield for stresses resulting from strains after multiplication with elastic modulus E (**c**); simplified analysis according to Nikolaev in Ref. [8]

a computer. The total non-linear, temperature-dependent and transient behaviour in the area surrounding the moving heat source can be simulated. The plate strip model may be extended three-dimensionally to bars with a general cross-section contour. Only the assumption that the stresses are uniaxial and that the cross-sections remain plane is retained in this case.

The component cross-section is considered to be discretized into a large number of rod elements with the initially fictitious small length Δl, depicted in Fig. 131 on the plane model. The rod elements subjected to temperature increments ΔT are connected in the gravity centre of their two end cross-sections with a straight and rigid transverse bar, which is the means of simulating that the cross-sections of the component remain plane during the welding process. The rod elements may move freely in transverse direction so that varying transverse contraction may take place without suspending the transverse connection of the rod elements and without giving rise to transverse forces. The transverse bar on the right side may also shift in the rod element direction

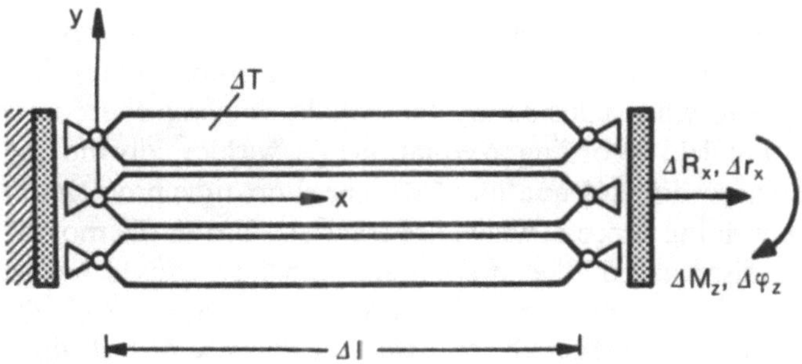

Fig. 131. Rod element model, increments of temperature, force and displacement, rigid transverse bars

x (translation Δr_x) and rotate around the cross-section axis z (rotation $\Delta\varphi_z$). In addition to the longitudinal force ΔR_x, the bending moment ΔM_z may also act on the transverse bar.

The plane rod element model can be expanded three-dimensionally to the x–y plane in order to represent the residual stresses in longitudinally welded section bars. The coordinate system should then be positioned in the section bar in such a way that the x-axis coincides with the centre of gravity line and the y- and z-axes coincide with the principal axes of inertia. The rod elements may be of different lengths before connection to the rigid transverse plates, which, following connection, may simulate the residual stress state prior to welding.

The inhomogeneous and transient, measured or calculated changing temperature distribution in the cross-section during welding is imposed on the rod elements. In line with its position in the cross-section, each rod element thus generally experiences a different temperature cycle, consisting of a rise and fall which is marked to a different extent depending on the closeness to the weld seam, the respective maximum of which occurs at different times. Thermal and transformation strains occur in line with the temperature course. The elastic-plastic, or as the case may be also elastic-viscoplastic, mechanical behaviour of the rod elements is characterized by temperature-dependent (as the case may be also time-dependent) stress-strain relations. The temperature course and the course of the resulting exterior force, including moment, are applied in steps through to complete cooling down. Following each step, compatibility of the rod element longitudinal displacements in the sense of plane cross-sections and equilibrium of the rod element longitudinal forces is assured. The rod elements undergo mutual counterstressing. The counterstressing state which remains after cooling is the residual stress state sought.

The change in longitudinal residual stresses as a result of subsequent tensile or bending load can likewise be simulated with the rod element model. Such loads are intentionally applied for reducing residual stress (see Section 4.4.3.3.1).

There follows a more profound presentation of the principles of the rod element model in its application to butt welding of membrane plates. The starting point is the quasi-stationary temperature field (i.e. temperatures dependent on position but independent of time) of the uniformly moving line source in the infinite plate when considering the equally moving coordinate system x, y, z (temperature field according to equation (33), with rapidly moving high-power source also according to equation (48)). The short-time processes of the initial phase of the welding process, which are transient also in the moving coordinate system, are disregarded. The stress and strain field caused by the temperature field is likewise assumed to be quasi-stationary, apart from the initial phase. This assumption includes the fact that, in the case of restrained components, the reaction forces which develop in the initial phase also remain unchanged during the welding operation. The cross-sections perpendicular to the weld are assumed to remain plane during welding. The shear-related warping of the cross-section which has a stress-reducing effect and occurs in reality as a secondary effect remains ignored. The shear stress distribution in the cross-section of a purely elastic plate with moving heat source is given by equation (114). A statement with regard to shear deformation closer to reality is offered by Ref. [8] (ibid., p. 66).

The plane temperature and stress field of the membrane plate is retained for the purpose of presentation. Proceeding from the quasi-stationary temperature field, the temperature increments $\Delta T(y)$ are determined in distances Δx, which identify the centres of plate strips perpendicular to the weld (width Δl) to be considered one after the other between plane cross-sections (Fig. 132). The process can also be described in such a way that the perpendicular plate strips are in time intervals Δt at one and the same position in the stationary coordinate system while the heat source has been moving on at speed v. It is then:

$$\Delta x = v\Delta t. \tag{135}$$

The temperature and stress states in the same cross-section at different times are therefore equivalent to the states in different cross-sections at the same time. The above subdivision of the plate into perpendicular strips therefore provides the finite model for the quasi-stationary stress distribution.

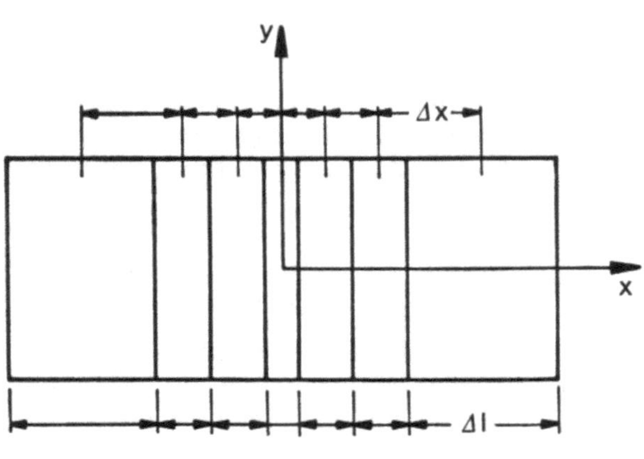

Fig. 132. Plate strips (in direction of y) transverse to weld (in direction of x)

The temperature increment $\Delta T(y)$ now has to be imposed on the plate strips in the time intervals Δt simultaneously enforcing that the strip edges remain plane. The stress reaction to this has to be calculated. For the calculation, the strip is subdivided into the rod elements described. The rod elements are subjected to the position-dependent sequence of temperature increments, the tensile and compressive stresses being built up under the restraint that the cross-sections remain plane. On the basis of the hypothesis of plane cross-sections, it would actually be reasonable to assume a linear distribution of stresses over the rod cross-section, i.e. tension and compression with superimposed bending. In fact, the bending deformation superimposed on the tensile deformation is ignored. The bending stiffness is set equal to zero. The error resulting as a consequence in the equilibrium condition for the total cross-section is small if the rod element cross-sections are selected adequately small. The peculiarities of the rod element model include the fact that there is a slight difference in the edge strains of adjacent rod elements (in accordance with the stepped deformation profile over the cross-sections) and that the cross-section centre line does not deflect (Fig. 133). The above simplification is avoided if the computationally more expensive bending-tension bar elements take the place of the pure tension rod elements. A further even more expensive step is the use of finite plate or solid elements in place of rod elements (see Section 3.2.5).

The equations for the incremental loading sequence in the rod element model required for the numerical analysis are stated in Ref. [209, 210]. The temperature-dependent stress-strain curves for mild steel are characterized there by the temperature dependence of elastic modulus E, yield limit $\sigma_{0.1}$, and hardening module H (Fig. 134). The reference parameters introduced are the values of the stated parameters at 0 °C, E_0, $\sigma_{0.1\,0}$ and H_0. In addition, the linear thermal expansion coefficient α, referred to its maximum value α_1 at elevated temperature, is included. The reduction of yield limit which occurs during unloading from the elastic-plastic state (Bauschinger effect) is taken into account.

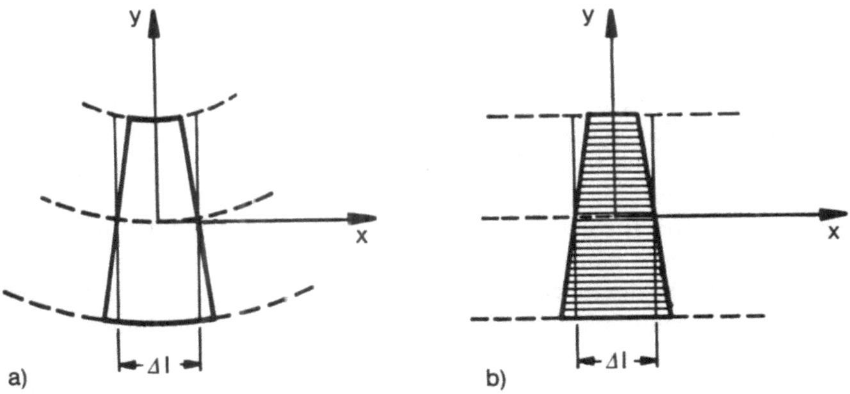

Fig. 133. Bending deformation of bar element (**a**) approximated by (pure tension) rod elements (**b**)

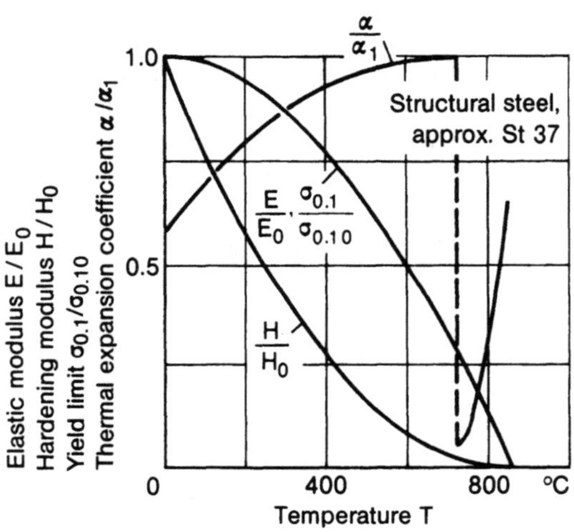

Fig. 134. Mechanical material characteristic values of structural steel schematized and normalized for the purpose of analysis

As an example of application, the longitudinal residual stresses during submerged arc welding as well as during electron beam welding of plates made of mild structural steel have been calculated using the rod element model. The plate width in this calculation is at first assumed to be large, and then relatively small. In Fig. 135, temperature and stress distribution are presented at different times *t* after passage of the heat source or at various distances *x* from the heat source. Following complete temperature equalization, the residual stresses shown with curve VIII remain. The residual stresses after submerged arc welding are drawn as a continuous curve, while those after electron beam welding are shown as a dashed curve, in each case for large plate width. The residual stresses after submerged arc welding with small plate width are shown as a dot-dash curve. The maximum residual stresses at the level of the strain-hardened yield stress occur in the centre of the weld.

3.2.3 Ring element model

The tangential, radial and axial residual stresses of the following groups of welded joints can be modelled as axisymmetrical: spot weld, plug weld, circular patch weld, stud weld, circular rod butt weld, circumferential weld on cylindrical and spherical shell. The residual stresses at flame cut holes are also approximately axisymmetrical.

The axisymmetrical solutions reported on below are combined under the heading "ring element model" irrespective of whether the procedure is completed through to the discretization into finite elements or the result is achieved already with the continuum model. In the simpler form of the finite element solution, the thermomechanical states are dependent, apart on time, only on the radial coordinate (single-layer membrane ring model), while in the more complex form allowance is made additionally for the axial dependence (single- or multi-layer solid ring model). In the former case, the solution is mathematically

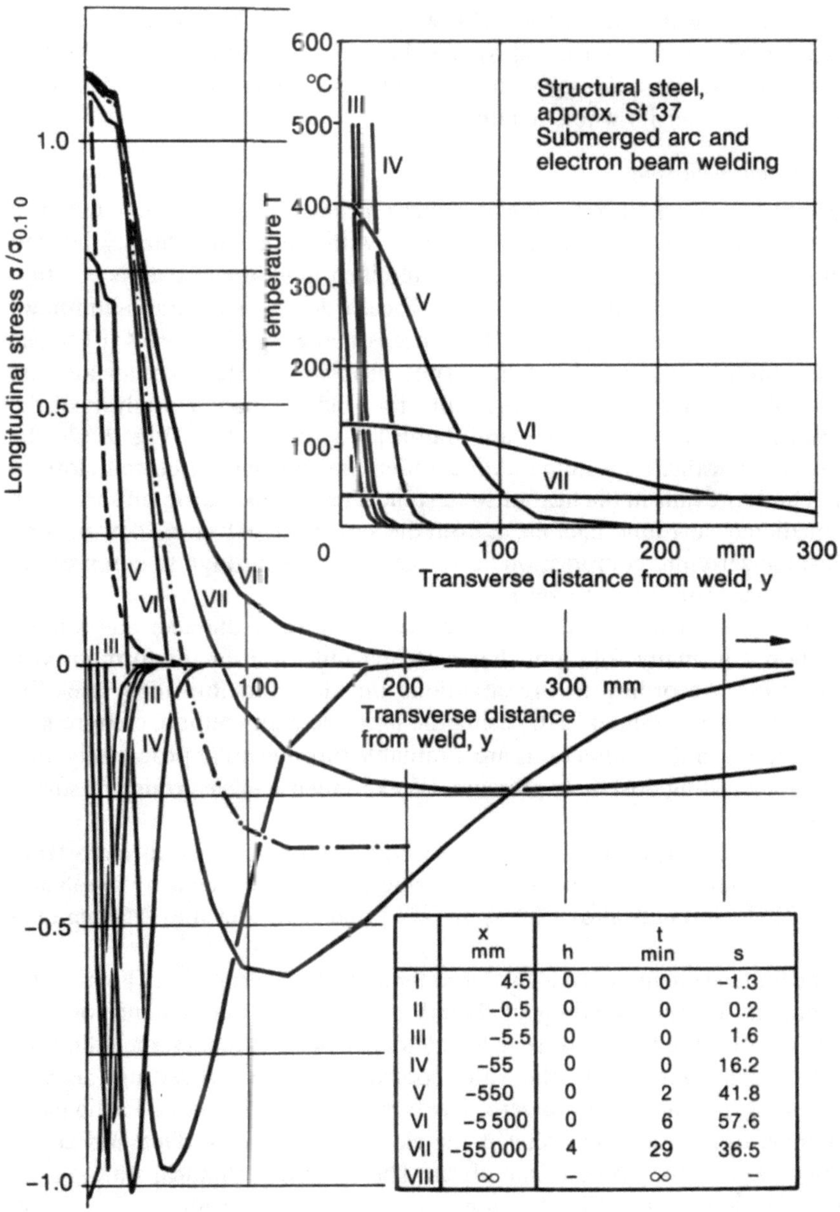

	x mm	h	t min	s
I	4.5	0	0	-1.3
II	-0.5	0	0	0.2
III	-5.5	0	0	1.6
IV	-55	0	0	16.2
V	-550	0	2	41.8
VI	-5 500	0	6	57.6
VII	-55 000	4	29	36.5
VIII	∞	-	∞	-

Fig. 135. Temperature and stress profiles transverse to butt weld during and after welding, structural steel (approx. St37), infinite plate and plate strip submerged arc welded (solid curves and dot-dash curve), infinite plate strip electron beam welded (broken curve); after Radaj [209, 210]

one-dimensional, and in the latter it is mathematically two-dimensional. In physical terms, however, two- and three-dimensionality exists, which necessitates the introduction of general yield and hardening laws. The condition that the (radial) cross-sections remain plane is exactly met. The starting basis of the analysis is the axisymmetrical transient temperature distribution in the area of the weld, which can be determined by calculation or measurement. To preserve the axisymmetry, the weld seam in the model is considered to be deposited instantaneously over its whole length.

Heat spot and weld spot

The result of the residual stress analysis after Gurney [212] for the circular steel plate with heat spot using a lattice model with radial and tangential rod elements is shown in Fig. 112. A greatly simplified functional-analysis solution for the circular plate with heat spot in the case of heat softening aluminium alloys is offered by Pelli [52, 214]. Reference is made by Vinokurov [8] to the early non-finite solutions of the heat spot problem by the Russian authors Leikin, Bakshi and Zolotarev. The result of an older Japanese analysis of the plug weld problem by Watanabe and Satoh [213] is shown in Fig. 9. On the basis of the investigations stated above, radial and tangential tension close to the yield limit prevails in the heat or weld spot. These stresses drop off sharply in the immediately surrounding area, σ_r from the above value to zero, σ_t by passing first through zero into compression (as the case may be through to compressive yield) and then from there to zero.

According to the investigations of the Russian authors, the electrode contact force, ignored in many solutions, has a strong influence on the formation of residual stress. According to investigations on mild and low-alloy steels by Popkovskii and Berezienko [220] using the finite element method, compressive residual stresses in three directions are attainable through an appropriately high level and long lasting post-weld pressure. Weak influence of electrode pressure is stated in Ref. [215, 216].

The residual stresses at spot welds, without taking account of microstructural transformation, have been analysed using solid ring elements by Lindh and Tocher [215] (titanium alloy), Nguyen [216] and Schröder and Macherauch [217].

In the finite element analysis by Schröder and Macherauch [217] with solid ring elements for spot welding of 2 and 10 mm thick plates made of steel X4CrNiMo1913, the temperature field calculated close to reality, including the strong heat propagation into the water-cooled electrodes but setting the electrode pressure unrealistically to zero, has been taken as the basis. Particularly steep temperature gradients occur at the electrode contact surface and at the weld spot edge on the inside of the plates. The thermomechanical state on the inside of the plates at the end of heating is characterized according to Fig. 136 by a high temperature in the weld spot centre, a sharp temperature drop at the weld spot edge, temperature-related yield stress reduction in the weld spot, and a surface stress maximum somewhat outside the weld spot edge. Temperatures and stresses vary widely in the axial direction. After complete cooling, the stress curves for the inside and outside of the plate, shown in Fig. 137, are relatively less variable in the axial direction. The basic pattern of the stresses known from the corresponding membrane ring models is confirmed. The effects resulting

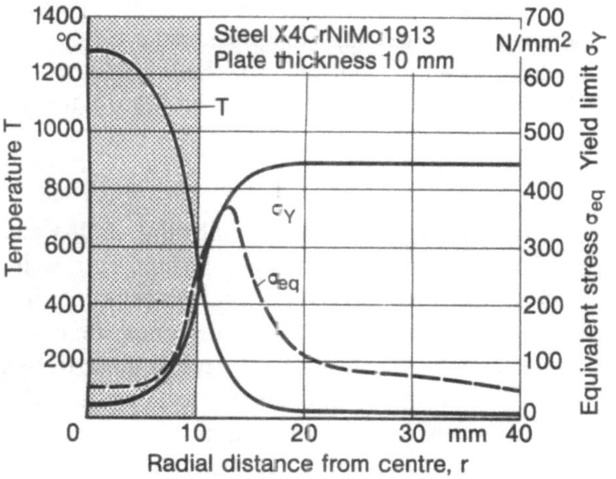

Fig. 136. Temperature T, yield limit σ_Y and equivalent stress σ_{eq} (after von Mises) on inner surface of spot welded joint at end of heating; after Schröder and Macherauch [217]

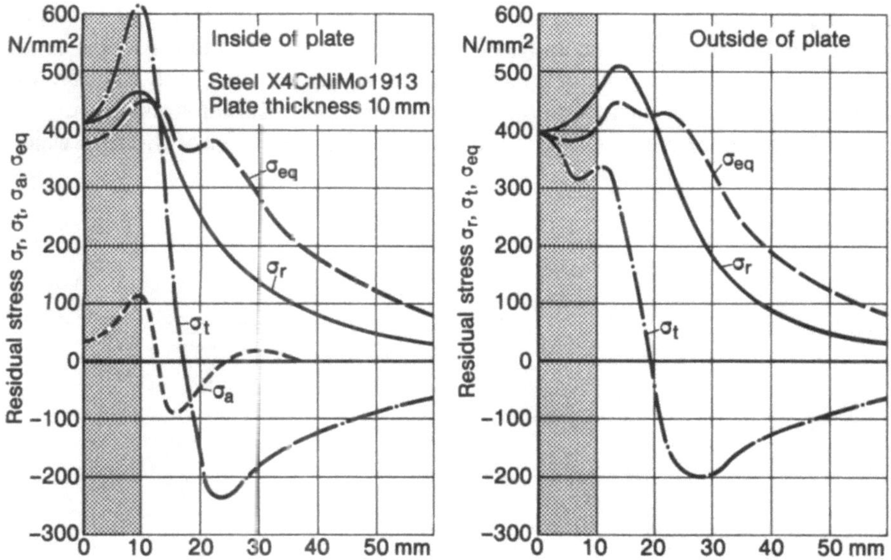

Fig. 137. Residual stresses σ_r, σ_t, σ_a, σ_{eq} on inner surface (**a**) and outer surface (**b**) of spot welded joint; after Schröder and Macherauch [217]

from the three-dimensionality of the model should, however, be emphasized. A triaxial tensile stress state occurs at the weld spot edge. The stress σ_a in the axial direction on the inside of the plates occurs with the condition that slot opening is suppressed. On account of the triaxial nature of the tensile stresses, the tangential stress σ_t at the weld spot edge is raised considerably above the uniaxial yield limit σ_Y.

The computational approach by Gorissen [218], to determine welding residual stresses for spiral-type welds in large-diameter pipes using a (locally) axisymmetrical model, led to false results, contrary to the arguments in Ref. [218]. Model, measurement and residual stress axiom used are obviously incorrect [219].

Ring weld

A functional-analysis approach on the basis of the equilibrium and yield condition is discussed in Ref. [8] for the instantaneously deposited ring weld in the circular plate made of mild steel (an approximation, for example, for the patch weld) (Fig. 138). A distinction is made between three zones, the inner zone I, the weld zone II, and the outer zone III. Tangential stresses close to the yield limit and lower radial stresses, increasing to the outside, occur in the weld zone. Fig. 138 shows three possibilities of stress distribution in this connection. Low biaxial tension or compression prevails in the inner zone. Radial tension and tangential compression (decreasing to the outside) prevail in the outer zone. The actual stress profile depends on the stiffness of the inner and outer zones, on the ring weld diameter, on the welding and material parameters. For example, with a small, stiff inner zone and wide outer ring, tensile residual stresses occur in the inner zone as a consequence of high heating-up whereas, with a larger, less stiff inner zone with narrow outer ring, compressive residual stresses are produced in the inner zone as a consequence of the contraction of the ring weld.

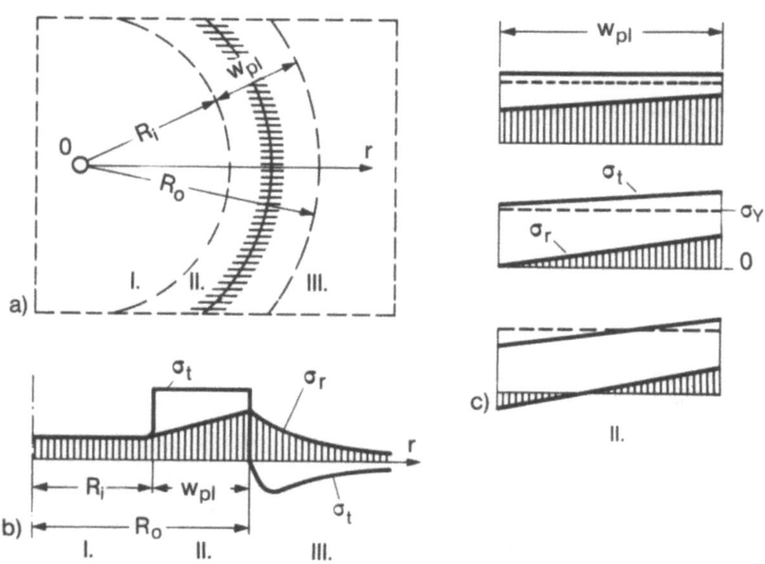

Fig. 138. Residual stress profiles transverse to ring weld in circular plate of mild steel; plate areas I to III (**a**), radial and tangential residual stress profile (**b**), three variants of residual stress profile in area II (**c**); after Vinokukrov [8]

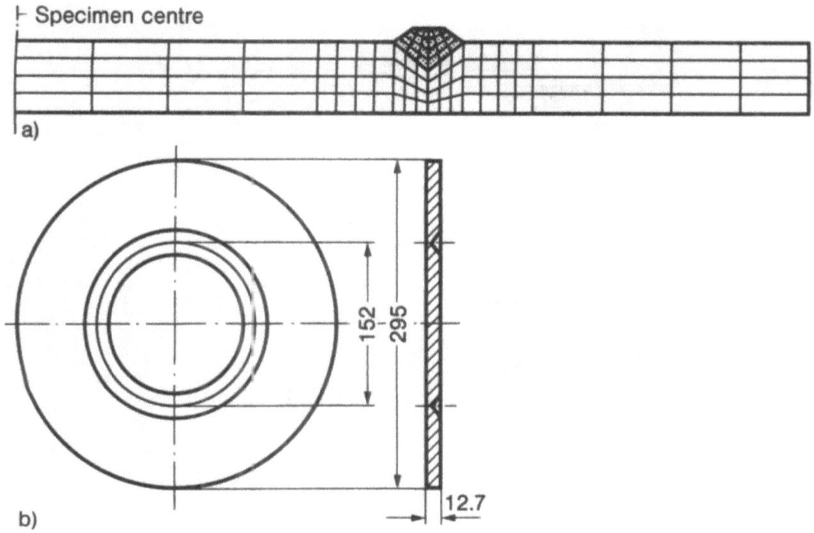

Fig. 139. Solid ring element model (shown with surfacing weld) (**a**) of ring groove weld specimen (shown without surfacing weld) (**b**); after Hibitt and Marcal [221, 222]

Hibbitt and Marcal [221, 222] analysed the transient temperature field, the welding residual stresses, the plastic strains and the residual stress reduction as a result of hot stress relieving for the instantaneously deposited ring weld on the circular plate of the ring groove weld test using solid ring elements (Fig. 139). The poor correspondence between analytical and experimental results was correctly attributed to the fact that microstructural transformation processes were disregarded in the analysis.

Nickel and Hibbitt [37] analysed the residual stresses relating to the first pass of a ring-shaped butt weld produced with a total of three passes on the torus-shaped connecting ring of a nuclear reactor component made of the NiCrFe alloy Inconel 600.

In order to simulate the strong influence of the different cooling conditions and microstructural transformation processes on the surface and in the interior of weld joints on the residual stresses, annular plates with a centric weld seam have been analysed by Yu [223] using solid ring elements. The annular plate is to simulate the infinite weld seam in a straight plate strip with unrestrained transverse deformation. Therefore, the annular plate is only slightly curved; plate width is 80 mm with 1000 mm outer diameter. The plate thickness is 6 mm.

Transverse residual stresses equal to zero should be expected in the mean over the plate thickness. Considerable surface stresses are, however, actually measured in these and comparable specimens. Numerical analysis shows that the residual stresses on the surface and in the interior add up to a zero resultant. The fundamentally different stress distributions in a non-transforming steel

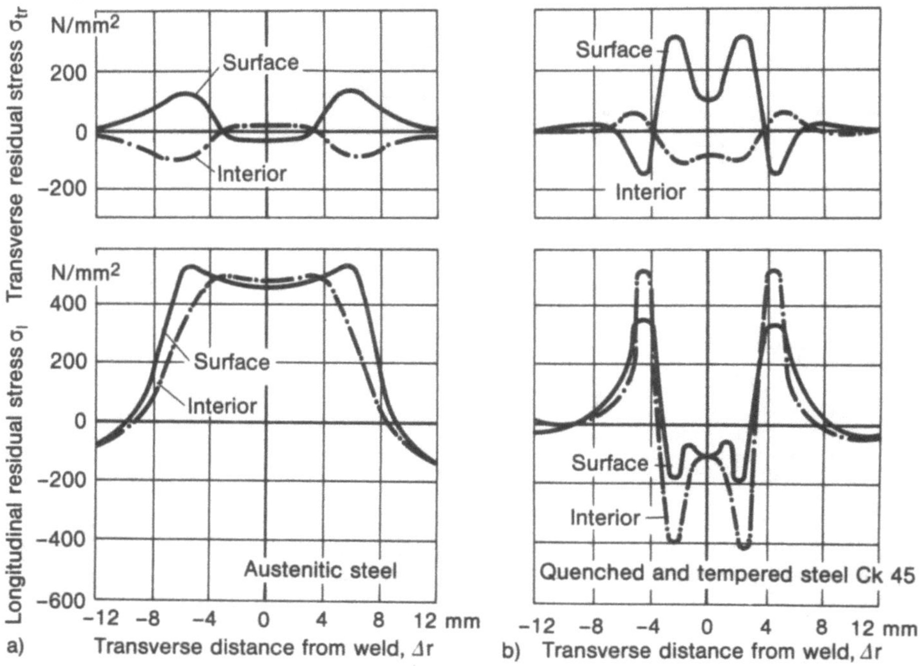

Fig. 140. Weld transverse and weld longitudinal residual stresses at centric butt weld in annular plate (outer diameter 1000 mm, width 80 mm, thickness 6 mm) of non-transforming austenitic steel (**a**) and transforming quenched and tempered steel CK 45 (**b**), on surface and in interior; after Yu [223]

(austenitic steel) and in a transforming steel (quenched and tempered steel Ck45) are shown in Fig. 140. The differences between surface and interior are particularly pronounced in the transforming steel. The reasons for the differences are the cooling and transformation processes, which do not occur simultaneously on the surface and in the interior. Maximum transverse stress values generally occur at points of greatest difference in longitudinal stress between surface and interior. Without transformation, maximum transverse tension occurs on the surface next to the weld seam; with transformation, maximum transverse compression occurs.

Butt welding of rods with circular cross section

Pressure butt welding of a bolt to a base plate of beryllium, including subsequent machining of the opposite side of the base plate and loading of the base plate edge, has been analysed by Cyr, Teter and Stocks [224] using solid ring elements (Fig. 141).

The quasi-elastic solution for the three-dimensional welding residual stress state in the interior of a resistance butt-welded circular section rod, proceeding solely from the temperature drop in the axial direction of the rod, is presented in Ref. [8] (Fig. 142). Whereas uniaxial compressive stresses in axial direction are determined for the surface, the crack-initiating triaxial tensile stress state with

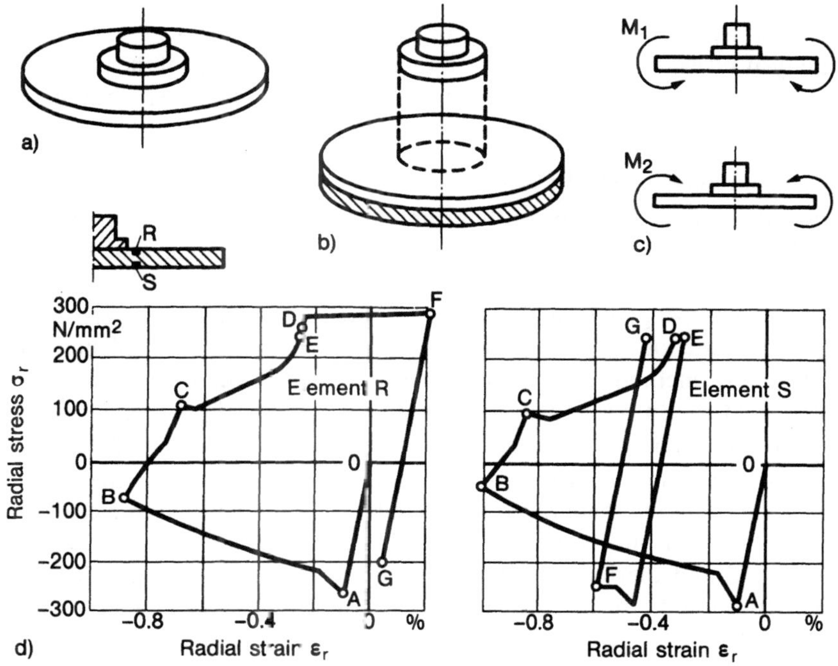

Fig. 141. Bolt on plate of beryllium (**a**), pressure butt welding followed by machining of the hatched layer (**b**), loading of plate edge by alternating bending moments M_1 and M_2 (**c**), stress-strain cycle in solid ring elements R and S (**d**): heating OAB, cooling BCD, machining DE, loading EFG; after Cyr, Teter and Stocks [224]

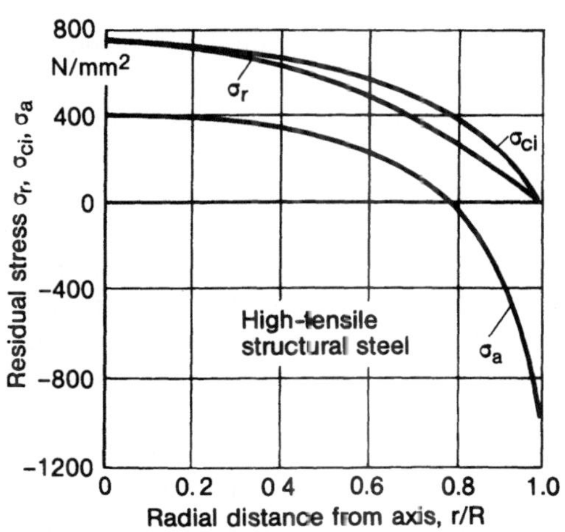

Fig. 142. Welding residual stress state in centre cross-section of resistance butt welded circular rod (outer radius R) of high-tensile structural steel; quasi-elastic analysis after Vinokurov [8]

a maximum value at the yield limit occurs in the interior. It is therefore recommended to perform a post-weld heat-treatment to reduce residual stresses.

The finite element analysis of flash forming during upset welding conducted by Nied [225] refers to large displacement, large strain and viscoplastic metal behaviour at elevated temperatures preceded by thermoelectric heating. This is a flow model neglecting elastic stresses (including residual stresses). The results of the analysis reveal that the material originally on the interface of the two parts to be welded is heavily deformed while expelling into the flash (but never substituted by material from the nearby inside).

Circumferential weld in cylindrical or spherical shell

The thin-walled cylindrical shell with instantaneously deposited circumferential weld (and unrestrained shell edges) and the corresponding spherical shell are dealt with by Fujita [226, 227]. He used a mathematically one-dimensional approach for heat propagation and stress formation, i.e. series expansion using the deflection function of the elastic cylindrical and spherical shell bending theory, the Ritz method applied per load increment, initial strain method, von Mises yield condition, improved Prandtl and Reuss yield law, slight strain hardening (with $E/100$) and temperature-dependent yield limit. The parameter $\lambda = \sqrt[4]{3(1 - \nu^2)/R^2 h^2}$ according to equation (160), known from the elastic cylindrical shell bending theory, is revealed to be the dominant parameter of the process (this results from the theoretical basis chosen). Systematic variation of the geometrical and thermal parameters results in largely identical deflection (i.e. constriction) and stresses with identical as well as largely identical stress maximum values with identical product $q_w \lambda$ in which q_w is the heat input per unit length of weld. Consequently, an equivalence of geometrical and thermal influencing parameters exists in respect of the residual stresses. Comparative calculations without and with incorporation of the heating-up phase reveal a strong influence of this phase, at least on the deflections. A high degree of correspondence between calculated and measured deflections and stresses was verified. Reference should additionally be made to the solution on the basis of the shrinkage force model [264] (Fig. 174). This solution is similar in respect of the deflection function used.

Figure 143 shows the pattern of the circumferential and axial residual stresses on the inner and outer surface of the cylindrical shell with circumferential weld. Tensile bending stresses occur on the inside in axial direction transverse to the weld, while corresponding compressive bending stresses occur on the outside. These stresses can be explained as being caused by a constriction of the cylindrical shell resulting from the longitudinal contraction of the circumferential weld. The maximum values occur characteristically alongside of the weld (i.e. not in the weld). Further away the sign of the stresses is reversed. A stress pattern similar to that known from the plane butt-welded plate occurs on the inside of the shell as a circumferential stress in the direction of the weld, high tensile stress in the weld and low compressive stress alongside it. On the outside of the shell, the longitudinal stress in the weld is largely reduced. A pronounced compressive stress maximum occurs alongside the weld. The considerable differences in the stress pattern between curved shell and plane plate needs to be emphasized. The stress pattern which occurs for the spherical shell with circumferential weld is similar to that of the corresponding cylindrical

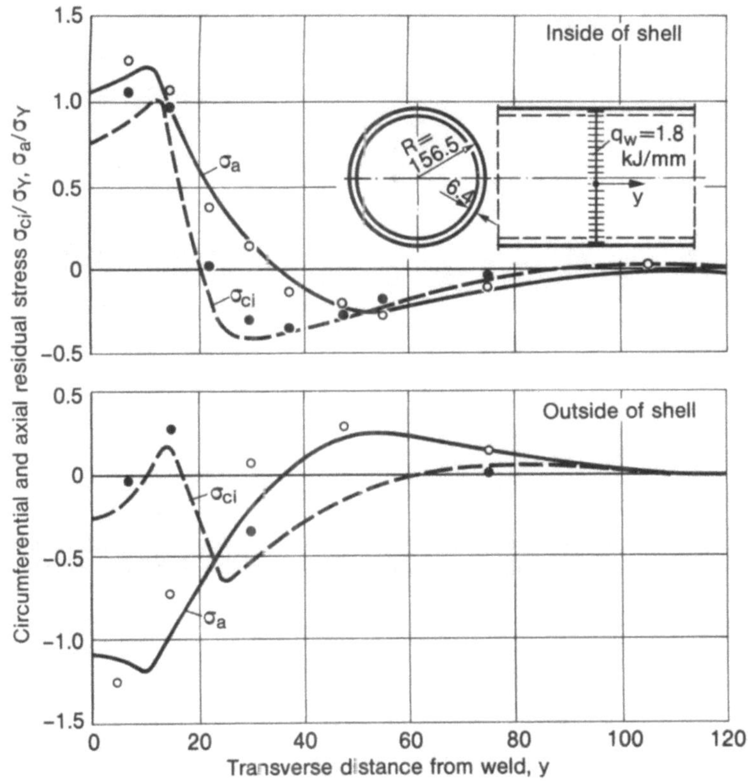

Fig. 143. Circumferential and axial residual stresses in cylindrical shell with circumferential weld, comparison of analytical and experimental results; after Fujita, Nomoto and Hasegawa [227]

shell (Fig. 144). Finally, it was possible to analyse the welded joint between cylindrical shell and semi-spherical shell in respect of residual stresses by means of the above computational method (Fig. 145).

In the contributions of Rybicki et al. [228, 229] (see also Ref. [230]), solid ring elements are used in combination with the quasi-stationary temperature field equation of seam welding for analysing the longitudinal and transverse residual stresses of a pipe circumferential weld (stainless austenitic steel, one-sided multi-pass weld) (Fig. 146). High tensile stresses, which drop to compression in the heat-affected zone, are determined in the axial and circumferential direction on the inner surface of the pipe with the weld root. The circumferential shrinkage force causes a contraction of the circumferential weld to a reduced diameter, i.e. a constriction of the pipe. It was possible to confirm these results by measurement, at least qualitatively. A further investigation [423] deals with the stress reduction as a result of inductive post-weld heat-treatment (see Section 4.4.3.3.4).

Further finite element analysis results from Josefson [379, 380] relating to weld residual stress distribution at single-pass and multi-pass circumferential

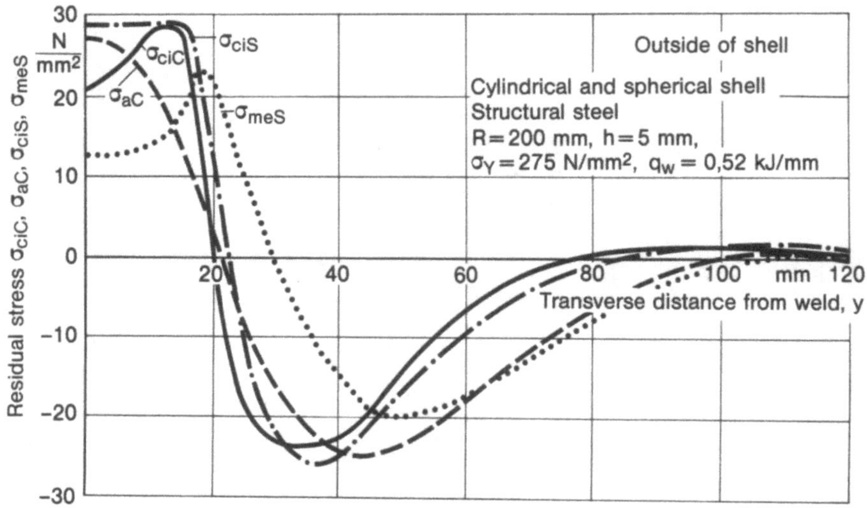

Fig. 144. Circumferential and axial or meridional residual stresses in cylindrical and spherical shell (indices *C* and *S*) with circumferential weld, comparison of analytical results; after Fujita, Nomoto and Hasegawa [227]

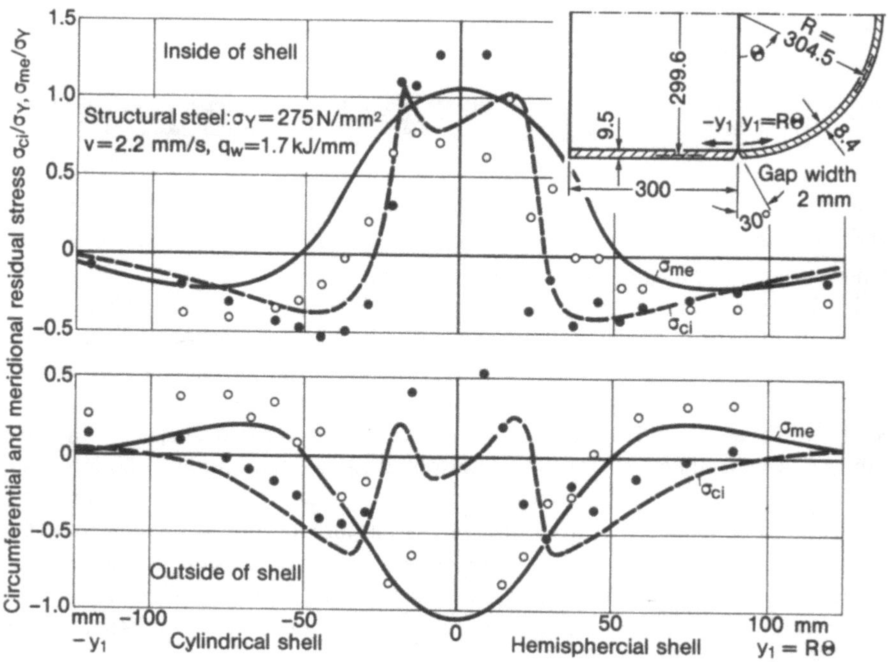

Fig. 145. Circumferential and meridional residual stresses in cylindrical shell jointed to hemispherical shell by circumferential weld, comparison of analytical and experimental results; after Fujita, Nomoto and Hasegawa [226]

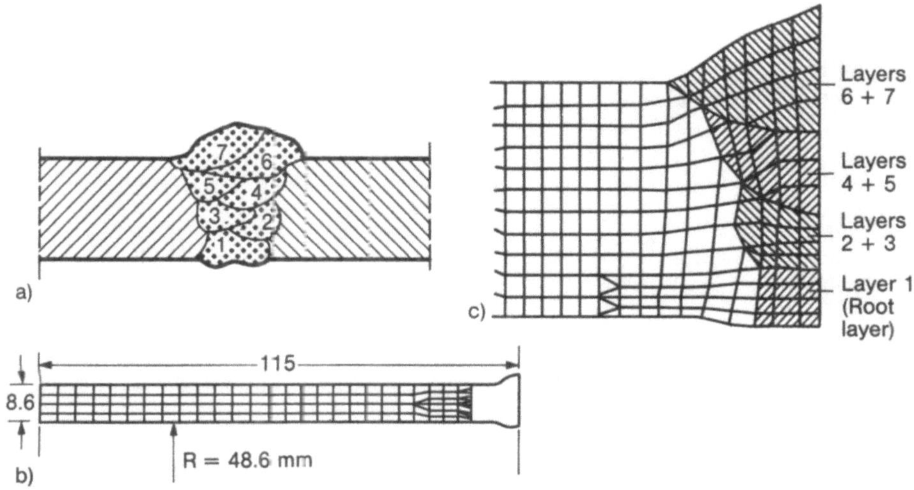

Fig. 146. Discretization into solid ring elements of multilayer circumferential weld in pipe; layer structure in reality (**a**), solid ring element model with enlarged detail (**b**, **c**), symmetry half with combined weld layers; after Rybicki and Stonesifer [229]

welds are presented in combination with hot stress relieving in Figs. 245 and 247. Particularly high residual stresses are observed in the martensitic part of the heat-affected zone. The strength-reducing triaxial tensile residual stress state occurs for the CMn-alloyed steel in question neither in the case of single-pass welding nor in the case of multi-pass welding. The three-pass circumferential weld is analysed in Ref. [231].

3.2.4 Membrane plate element model in the plate plane

It is an obvious step to determine the longitudinal and transverse residual stresses, as well as the appertaining strains and deformations, in relatively thin plates with butt welding by means of a finite membrane plate element model (elements arranged in the plate plane), assuming a plane temperature and stress state. Such investigations have been carried out with respect to hot cracking at the ends of long butt welds in plate assemblies, to their groove gaping during welding and to the weld longitudinal residual stresses.

The finite element discretization for longitudinal seam welding of large plates of a fine-grained structural steel, such as is used for example in ship-building, is shown in Fig. 147. The mesh refinement from the load-free plate edge to the weld is solved in an exemplary manner. The transverse strains and stresses at the start and end of the weld were analysed with this model in order to provide a basis for explaining the hot crack formation observed here (see Ref. [232–238, 194]. The temperature-dependent material characteristic values of the applied fine-grained structural steel, which is transforming and suitable for welding, is shown in Fig. 114. The non-thermal transverse strain

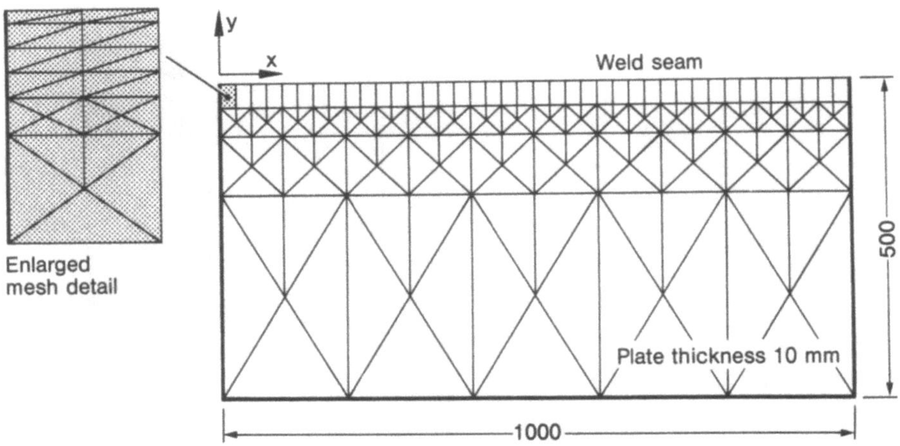

Fig. 147. Finite element mesh for rectangular plate with continuously deposited centric weld, triangular elements with cubic displacement approximation; after Jonsson, Karlsson and Lindgren [238]

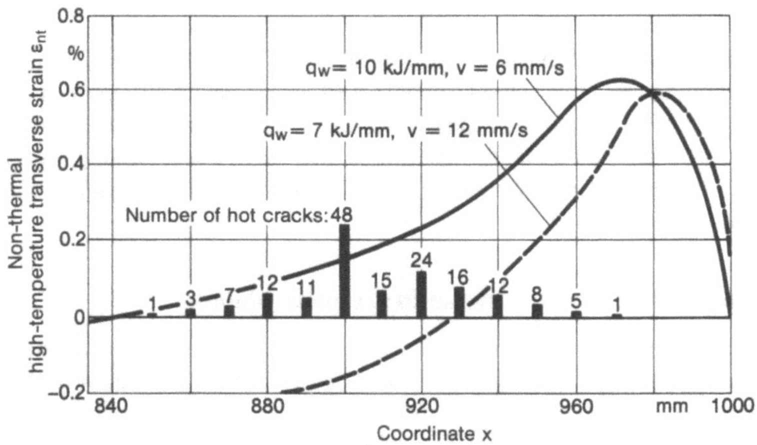

Fig. 148. Non-thermal high-temperature transverse strain ε_{nt} in temperature range between 1400 and 1100 °C determined by finite element analysis at end of weld, rectangular plate of fine-grained structural steel, different heat input per unit length of weld, number of hot cracks; after Andersson and Karlsson [236]

$\varepsilon_{nt} \geq 5 \times 10^{-3}$ in the cooling interval between 1400 and 1000 °C is regarded as the cause of hot crack formation. With increasing heat input per unit length of weld and decreasing welding speed, its maximum value shifts from shortly within the weld end in the direction of the centre of the weld length (Fig. 148). Multi-pass welding is advantageous. The correlation with the number of hot cracks is, however, only qualitatively satisfactory. The number of hot cracks was reduced by a factor of ten by extra local heating. The propagation of the hot cracks, as cold cracks, in the direction of the centre of the weld length is controlled by the transverse tensile stress (Fig. 149), whereas the weld ends are protected by transverse compressive stress (a risk of warpage occurs here). The transverse movement of the groove edges for the pre-tacked weld was also calculated. The groove gap first closes, and then, after overwelding of the centre tack weld, opens again. The maximum gap change is $\pm$ 0.15 mm. These results are typical of transforming structural steels.

The residual stresses longitudinal and transverse to the weld (instantaneously deposited in the model) were analysed using a membrane model and compared with measured values for a rectangular plate with slot weld tested in respect of hot and cold cracking (Figs. 150 and 151) (see Fujita et al. [239, 240]). The incremental initial strain method with improved yield law was applied. Of particular note are the favourable transverse compressive residual stresses in front of the weld end and the unfavourable biaxial tensile residual stresses within the weld end.

In the same way (see Ref. [240]), the residual stresses in sheet metal plates with welded continuous rib stiffeners were also analysed. The width of the weld zone with high tensile longitudinal residual stresses and the level of the lower compressive longitudinal residual stresses in the adjacent plate and rib areas could be correlated within a limited parameter domain, independent of dimensions, with the global post-weld temperature increase ΔT_0 (Fig. 152). This heating-up results from the heat input Q per deposited fillet weld, the volume-

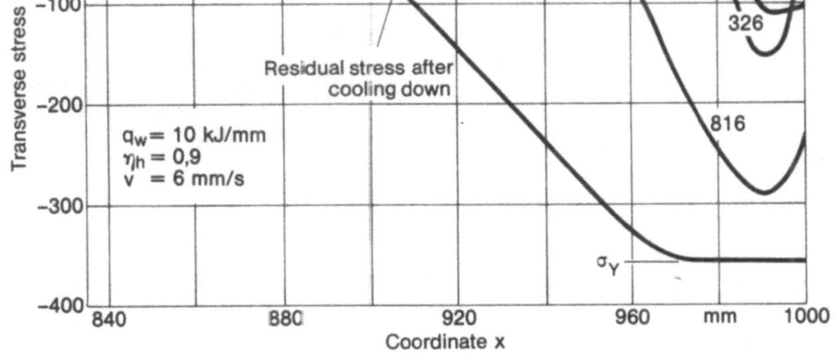

Fig. 149. Transverse thermal and residual stress at end of weld, rectangular plate of fine-grained structural steel different time intervals t_e after end of welding; after Andersson and Karlsson [236]

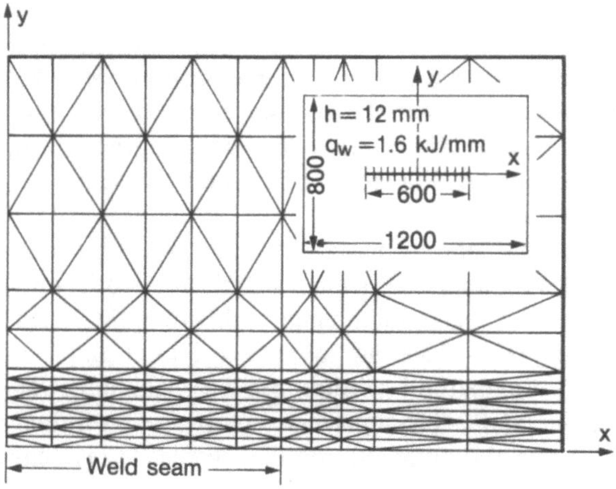

Fig. 150. Finite element mesh for plate with slot weld, symmetry quarter; after Fujita, Nomoto and Hasegawa [240]

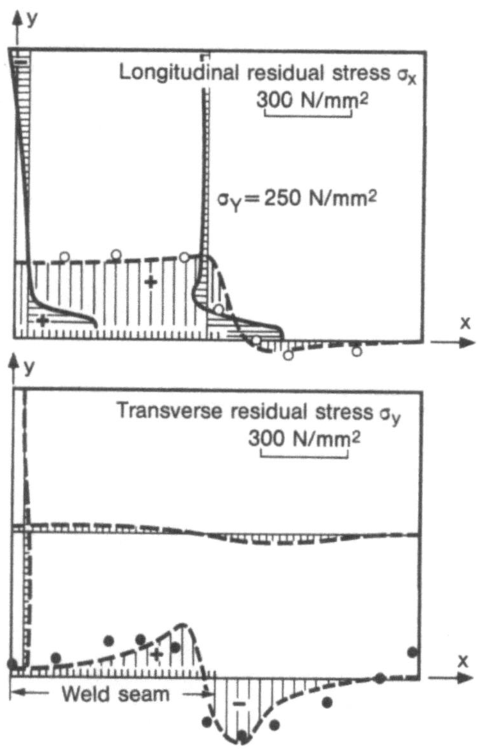

Fig. 151. Longitudinal and transverse residual stress in rectangular plate with slot weld, comparison of analytical and experimental results; after Fujita, Nomoto and Hasegawa [240]

heat $c\varrho$, the plate width w and the plate thickness h (neglecting the ribs):

$$\Delta T_0 = \frac{2Q}{c\varrho h w}. \tag{136}$$

Slightly curved thin shell elements, defined as degenerated solid elements, have

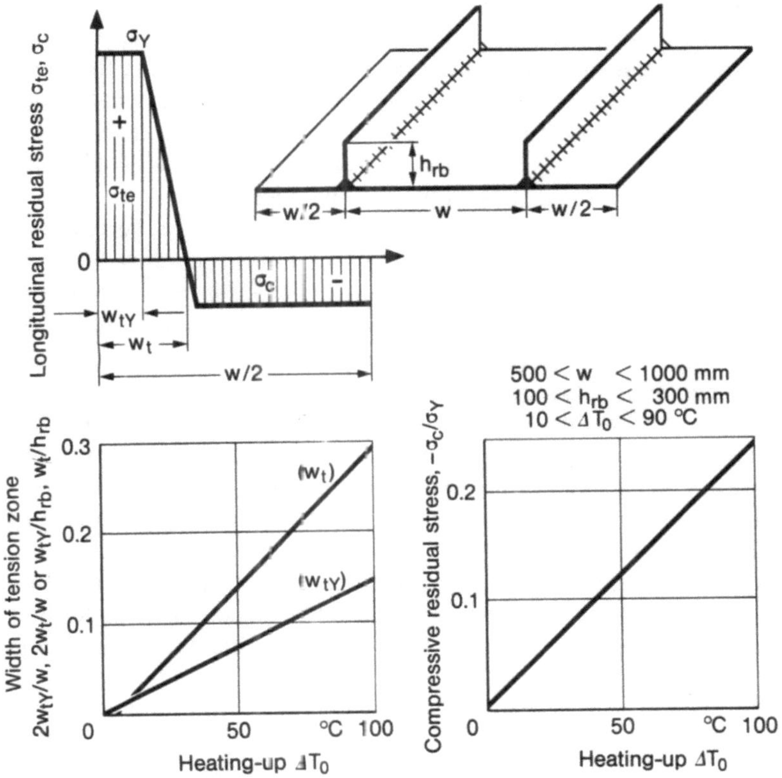

Fig. 152. Longitudinal residual stress σ (tensile stress σ_{te}, compressive stress σ_c) in plate with ribs over global temperature rise ΔT_0 of plate or rib, tension zone widths w_{tY} and w_t, result of finite element analysis based on plate model; after Fujita, Nomoto and Hasegawa [240]

been used by Lindgren and Karlsson [241] for analysing the residual membrane stresses in a thin walled pipe made of CMn steel and circumferentially GMA welded. The out-of-plane deformations are restrained in the model which thus behaves similar to a plane membrane plate element model.

3.2.5 Membrane plate element model in the cross-section

A second possible application of membrane plate element models, this time with a plane deformation state, is to consider the transverse contour model of the welded joint (i.e. the cross-section transverse to the weld seam). The biaxial stress state in the cross-section plane is obtained proceeding from the temperature field in this plane. The normal stress perpendicular to the cross-section plane, i.e. in the direction of the weld, follows from the condition of plane deformation. This corresponds to Bernoulli's hypothesis of the cross-sections remaining plane. The additional condition of completely suppressed normal displacements applies to symmetrical problems with a relatively large welding-unaffected cross-section. The less extreme assumption of plane cross-sections which can be displaced and rotated is recommended in the case of non-symmetrical problems, e.g. problems with bending shrinkage. Solid elements with an appropriate restriction of the degrees of freedom are also suitable for modelling the plane strain state. In the special case of element-constant state variables, what results is the rod element model. Such finite element analyses using the cross-section model have been performed, particularly in respect of hot and cold crack formation as well as of residual stresses in multi-pass welding.

The finite element discretization for the cross-section model of the specimen of the rigid restraint cracking test is shown in Fig. 153. Proceeding from the plane in-stationary temperature field (calculated using the finite difference method), the model is applied to determine the stress and strain course at the points A to D, which are considered to be particularly prone to hot and cold cracking (see Ueda et al. [242–246, 252]). Corresponding analyses have been conducted for the double fillet weld specimen as well as the edge fillet weld specimen. They permit a rough quantification of these tests. These self-restraint tests are, however, increasingly being abandoned in favour of tests with force-controlled global stress. The latter likewise permit more meaningful statements based on local stress analyses.

The dependence of the reaction force and reaction stresses in multi-pass welding of a transverse butt weld between rigidly restrained tensile plates (hot crack specimen in the rigid restraint cracking test) with different preheating in

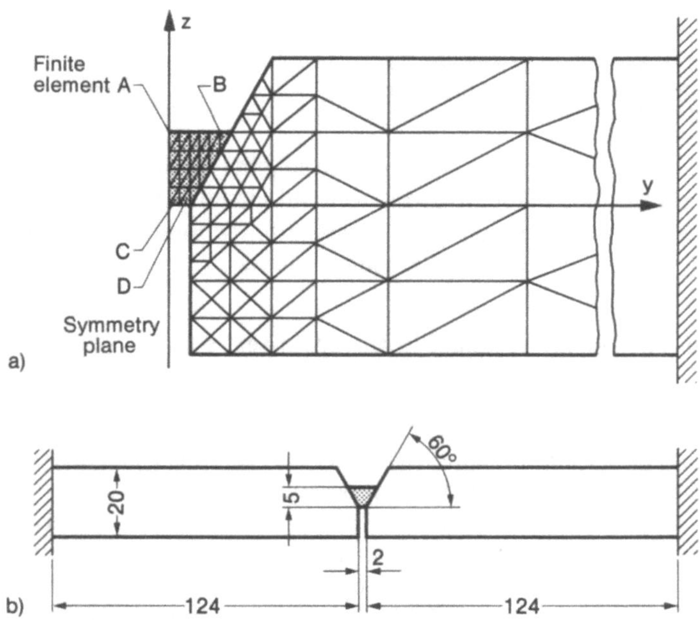

Fig. 153. Finite element cross-section model (**a**) of specimen (**b**) of rigid restraint cracking test; after Ueda and Yamakawa [242]

the weld area has been examined by Satoh et al. [244]. No proportionality exists between reaction force and reaction stress because the load bearing cross-section is continuously increased by the deposited weld passes. The reaction force directly after deposition of the respective weld pass increases with the number of passes, the reaction stress, by contrast, decreases (see also Fig. 185). Preheating of the weld area reduces reaction force and reaction stress during welding. Following welding and cooling down, by contrast, an increase occurs.

The finite element residual stress analysis after Andersson [235] using the cross-section model of a single-side submerged-arc butt weld in a fine-grained structural steel plate (material characteristic values according to Fig. 114 and similar to Fig. 123) resulted in weld-transverse and weld-longitudinal residual stresses in the top surface as shown in Fig. 154. Base and filler metal have different characteristic values (base metal $\sigma_Y = 350\ \mathrm{N/mm^2}$, filler metal $\sigma_{Yw} = 450\ \mathrm{N/mm^2}$, hardening not yet taken into account). The transformation influence is marked in the weld and heat-affected zone in accordance with the different yield limits by separate troughs in the stress curves.

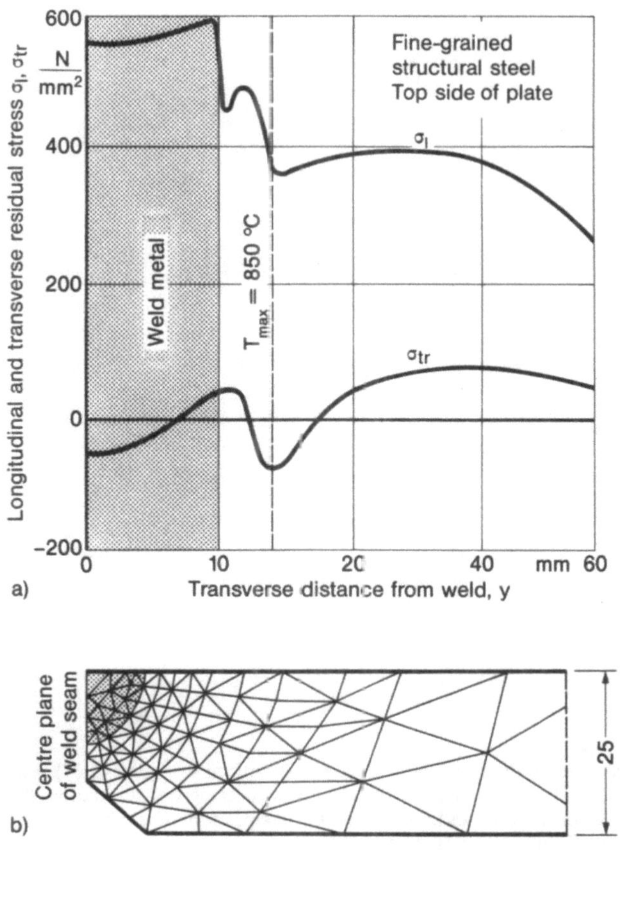

Fig. 154. Longitudinal and transverse residual stresses in top surface of structural steel plate with single-side submerged-arc butt weld (**a**) and appertaining finite element cross-section model (**b**); after Andersson [235]

The residual stresses of the first pass of a gas metal arc weld in the double-V groove of a 25.4 mm thick plate of the high-strength steel HY130 (with microstructural transformation) were determined by Papazoglou and Masubuchi [250] proceeding from the cross-section model. The butt weld with square groove in a 2.54 mm thick plate made of a NiCrFe alloy was analysed in a similar manner by Friedmann [249]. The maximum plastic strain (4.5%) occurs at the weld root at the transition from fusion to heat-affected zone [249]. The longitudinal residual stresses in the work-hardened weld area exceed the yield limit by as much as 30% [249].

In the finite element residual stress analysis by Argyris et al. [248, 193] for the double-V groove weld in structural steel St37 (material characteristic values according to Figs. 117 and 127), the two passes deposited simultaneously from both sides, produced the longitudinal and transverse residual stresses in the surface shown in Fig. 155. A microstructural transformation to the martensitic phase takes place in the fusion and heat-affected zone. The transformation strain is introduced, greatly simplified with constant magnitude, in all regions with

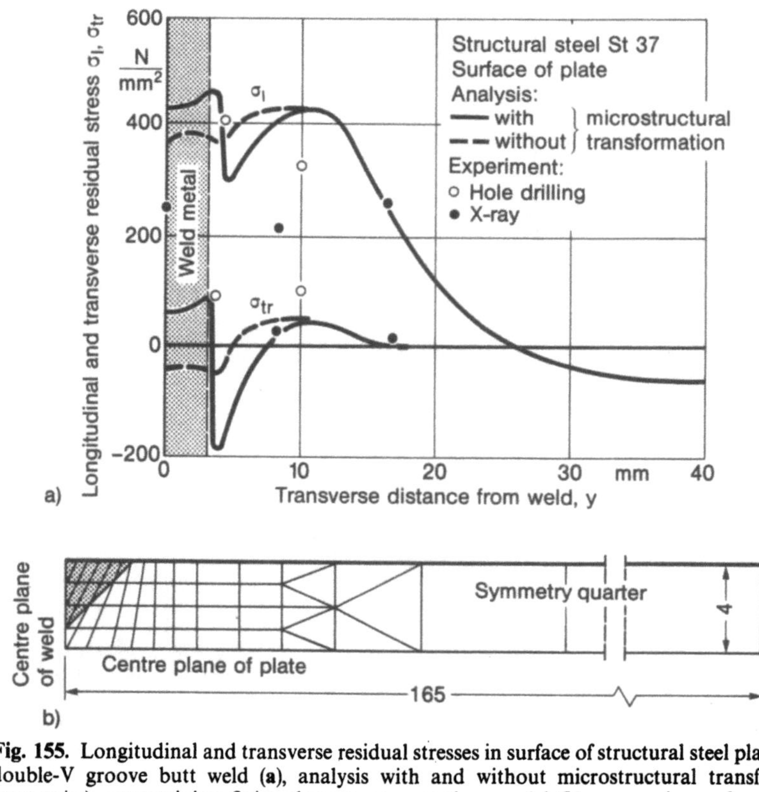

Fig. 155. Longitudinal and transverse residual stresses in surface of structural steel plate (St37) with double-V groove butt weld (**a**), analysis with and without microstructural transformation (to martensite), appertaining finite element cross-section model (**b**), comparison of analytical with experimental results; after Argyris, Szimmat and Willam [248]

more than 50% martensitic transformation, which is determined depending on cooling rate from the TTT diagram in Fig. 86. Analysis results with and without transformation as well as measurement results (X-ray stress measurement and hole drilling method) are compared. The effect of microstructural transformation is expressed in the curves as a stress peak in the fusion zone and a stress trough in the heat-affected zone, when taking the transformation strain and the increase of the yield limit by martensitic transformation into account. These investigations were continued in Ref. [247] in respect of a multi-pass weld in single-side groove (fine-grained low-alloy steel 22NiMoCr37 with microstructural transformation).

The root pass of a submerged arc butt weld in fine-grained low-alloy steel 22NiMoCr37 has been analysed by Kussmaul and Guth [254] based on the cross-section finite element model taking both transformation expansion and transformation plasticity into account. The transformation causes high longitudinal compressive stresses in the heat-affected zone.

The finite element residual stress analysis after Ueda et al. [242–246, 252] (Fig. 156), for multi-pass narrow-gap welding (20 passes) of structural steel with $\sigma_Y = 400\ \text{N/mm}^2$ for the base metal and $\sigma_{Yw} = 460\ \text{N/mm}^2$ for the filler metal, with $\sigma_{Y\,max} = \sigma_U = 550\ \text{N/mm}^2$, calculated without microstructural transformation, with the lower edge of the cross-section model supported bending-free and bending-rigid respectively, resulted in the longitudinal and transverse residual stresses shown in Fig. 157. The bending-rigid support increases the area of high longitudinal residual stress compared with bending-free support. The transverse residual stresses at the lower edge of the cross-section model show a tension or compression peak depending on the support condition. Comparative analyses show that useful results are already obtained with a coarser mesh, with two weld passes each being deposited together and with the analysis performed only for the last layers (not suitable for generalization; see Ref. [251]).

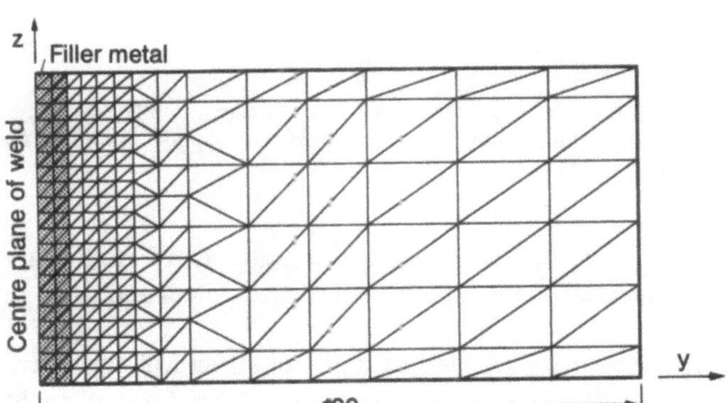

Fig. 156. Finite element cross-section model for residual stress analysis of multi-layer narrow-gap welding (20 layers), fine mesh; after Ueda and Nakacho [246]

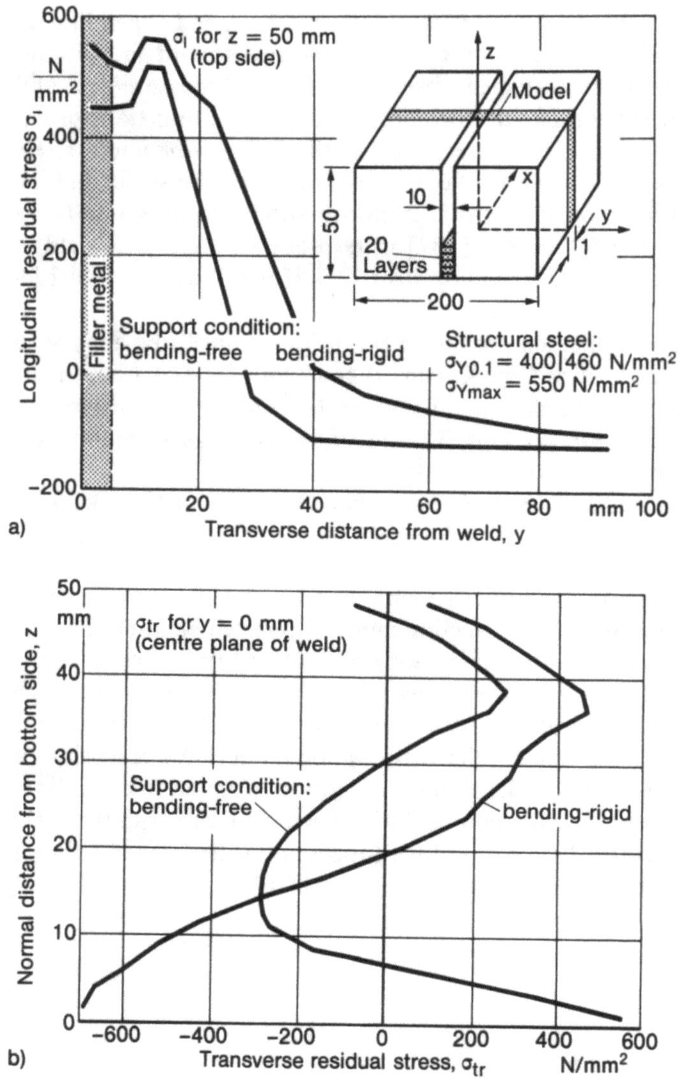

Fig. 157. Longitudinal residual stress (**a**) and transverse residual stress (**b**) after multi-layer welding, finite element analysis without microstructural transformation, two different support conditions on bottomside; after Ueda and Nakacho [246]

Further analysis results relating to cross-section models of multi-pass welds are presented in connection with post-weld heat-treatment methods in Section 4.4.3.2.6.

Longitudinally electron-beam welded vessels made of structural steel have been analysed by Carmet et al. [253] using the cross-section finite element model of the welded joint (Fig. 158), comprising the interaction between transformation plasticity and strain hardening phenomena. The investigation is

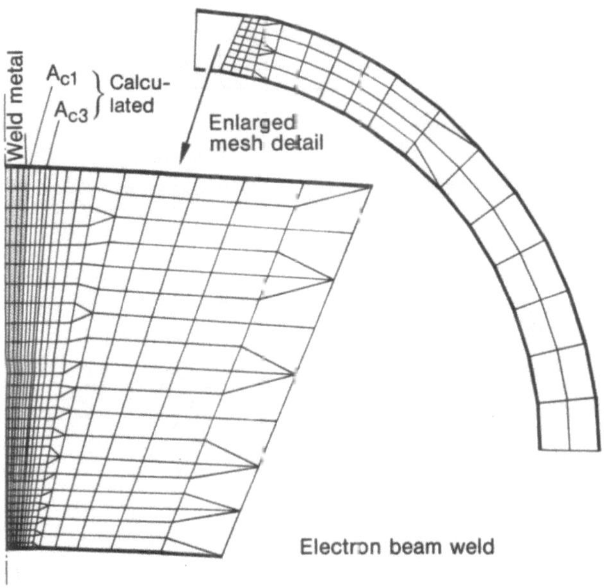

Fig. 158. Finite element cross-section model of longitudinal electron beam weld in cylindrical vessel made of structural steel, calculated A_{c3} and A_{c1} temperatures; after Carmet et al. [253]

extended to the fracture mechanics analysis of a semi-elliptical longitudinal crack subjected to the calculated residual stresses.

3.2.6 Solid element model

In recent years, the three-dimensional finite element models based on solid elements for transient heat diffusion in welding and related mechanical response have been developed and successfully applied [255–260]. The advances in computer technology made this possible but the expense in computer time for one analysis is extremely high.

The solid element model according to Fig. 159 has been used by Ueda et al. [255] to calculate the temperatures and stresses in a thick rectangular mild steel plate with a continuously deposited short surfacing weld (typical for repair welds). The temperature distribution on the surface of the plate in the weld centre line at times t_e after end of welding is shown in Fig. 160. The temperature maximum shifts from the end to the centre of the weld during cooling down. The residual stresses after cooling are plotted in Fig. 161 for the border line of the heat-affected zone 8 mm below the weld centre line (a) and for the vertical line underneath the weld centre (b). Rather high triaxial tensile residual stresses occur in the heat-affected zone and in the fused regions nearby. The stress distribution ahead of the weld end is similar to the axisymmetric heat spot behaviour (radial tension, tangential compression).

A long surfacing weld on a large rectangular steel plate (steel type HY80, size $500 \times 400 \times 12\ mm^3$) has been investigated by Oddy, Goldak and Dill [257, 260]

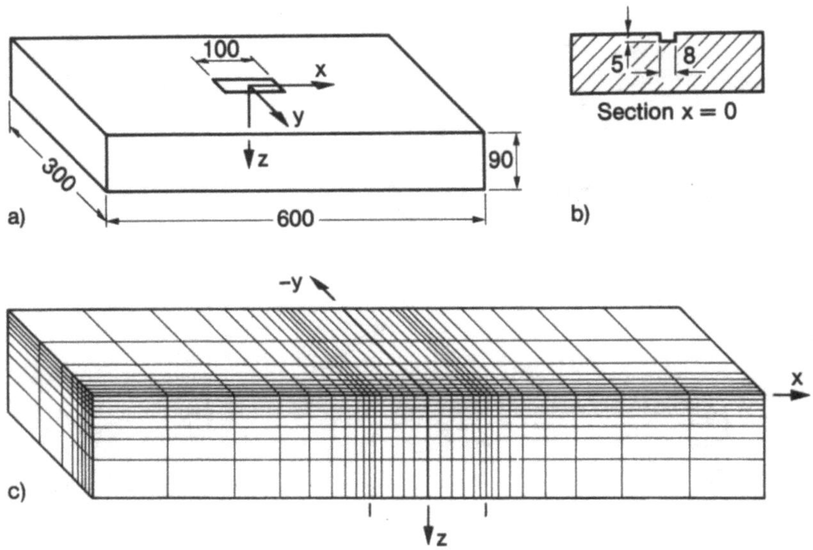

Fig. 159. Solid element model (**c**) of rectangular mild steel plate (**a**) with surfacing bead groove (**b**); after Ueda et al. [255]

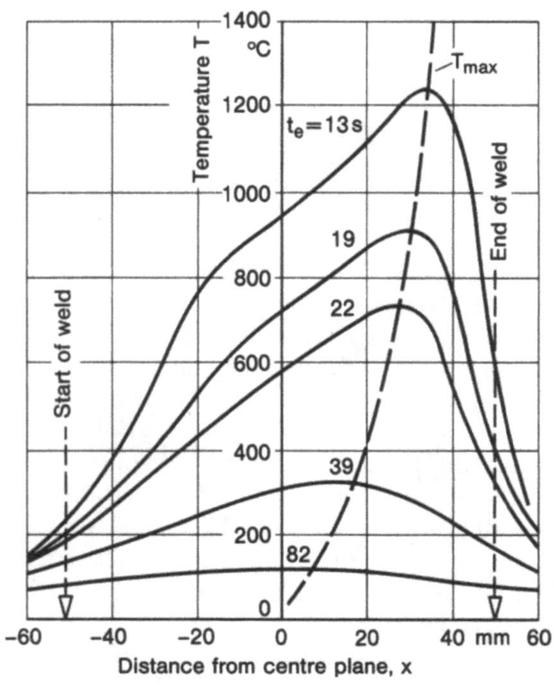

Fig. 160. Temperature profiles on surface of plate in weld centre line at times t_e after end of welding, surfacing bead on steel plate; after Ueda et al. [255]

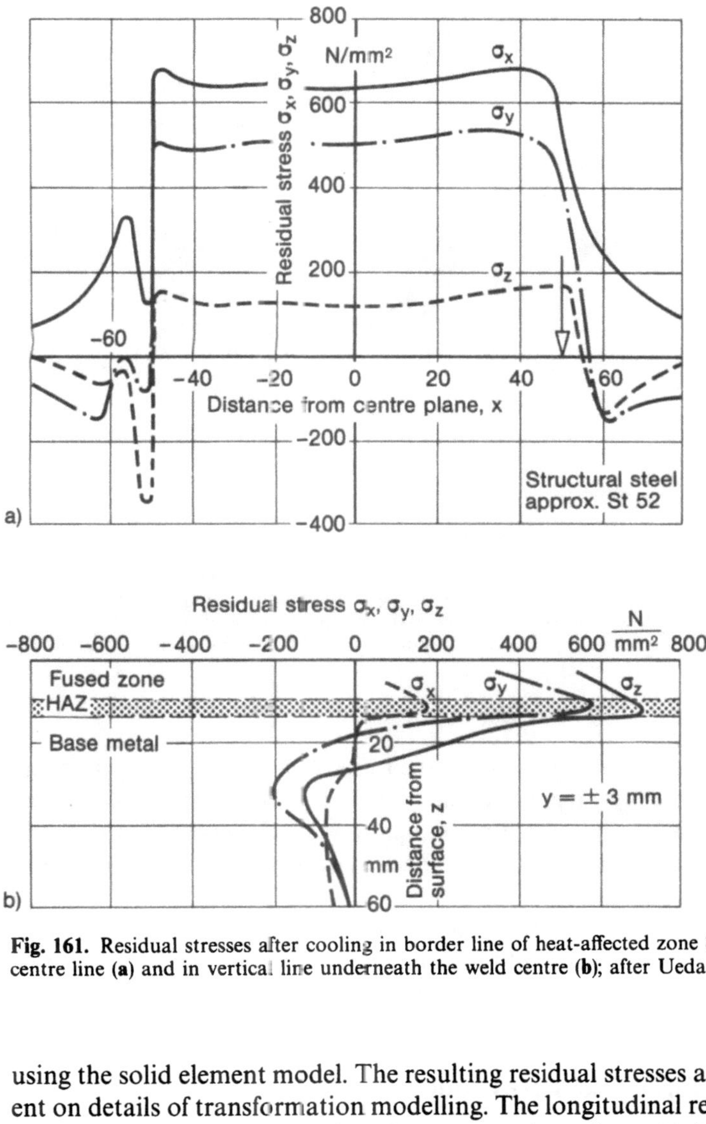

Fig. 161. Residual stresses after cooling in border line of heat-affected zone 8 mm below the weld centre line (**a**) and in vertical line underneath the weld centre (**b**); after Ueda et al. [255]

using the solid element model. The resulting residual stresses are highly dependent on details of transformation modelling. The longitudinal residual stresses on the plate surface plotted in Fig. 162 transversely to the weld depend on whether or not volume change, yield strength hysteresis and transformation plasticity are included.

The single pass transverse butt weld in a plate strip specimen (gas metal arc welded, V groove, mild steel) has been analysed by Tekriwal and Mazumder [256] based on a solid element model. The largest stresses occur transverse to the weld near the heat-affected zone boundary. The analysis comprises models with different cut-off-temperatures lower than melting temperature (no strong influence).

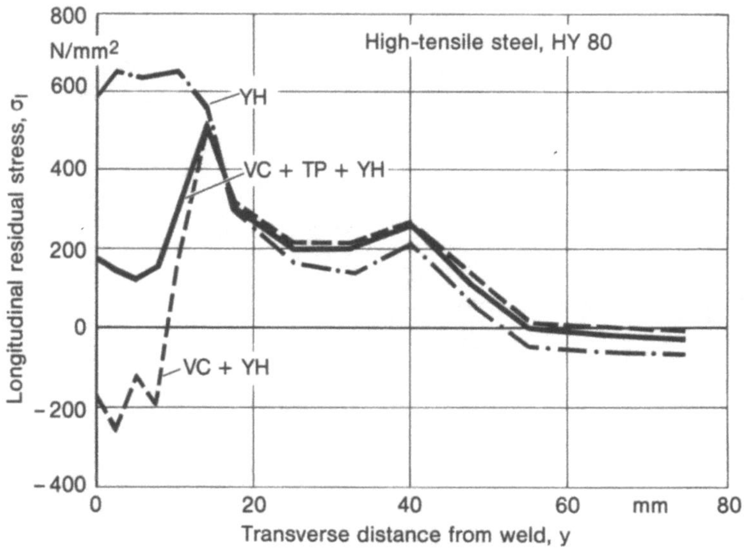

Fig. 162. Longitudinal residual stress on plate surface over transverse distance from weld, high-tensile steel HY80 plate with surfacing weld, finite element analysis including volume change by transformation, transformation plasticity and yield strength hysteresis (VC + TP + YH), the same without transformation plasticity (VC + YH) or additionally without volume change (YH); after Oddy, Goldak and Dill [257]

The circumferential weld in a thin walled pipe (CMn steel, outer diameter 203 mm, thickness 8.8 mm) has been analysed by Josefson and Karlsson et al. [258, 259] based on a fully three-dimensional model. The stresses in the weld metal and heat-affected zone are found to be heavily dependent on the details of transformation modelling.

Of course, the stated sensitivity of results in respect of transformation modelling is not a characteristic of three-dimensional modelling. It can also be demonstrated based on lower dimension modelling.

3.3 Shrinkage force and stress source models

3.3.1 Longitudinal shrinkage force model

The longitudinal residual stresses in oblong plate-like components, as well as the related weld deformations, can be approximately determined proceeding from the (longitudinal) shrinkage force of the weld and the zones close to it based on a unidimensional elastic approach. It is imagined that, in place of the welds, there are prestressed wires which apply a load on the initially stress-free component. Shrinkage force refers to the prestressing force of the tensioning element directly before it subjects the component to a load and is partially relieved as a result of the elasticity of the component. The shrinkage force can be interpreted as the prestressing force of the component assumed to be rigid. It is an initial force in the sense of the residual stress theory. A variant of the formula uses, in place of the shrinkage force on the component imagined to be rigid, the "active" (or "effective") shrinkage force at the elastic component, which is in equilibrium with the "reactive" compressive forces of the cross-section. Unfortunately, a clear distinction is not always drawn between the two differently defined shrinkage forces.

If the shrinkage force of the weld is known, it is possible, for example, to determine the mean compressive prestress for a box girder with longitudinal welds, which is required for stability studies [261]. The compressive prestress is obtained from referring the total of the shrinkage forces of all the welds to the rod cross-section area. Or, it is also possible to calculate the bending shrinkage of girders with eccentric longitudinal weld, proceeding from the shrinkage force.

The quantitative relations in the simplest case of the plate strip unrestrained from all sides and with centric longitudinal weld are presented in Fig. 163. The weld cross-section area is subjected to tension σ_{te} at the level of the yield limit σ_Y. Its width w_t is equated in the sense of an approximation to the width w_{pl} of the plastic zone. The compressive stress σ_c follows from the equilibrium condition:

$$\sigma_c = -\sigma_{te}\frac{w_t}{w - w_t} = -\sigma_Y\frac{w_{pl}}{w - w_{pl}}. \qquad (137)$$

The (longitudinal) shrinkage force F_s, which is assumed to be known, is reflected in the width of the plastic zone w_{pl}:

$$F_s = (\sigma_{te} - \sigma_c)w_t h = (\sigma_Y - \sigma_c)w_{pl}h. \qquad (138)$$

A comparison of areas in Fig. 163 results alternatively in:

$$F_s = -\sigma_c wh. \qquad (139)$$

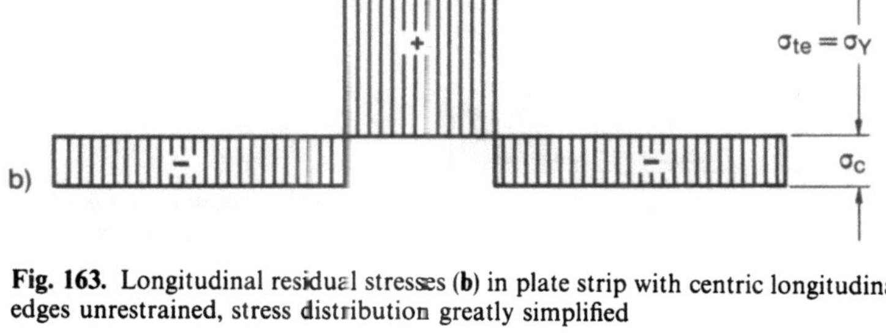

Fig. 163. Longitudinal residual stresses (**b**) in plate strip with centric longitudinal weld (**a**), all plate edges unrestrained, stress distribution greatly simplified

Consequently, the shrinkage force F_s is supported by the entire cross-section (not so the active shrinkage force).

Only in the case of the (longitudinally) rigidly supported (or also infinitely wide elastic) plate strip, is F_s identical with the resultant of the tensile zone, i.e. with the active shrinkage force:

$$F_s = \sigma_{te} w_t h = \sigma_Y w_{pl} h. \tag{140}$$

For the plate strip with the weld arranged eccentrically, the stress distribution which results with linear bending stress distribution (this means that the cross-sections remain plane), is presented in Fig. 164. Finally, the combined case of a T section bar with a web-to-flange weld is shown in Fig. 165. The flange plate is jointed centrically and the web plate eccentrically. The content of Figs. 164 and 165 is reflected in equations which are similar to equations (137) to (139) but more complex.

The shrinkage force and the size of the plastic zone which essentially determines the former depend primarily on the welding process parameters as well as on the thermomechanical material characteristic values and secondarily on the longitudinal stiffness of the component and on the heat flow in the welded joint.

The decisive parameters of the welding process are the heat input q and the welding speed v, combined into the heat input per unit length of weld q_w according to equation (3). The width of the plastic zone increases as heat input rises and welding speed drops, respectively. Approximation formulae therefore bring the shrinkage force F_s into relation with the heat input per unit length of weld q_w. For mild structural steels, for example,

$$F_s \approx 170 q_w \tag{141}$$

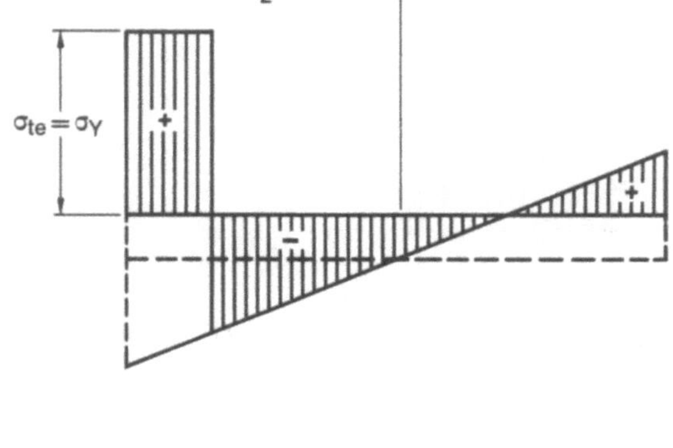

Fig. 164. Longitudinal residual stresses in plate strip with eccentric longitudinal weld, all plate edges unrestrained, stress distribution greatly simplified; after Hänsch [7]

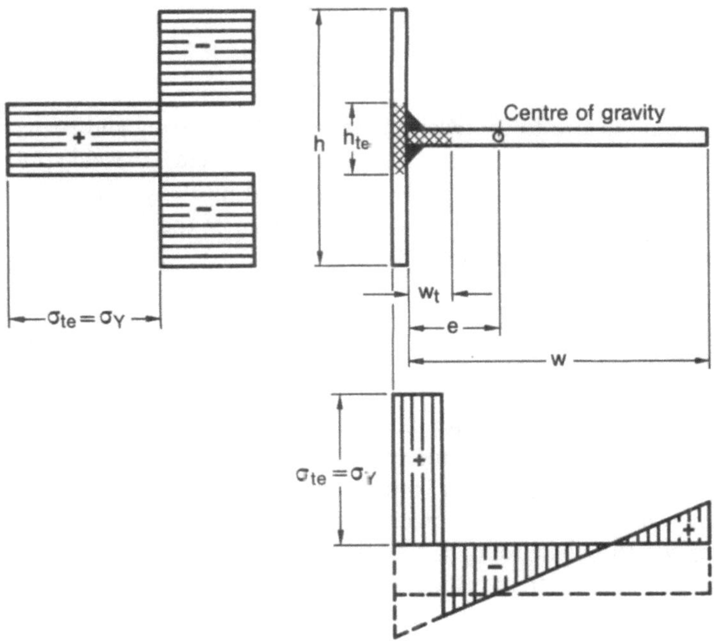

Fig. 165. Longitudinal residual stresses in T section bar with web-to-flange fillet weld, longitudinal edges of bar unrestrained, stress distribution greatly simplified; after Hänsch [7]

is stated in Ref. [7, 8] proceeding from equation (149). The heat input q_w should be inserted in [J/mm] in order to obtain F_s in [N]. The width w_{pl} of the plastic zone follows from equations (140) and (141):

$$w_{pl} \approx 170 \frac{q_w}{h\sigma_Y}. \tag{142}$$

The heat input per unit length of weld q_w in turn is set proportional to the volume of the molten filler metal per unit of weld length or to the weld cross-section area A_w.

$$q_w = kA_w. \tag{143}$$

The proportionality factor k is stated to be $k = 61$ for covered electrode welding, $k = 41$ for gas metal arc welding and $k = 72$ for submerged arc welding (see Ref. [7]) if A_w is inserted in [mm^2] and q_w is expected in [J/mm].

In multi-pass welding (number of passes n), the longitudinal shrinkage force according to equation (141) should be multiplied by the correction factor k_n:

$$k_n \approx n^{-2/3} \tag{144}$$

In single-pass and multi-pass double fillet welding, k_n according to equation (144) should be further modified depending on the welding sequence [7].

In intermittent (i.e. string) welding with the string weld length l_{st} and the related string weld interspace length l_i, on the other hand, the correction factor

k_i should be introduced:

$$k_i \approx \frac{l_{st}}{l_{st} + l_i}. \tag{145}$$

This factor applies roughly also to spot welds if, in place of the string and string interspace length, the corresponding parameters of the spot weld, i.e. weld spot diameter and weld spot interspace (the pitch), are inserted into equation (145). A more exact formula for the longitudinal shrinkage force of the spot weld is stated by Okerblom for mild steel in Ref. [8]:

$$F_s = 184 \frac{h_{tot}}{e^*} d_{150}^2. \tag{146}$$

The formula comprises the welded total plate thickness h_{tot} joined by welding, the weld spot interspace e^* and the diameter d_{150} of the plate area heated to about 150 °C (d_{150} should be inserted in [mm] in order to obtain F_s in [N]).

Only the mean value of the tensile longtitudinal stresses is determined for string and spot welds with F_s according to equations (141), (143), (145) and (146), respectively. The more precise stress profile is wave-shaped with an increase up to the value of the continuous weld in the centre of the string or spot weld and with a reduction in between down to zero.

The decisive material parameters in respect of the shrinkage force or of the width of the plastic zone are the yield limit σ_Y, the elastic modulus E and the thermal expansion coefficient α. These parameters determine the unloading as a result of cooling, proceeding from the compressive yield limit until the tensile yield limit is reached. This is presented in Fig. 166 for a rod made of low-strength structural steel rigidly restrained at both ends [261]. The temperature difference ΔT_{el}, which can be sustained with elastic unloading, is all the greater, the plastic zone consequently all the narrower; the larger σ_Y and the smaller E and α are:

$$\Delta T_{el} = \frac{2\sigma_Y}{E\alpha}. \tag{147}$$

Equation (147) gives $\Delta T_{el} \approx 180$ °C for the mild structural steel in Fig. 166. Consequently, for this material the plastic zone can be approximately set equal to the area heated to above 200 °C, assuming the weld is rigidly supported in longitudinal direction. The required temperature conditions in the weld area can be determined by measurement or by calculation.

In a simpler form (yield limit σ_Y temperature-independent and no hardening), the above procedure is used in Ref. [262] for approximately determining the longitudinal residual stresses in the entire cross-section under the condition of rigid supports, whereby the post-weld global rise in temperature, ΔT_0, of the component is also taken into account (Fig. 167). In the weld area, the tensile yield limit is reached via OACD, in the outer area the compressive stress σ_c via OZ and in the intermediate area a transition state which depends on the

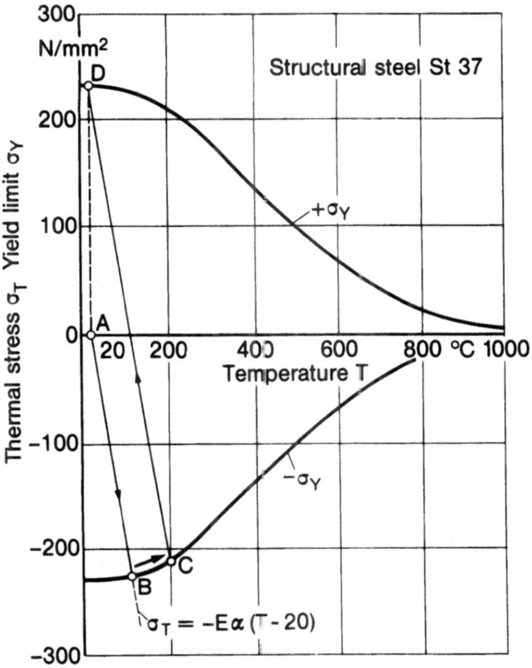

Fig. 166. Stress-temperature cycle in rod rigidly restrained on both ends, heating to 200 °C with subsequent cooling, residual stress equal to tension yield limit; after Buttenschön [261]

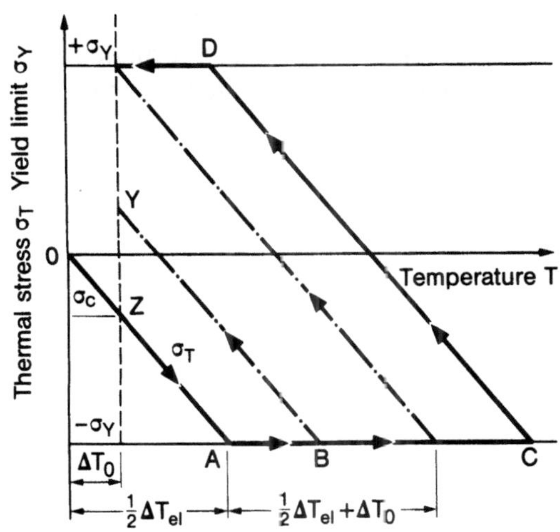

Fig. 167. Simplified scheme for determining residual stress in rigidly restrained rod for different intensities of heating with subsequent cooling to ΔT_0; after Wells [262]

locally attained maximum temperature via OABY. Hence, with rigid supports:

$$\sigma_c = -E\alpha\Delta T_0. \tag{148}$$

The material influence can also be determined on the basis of the maximum temperature profile according to equation (49) for the rapidly moving high-performance source and from an approximation (obviously) in accordance with

Fig. 129 for the elastic compressive shortening which accompanies the shrinkage force (according to Ref. [2], equation (52) in Ref. [7], derivation unclear; shrinkage strain ε_T^*, longitudinal stiffness factor $\mu_l = 0.335$):

$$\varepsilon_T^* = \mu_l \frac{\alpha}{c\varrho} \frac{q_w}{hw}. \tag{149}$$

This expression results in the shrinkage force according to equation (141) for mild structural steel [7].

In addition, the (longitudinal) stiffness of the component or plate strip has an (in general secondary) influence on the shrinkage force or on the width of the plastic zone. A reduction in stiffness compared to the rigid support results in a narrowing of the plastic zone and thus in a reduction of the shrinkage force. This occurs, assuming unrestrained plate strip ends, with reduced plate width or eccentric weld (as a consequence of bending deformation). The different conditions in the narrow longitudinally unrestrained plate strip compared to the infinitely wide longitudinally rigidly supported plate strip are presented in Fig. 168 in continuation of the model according to Fig. 129. In the longitudinally rigidly supported wide plate strip, the part of the thermal residual strains ε_R, shown shaded in Fig. 168a, occurs as residual stress. In the longitudinally unrestrained narrow plate strip, the zero stress level is raised, on the one hand as a result of the unrestrained formation of the averaged cooling strain, on the other hand as a result of the compressive elasticity of the plate strip. The width w_{t0} of the tensile zone in the wide plate strip is thus reduced to the width w_t of the tensile zone in the narrow plate strip. Elaboration in terms of a formula [8] of the relation between w_t and w_{t0}, which is more clearly presented graphically in Fig. 168, is dispensed with here.

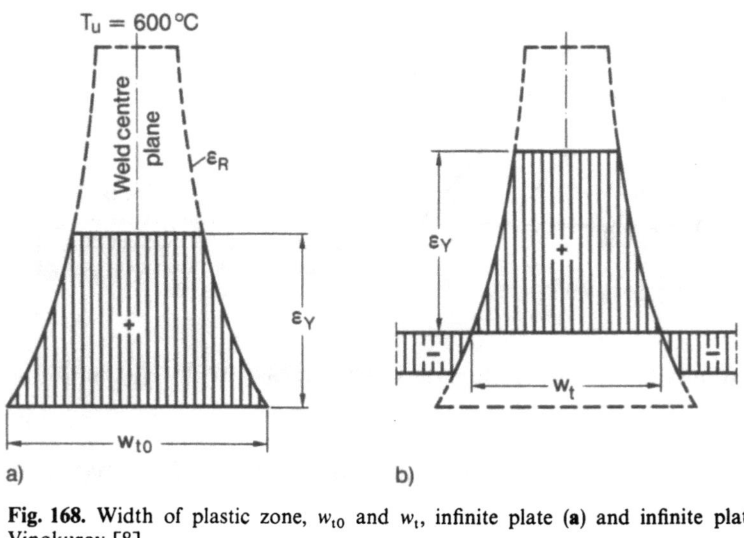

Fig. 168. Width of plastic zone, w_{t0} and w_t, infinite plate (**a**) and infinite plate strip (**b**); after Vinokurov [8]

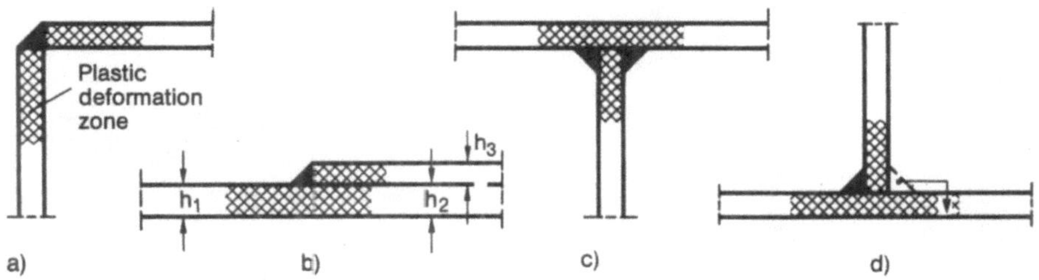

Fig. 169. Plastic zones in different joints (**a**) to (**d**), and superposition effect of successively deposited fillet welds (**d**)

The allocation of the heat input to the plates connected in the joint should be estimated for welded joints which deviate from the butt or corner weld between two plates of equal thickness in order to determine position and, also to a certain extent, size of the plastic zone. This should always be performed if the thicknesses of the jointed plates are different, if the plates are staggered (e.g. lap joint) or if more than two plates are connected in the joint (e.g. T joint).

The heat input is allocated according to the "reduced thicknesses" of the plates connected in the joint, through-plates being treated as two plates [2]. The heat input q_i in plate i of n plates results from the (effective) total heat input q and the plate thicknesses h_ν according to:

$$q_i = \frac{h_i}{\sum_{\nu=1}^{n} h_\nu} q. \tag{150}$$

To illustrate this, the welded joints of Fig. 169 are considered. The corner joint with a uniform plate thickness behaves like a butt joint with equal heat flows. Lap joint and T joint subdivide the heat flow in the ratio of the plate thicknesses. If the two fillet welds at the T joint are deposited one after the other, with intermediate cooling, the relevant plastic zones are superimposed. The total plastic zone is then only approximately 15% larger than the plastic zone of the single weld [8]. Reference is made in Fig. 152 to the correlation between the size of the plastic zone and the post-weld rise of temperature ΔT_0 (to be understood as related to the considered joint part).

The representation of the shrinkage force or plastic zone size as a function also of secondary parameters such as longitudinal stiffness and heat flow does not, however, constitute the actual contents of the shrinkage force model, namely the shrinkage force being dependent on process and material only and independent of type, geometry and dimensions of the welded joint and component, for example in accordance with equations (141) or (142).

3.3.2 Transverse shrinkage force model

Whereas the weld-longitudinal residual stresses as "constraint stresses" depend only secondarily on the degree of the (longitudinal and transverse) restraint of

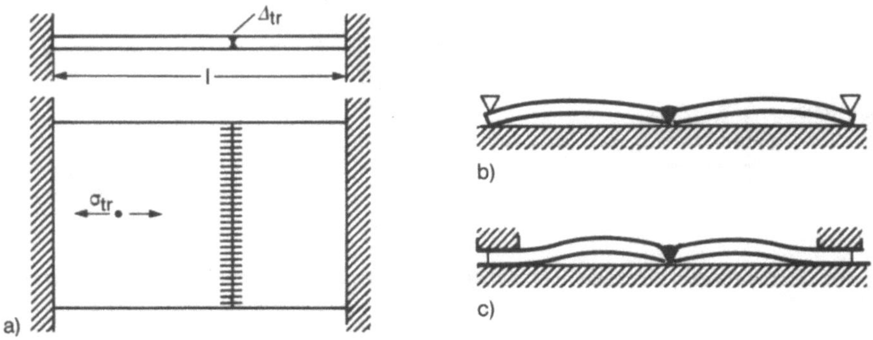

Fig. 170. Residual stresses transverse to weld, σ_{tr}, in the form of membrane reaction stress caused by suppressed transverse shrinkage (**a**) and of bending reaction stress caused by suppressed angular shrinkage (**b**, **c**)

the parts connected by welding, the weld-transverse residual stresses depend decisively thereon. The transverse shrinkage Δ_{tr} of the weld and of the heated area close to the weld causes all the higher transverse stresses the more rigidly and the more shortly the parts are fixed transversely. For the butt weld between two rigidly restrained plates (straining length l), (Fig. 170a), this results within the framework of a one-dimensional consideration, in the transverse stress σ_{tr}, as an elastic reaction stress (elastic modulus E) originating from the transverse shrinkage Δ_{tr}:

$$\sigma_{tr} = \frac{\Delta_{tr} E}{l}. \tag{151}$$

The related (active) transverse shrinkage force F_{st} follows with the cross-section area A of the plate:

$$F_{st} = \sigma_{tr} A. \tag{152}$$

The transverse shrinkage Δ_{tr} is determined with unrestrained (i.e. free) plate edges (see Section 3.5.2) and transferred to the restrained plate in accordance with equation (151), whereby it is assumed that the yield limit σ_Y is not exceeded. In case that plastic transverse shrinkage superimposes, Δ_{tr} should be reduced accordingly.

The transverse residual stresses in string and spot welds in transversely unrestrained plates consist of transverse tension in the string or spot weld areas and transverse compression in the areas between them (in the case of gap-free welding). These stresses are shifted in the direction of higher tension in transversely restrained plates.

The transverse bending stresses arising from restrained angular shrinkage $\Delta\beta$ can be determined using the engineering beam theory in an appropriate manner, e.g. for restraint conditions according to Figs. 170b and 170c, provided $\Delta\beta$ is known (approximations of $\Delta\beta$ in Section 3.5.4).

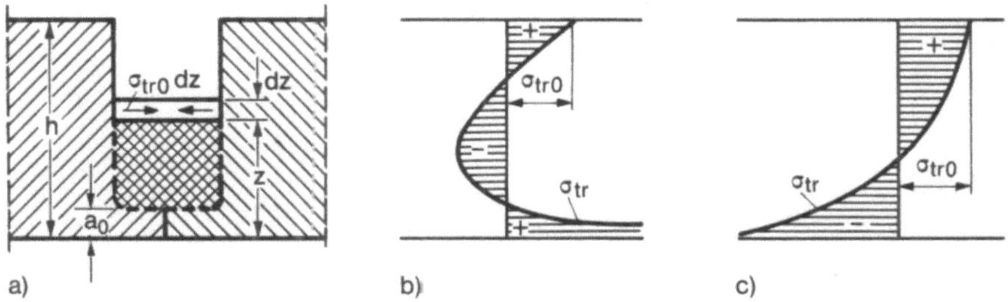

Fig. 171. Groove model for determining transverse residual stresses in multi-layer welding (**a**), angular shrinkage allowed (**b**) or suppressed (**c**); after Vinokurov [8]

The transverse residual stresses during and after multi-pass welding can also be approximately determined on the basis of shrinkage forces. The groove filling model according to Fig. 171a is used for this purpose [8]. The constant surface shrinkage stress σ_{tr0} is moved together with increasing groove filling, giving rise to membrane and bending stresses in the already filled groove longitudinal section. The root pass with thickness a_0 is introduced as being initially free of shrinkage stress. The integration following joint filling in the case of unrestrained plate edges, and, accordingly, severe angular shrinkage, results in the transverse stress distribution according to Fig. 171b with tension at the weld root. If, by contrast, the angular shrinkage is completely suppressed, which can occur only incompletely in practice, this results in the transverse stress distribution according to Fig. 171c with compression at the weld root.

The transverse shrinkage force model has been further developed in respect of underbead, root and laminar cracks, as well as in respect of the related cold crack testing (rigidly restrained plate with transverse weld, i.e. rigid restraint cracking (RRC) test, or also unrestrained rectangular plate with slot weld) [265–269, 297, 298]. Of interest as the least favourable case is the thin weld between thick plates as a simulation of tack welding or of the first pass of a multi-pass weld. The conditions are relatively clear only at the rigidly, or also elastically, restrained plate with transverse weld.

The (unrestrained) transverse shrinkage $\varDelta_{\text{tr}}$ is introduced as the parameter independent of the degree of restraint of the weld. It can be determined by measurement or by approximative calculation according to equations (166), (168), (169) or (172). The reaction forces which are obtained for restrained transverse shrinkage depend on the stiffness of the weld and of the restrained plates. This is presented in Fig. 172 for clamping a plate strip with short and long straining length for the purpose of comparison, with the elastic curve of the clamped plates in both cases in the left part of the diagram, the elastic-plastic curve of the weld (subjected to transverse load) in both cases in the right part of the diagram. The transverse shrinkage force F_{st}^* per unit of weld length is plotted versus the plate elongation Δl and versus the weld transverse deformation Δw^* respectively. The mutual prestressing (or counter-loading) is obtained by proceeding from the transverse shrinkage $\varDelta_{\text{tr}}$. In the cases presented, the yield limit

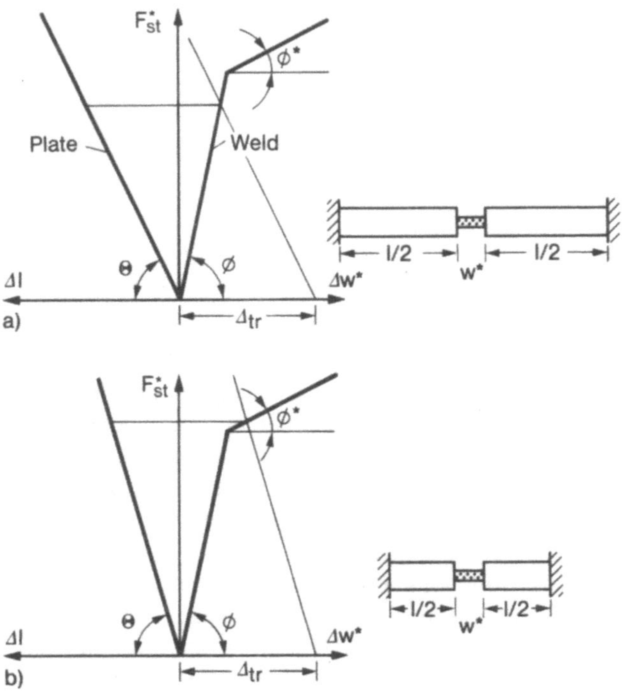

Fig. 172. Counter-loading diagram of tensile specimen with transverse weld clamped with long straining length (**a**) or short straining length (**b**), equal transverse shrinkage Δ_{tr}; after Satoh [267]

is not exceeded in the weld with long straining length, but it is with short straining length. The stiffness of the clamped plates characterized by the inclination angle θ of the force-displacement curve is designated as the restraint intensity R acting on the weld:

$$R = \tan\theta = \frac{F^*_{st}}{\Delta l} = \eta^* \frac{Eh}{l}. \tag{153}$$

The restraint intensity R may be interpreted as the force per unit of weld length which produces the unit of transverse elongation $\Delta l = 1$. The shape factor η^* according to Fig. 173 takes into account the reduction in stiffness of the clamped plates caused by inhomogeneous stress distribution in areas close to the weld while the transverse stiffness of the weld is accounted for separately by means of Δw^* in the counter-loading diagram of Fig. 172.

In respect of the reaction stress in the plate, $\sigma_{tr} = F^*_{st}/h$ (h plate thickness), or in the weld, $\sigma^*_{tr} = F^*_{st}/a$ (a weld thickness), this results with equation (153) and $\Delta l = \Delta_{tr}$ for $\Delta w^* \ll \Delta l$ in:

$$\sigma_{tr} = R\frac{\Delta_{tr}}{h}, \tag{154}$$

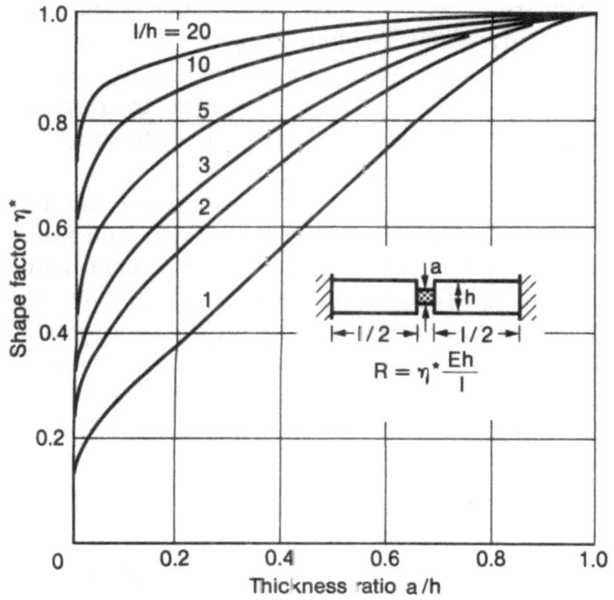

Fig. 173. Shape factor η^* of restraint intensity R of tensile specimen with transverse weld dependent on thickness ratio; after Ueda, Fukuda and Kim [269]

$$\sigma_{tr}^* = R\frac{\Delta_{tr}}{a}. \tag{155}$$

The transverse shrinkage force model is refined in Ref. [280] to the effect that a distinction is made, in evaluating appropriate strain measurements, between the deformation restraint of the weld area in itself, recognizable from the strain drop alongside the weld, and the deformation restraint as a result of the adjoining plates, recognizable from the constant level of strain. The model is extended in Ref. [269] to the slot weld in the rectangular plate. The transverse shrinkage Δ_{tr} can be introduced as constant only in the middle section of the weld length l_w in order to obtain a transverse stress profile with $\sigma_y = \sigma_Y$ over the length l_e at the weld ends and a parabolic drop over the length l_c to the centre of the weld length. The restraint intensity R turns out to be lowest in the centre of the weld and highest at the ends (singular behaviour).

3.3.3 Application to cylindrical and spherical shells

The shrinkage force model can be applied advantageously to determine the welding residual stresses in cylindrical and spherical shells. With cylindrical shells, longitudinal and circumferential welds as well as ring welds at nozzles or block flanges are of interest. Circumferential and ring welds are likewise found in the case of spherical shells. A further possible application is the ring weld at the flange of hemispherical caps.

If the shell (in reality or only in theory) is supported during welding and cooling by inserted rings as a shape-stabilizing measure, the welding residual stresses are generated in a similar way as in the plane unrestrained membrane plate. If the inserted ring is removed after welding, however, or is not even inserted at all, considerable bending deformations and related bending residual stresses occur. In the derivations below, the weld is assumed to be deposited instantaneously over its whole length, which, in the case of circumferential and ring welds, ensures axisymmetry of the model.

The longitudinal weld in the cylindrical shell causes bending deformation in the longitudinal and (thus also) in the transverse direction as a consequence of the longitudinal shrinkage force, with bending tension on the inside and bending compression on the outside (see the deformation patterns in Fig. 194). The (active) transverse shrinkage force F^*_{st} per unit of weld length is relatively small as a consequence of the low transverse stiffness of the cylindrical shell (radius R, thickness h) after being cut longitudinally on the one side (equation (188) in Ref. [8]):

$$F^*_{st} = \frac{\Delta_{tr} E h^3}{24\pi R^3}. \tag{156}$$

The (active) transverse shrinkage moment M^*_{st} per unit of weld length, which is additionally arising in the single-sided weld, follows from the angular shrinkage $\Delta\beta$ (equation (189) in Ref. [8]):

$$M^*_{st} = \frac{\Delta\beta E h^3}{24\pi R}. \tag{157}$$

The circumferential weld in the cylindrical shell causes a marked constriction of the shell as a consequence of the longitudinal shrinkage force. The deflections and residual stresses can be represented on the basis of functional-analysis using the shrinkage force model [8, 263, 264]. The starting point is the assumption of a constant circumferential initial stress σ_0, over the width $2w_{pl}$ covering both sides of the weld. The variables w_{pl} (corresponding here to the half width of the plastic zone) and σ_0 can be determined experimentally or numerically; empirical values exist additionally. In many cases, σ_0 can be set equal to σ_Y. The conversion of this "residual stress source" into the membrane and bending stresses of the cylindrical shell can be represented in terms of the engineering theory of elastic cyindrical shells. The load case of constant radial pressure p_0 acting on a $2w_{pl}$ wide circumferential strip of the cylindrical shell is considered with the equilibrium condition:

$$p_0 = \frac{\sigma_0 h}{R}. \tag{158}$$

This condition follows from the consideration that the circumferential strip with initial stress σ_0 and suppressed deflection produces the inside support pressure p_0, which is to be cancelled by an opposing outside pressure p_0 (or inside tension $-p_0$) if the support is eliminated. For the load case reduced initially to a circumferential line with pressure intensity p^*, the deflection w follows an oscillatory curve dying out exponentially with transverse distance y from the weld (edge disturbance solution of cylindrical shell theory):

$$w = \frac{p^* e^{-\lambda y}}{8\lambda^3 D} (\sin \lambda y + \cos \lambda y), \tag{159}$$

$$\lambda = \sqrt[4]{\frac{3(1 - \nu^2)}{h^2 R^2}} \tag{160}$$

$$D = \frac{E h^3}{12(1 - \nu^2)} \tag{161}$$

The circumferential strip solution is gained by integrating the circumferential line solution. The relations which result for the deflections w, circumferential stresses σ_{ci0} (in the middle plane of the cylindrical shell) and axial bending stresses σ_{a0} in the centre plane of the circumferential weld ($y = 0$) are as follows (Fig. 174) (the variables are designated as maximum values in Ref. [264]):

$$w_0 = \frac{\sigma_0 R}{E} (1 - e^{-\lambda w_{pl}} \cos \lambda w_{pl}) \tag{162}$$

$$\sigma_{ci0} = \sigma_0 e^{-\lambda w_{pl}} \cos \lambda w_{pl}, \tag{163}$$

$$\sigma_{a0} = 3\sigma_0 \sqrt{3(1 - \nu^2)}\, e^{-\lambda w_{pl}} \sin \lambda w_{pl}. \tag{164}$$

The deflection w_0 rises with λw_{pl} ($\lambda w_{pl} = 0$ corresponds to the plane plate) from zero up to a maximum value for $\lambda w_{pl} = 3\pi/4$. The (membrane) circumferential stress σ_{ci0} (in the middle plane of the cylindrical shell) drops from a (tension) maximum for $\lambda w_{pl} = 0$ to a (compression) minimum for $\lambda w_{pl} = 3\pi/4$. The (bending) axial stress σ_{a0} on the inside and outside, respectively, of the shell reaches a maximum for $\lambda w_{pl} = \pi/4$. Whereas w_0 and σ_{ci0} mainly drop with λw_{pl}, they rise and subsequently drop with σ_{a0} [264].

The circumferential and axial stresses for the circumferential weld in the cylindrical shell have been calculated in Ref. [263], proceeding from the shrinkage force of the corresponding plate longitudinal weld. The width of the plastic zone was determined in accordance with Ref. [262] (see also Fig. (167)). Figure 175 shows the longitudinal residual stresses in the plate, Fig. 176 the circumferential and axial residual stresses in the cylindrical shell, in both cases compared with the results of stress measurements. The circumferential stresses in the cylindrical shell are significantly lower than the longitudinal stresses in the plate. High axial bending stresses occur transverse to the circumferential weld in the cylindrical shell.

The analysis of the cylindrical shell with circumferential weld described above is based on the assumption of constant initial stresses σ_0 over the cross-section of the circumferential strip, which corresponds only approximately to reality. The model ignores bending effects which originate from the asymmetry of the welding groove or the welding operation. Consequently, they

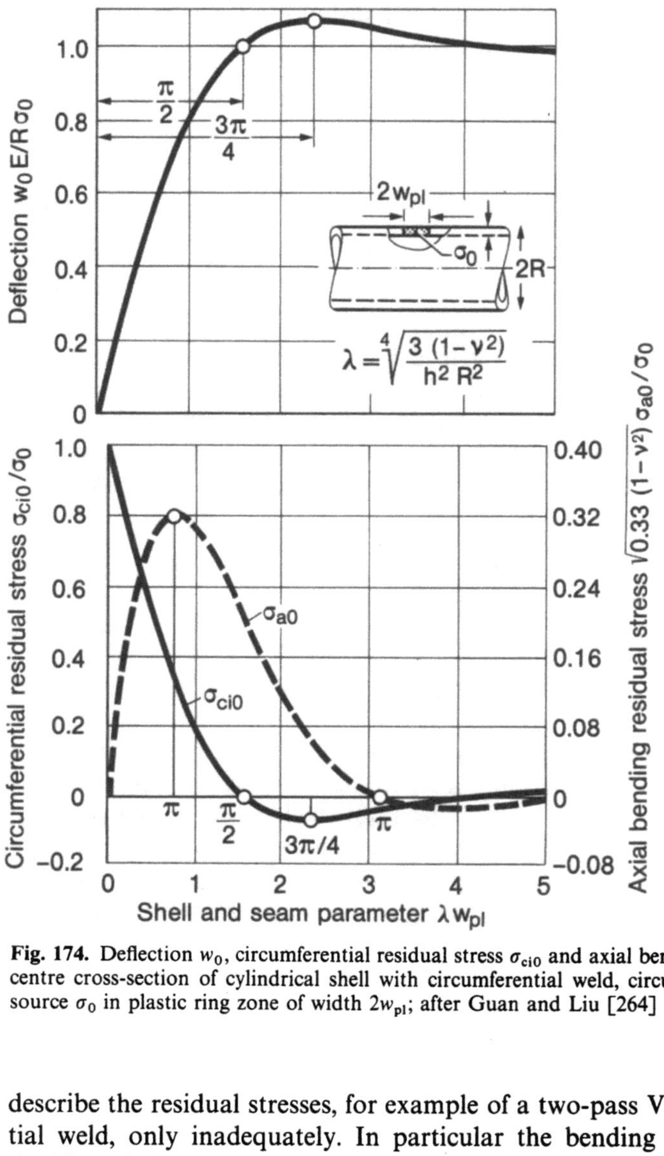

Fig. 174. Deflection w_0, circumferential residual stress σ_{ci0} and axial bending residual stress σ_{a0} in centre cross-section of cylindrical shell with circumferential weld, circumferential residual stress source σ_0 in plastic ring zone of width $2w_{pl}$; after Guan and Liu [264]

describe the residual stresses, for example of a two-pass V groove circumferential weld, only inadequately. In particular the bending stresses σ_{a0} may be significantly increased as a result of the asymmetry. The modelling is finally based on a supporting ring, which is not removed until after welding. The conditions which exist in the case of unrestrained shrinkage without supporting ring are admittedly similar, but not identical.

The analysis presented here for the cylindrical shell can be conducted in a corresponding manner for the spherical and toroidal shell.

In the case of aluminium alloys, the bulging may occur instead of constriction with circumferential welds. This reversal is explained by the wider heating-up zone transverse to the weld, which deflects elastic-plastically to the outside of

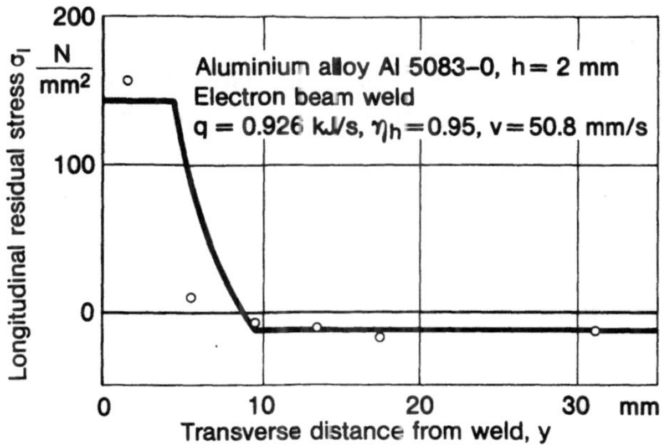

Fig. 175. Longitudinal residual stresses in plate strip with centric longitudinal weld, approximative calculation according to Ref. [262] and experimental results; after Vaidyanathan, Todaro and Finnie [263]

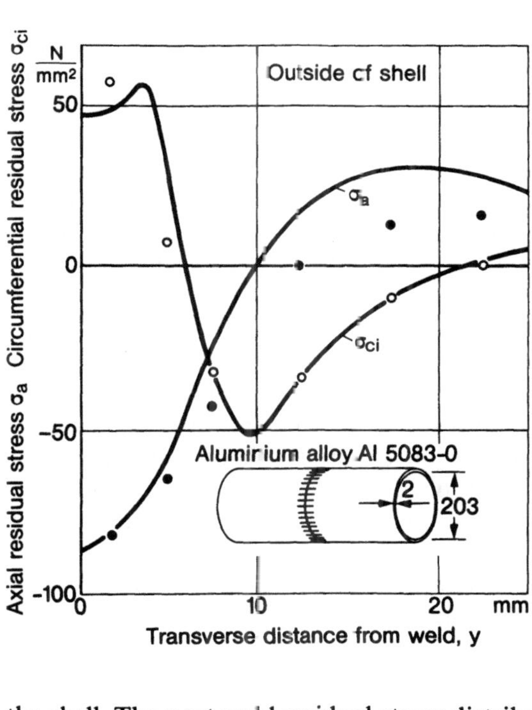

Fig. 176. Axial and circumferential residual stresses in cylindrical shell with circumferential weld, comparison of analytical and experimental results (experiments on spherical shell); after Vaidyanathan, Todaro and Finnie [263]

the shell. The post-weld residual stress distribution is not significantly changed as a result of this, however.

The residual stress distribution at ring welds (nozzle and block flange welds in cylindrical and spherical shells) is considerably complicated by the interaction of the longtitudinal and transverse shrinkage forces. The conditions in the case of a plane plate with a circular patch inserted by welding have been presented in Fig. 138. The conditions in this case are already very complex and manifold variants are possible (radial stresses for example between zero and the yield limit). The distribution changes when considering the curved shell instead of the plane plate, and bending effects occur additionally to the membrane stresses. The radial tension of the inserted part in the circumferential direction of the shell results in the inserted part being drawn into the shell, with the result that bending tension occurs on the inside and bending compression on the outside (approximation formulae for this are given in Ref. [8]). Heat spots in shells produce corrugation for the same reason, and not smoothing as in the plane case.

3.3.4 Residual stress source model

Residual stresses originate, as has been shown, as a consequence of inhomogeneous permanent deformations in components which, apart from this, behave elastically. In components with welding residual stresses, the permanent deformations are limited to the weld and its immediately surrounding area. This is where the large thermal and transformation strains occur in the course of the welding process and, in their train, the plastic deformations. Only in special cases do plastic deformations caused by welding also occur outside of the weld zones, for example at cracks and sharp notches. Welding residual stresses in the entire component can thus be attributed for the most part to permanent deformations in the weld zone.

The, in modelling terms, extremely complex explanation of the residual stresses based on the thermal and mechanical processes during welding can be substituted for the sake of simplicity by a fictitious, purely elastic correlation (which is independent of the stress and strain course during welding) of the residual stresses with local "extra strains" (these being incompatible strain components), "(macro) dislocations" (these being discontinuous displacement components) or "residual stress sources" (these being the non-equilibrium stresses which are assigned elastically to the extra strain). The possibility and single-valuedness of the purely elastic model requires to be examined in the individual case and assured by additional assumptions relating to the distribution functions (according to Reissner [270] single-valuedness is not given in the case of a general residual stress distribution).

Residual stress source distributions for weld seams also cannot be measured directly, but must be calculated back from measured residual stress parameters. The basis of this calculation is an elastic model for the component in question under the effect of residual stress sources, i.e. in simple cases a rod, beam, membrane, plate or shell model. This modelling and numerical processing is comparable with that which requires to be performed for the residual stress calculation based on the resilience values of the part from which material is removed in layers. The theoretical bases of the residual stress source method were developed by Reissner [270] and Rieder [271] and harnessed for welding residual stress problems by Schimmöller [272].

The solutions in Ref. [11, 273] according to the residual stress source method presented below have not proven to be generally practicable. The very successful practical application of the residual stress source method in combination with resilience measurements, on the other hand, is presented in Section 3.6.3.

According to Schönbach [273], a residual longitudinal stress source function with $\cos^2$-profile can be assumed over the width of the plastic zone for longitudinal welds on plate strips made of mild or low-alloy steel. The value $\varepsilon_{0\,max}$ equal 0.001 to 0.004 has proven to be a realistic maximum value of the extra strain in the weld longitudinal direction. The result of an analysis based on the residual stress source method in combination with beam bending theory for a plate strip manufactured from structural steel St37 with single-side edge surfacing weld is presented in Fig. 177.

A special form of the residual stress source method is the analysis method conceived in Japan and applied there to the cold cracking test specimen (rectangular plate with two slot welds in Y groove arranged in alignment one after the other, the Tekken weld test) [11]. Transverse strain dislocations ε_{0t} derived from transverse shrinkage measurements and longitudinal residual stress sources σ_{0l} derived from longitudinal residual stress measurements are combined. The integral equations which occur in treating the related elastic plate problem show an analogy to the aerodynamic wing theory and are treated according to the methods developed there.

The linearized residual stress source method, described above in its basic principles, has no practical significance for estimating unknown welding residual stresses in the absence of measurements. The residual stress source distribution is dependent to a complex extent on the material, dimensional, shape and restraining conditions. Questions concerning single-valuedness and accuracy remain open. The calculation of the residual stresses from the residual stress sources is, in general, not elementary. The residual stress source does not, therefore, lead to the desired information reduction.

This statement is not, however, intended to be generally disparaging of analyses which proceed from residual stress sources or extra strains. The

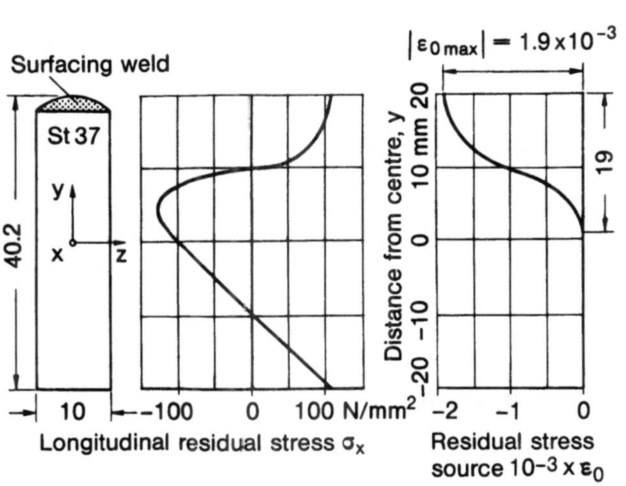

Fig. 177. Longitudinal residual stresses caused by residual stress source, plate strip with single-side edge surfacing weld, edges of plate strip unrestrained; after Schönbach [273]

shrinkage force models described are a useful special case of residual stress sources. The very successful determination of residual stresses from measured resilience parameters of the part from which material is removed in layers (see Section 3.6.3) is based on residual stress sources. And the incremental analysis algorithms relating to non-linear finite element structural models are likewise based on the concept of the residual stress sources or extra strains; in this case termed initial stresses or initial strains.

3.4 Overview of welding residual stresses

3.4.1 General statements

The results of analysis presented in Sections 3.2 and 3.3 serve primarily to demonstrate analysis methods without generally providing any coherent picture for practice. A brief overview of the principal statements of relevance for practice is possible, however, by including experimental findings.

The following statements apply irrespective of welding method, material selection and design detail:

- Particularly high multiaxial residual stresses occur in the fusion or pressure zone of the welded joint as well as in its heat-affected zone (zones of residual stress sources).
- The maximum residual stresses generally reach the yield limit (without hardening), although higher values are possible as a result of strain hardening as well as stress multiaxiality, and lower values as a result of superimposed transformation strains.
- The residual stresses may vary greatly from point to point in the fusion zone and heat-affected zone.
- Greatly differing residual stresses may occur on the surface and in the interior of parts particularly in the case of transforming alloyed metals.
- Considerable differences exist in respect of the residual transverse stresses between the longitudinal weld of a plate strip and the circumferential weld of a cylindrical or spherical shell, as a result of the constriction of the circumferential weld.
- Particularly unfavourable welding residual stress states occur in many cases at the weld ends.
- Considerable differences exist in respect of the residual stresses between the single-pass and the multi-pass weld; only the first and last pass of the multi-pass weld behaves similarly to a single-pass weld; the remaining passes are preheated by the preceding passes and post-weld heat-treated by the subsequent passes.

3.4.2 Weld-longitudinal residual stresses

Weld-longitudinal residual stresses arise as a result of the longitudinal contraction of the weld as it cools down, in certain cases superimposed by opposing

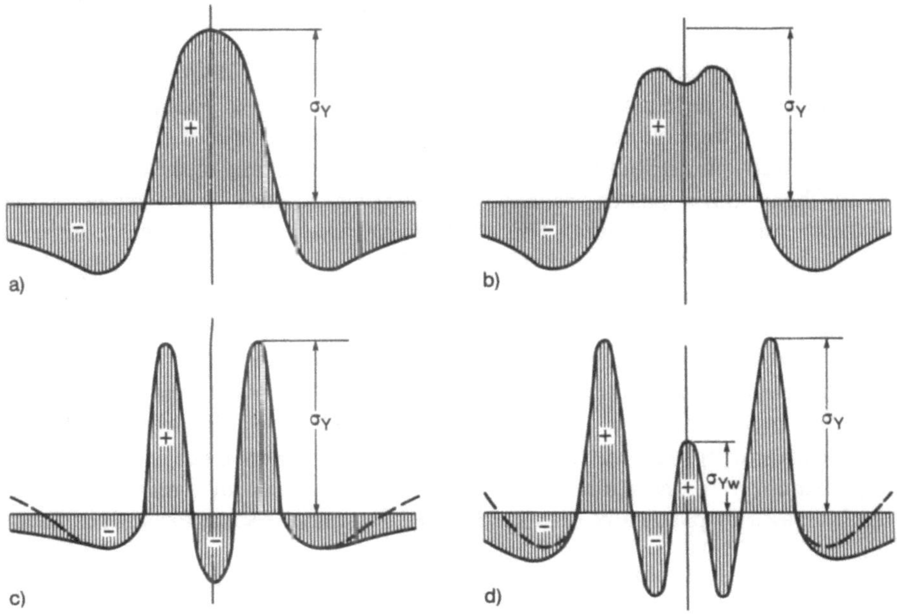

Fig. 178. Weld-longitudinal residual stresses, stress profile variants: mild steel (**a**), aluminium alloy (**b**), high-alloy structural steel with ferritic (**c**) and austenitic (**d**) filler metal

transformation processes. The different distributions presented in Fig. 178 can be observed. The basis of the presentation is a plate strip with centric longitudinal weld treated as a plane model. In mild and low-alloy steel, the simple distribution occurs with high tensile stresses (reaching the yield limit) in the weld and lower compressive stresses alongside to it (the W shape) (Fig. 178a). Such a distribution also occurs in the case of titanium, whereby the maximum stress is usually somewhat below the yield limit. In aluminium alloys, the maximum stress is likewise below the yield limit, but with a slight superimposed trough in the weld (Fig. 178b). In high-alloy steels with ferritic weld metal, the stress in the weld centre is displaced into the compression range (the M shape) as a consequence of the austenite-ferrite transformation at low temperature (Fig. 178c; see also Fig. 10). If, on the other hand, an austenitic electrode is used, the yield limit σ_{Yw} of the weld metal is reached in the weld. A compressive stress trough occurs to the right and left of that in the heat-affected zone, caused by transformation of the base metal after heating to above the transformation temperature A_{c1}. Further to the outside, the relatively high yield limit σ_Y of the base metal is reached as a consequence of heating to below A_{c1} (Fig. 178d). Even further to the outside, an additional reversal into the low tension range may occur as a consequence of the complex formation mechanism of residual stresses. The zones of particularly high tensile stresses prone to cold cracking occur in or close to the weld in accordance with the basic presentation of Fig. 178, depending on base and filler metal combination and process parameters. The great differences in stresses caused by the different cooling conditions on the surface and in the interior are shown for example in Fig. 140.

Longitudinal stress maximum values relative to yield limit based on practice-oriented experiments are summarized in Table 3. The reduction in maximum stress compared to the yield limit observed in the case of aluminium and titanium alloys is attributed in Ref. [8] to the shear warping of the cross-section as the heat source passes through. The warping, in turn, is the result of the elasticity of the component. This effect is more severe if the heat source is moving slower and the coefficient of thermal diffusivity of the metal is larger (it is particularly large in the case of aluminium and titanium alloys). Further results for aluminium alloys are presented in Ref. [286]. Stress levels up to the 0.2% offset yield limit have been measured on various welded joints.

The longitudinal residual stresses drop to zero at the transverse plate edge or the weld end. On the basis of elasticity theory, St Venant's Principle applies to such a stress drop, according to which the length of the drop range corresponds approximately to the width of the plate. The reduction in longitudinal residual stresses in plastically deformed zones, by comparison, takes place more rapidly.

Relatively short and narrow rectangular plates with a continuous or intermittent, through-welded or surfacing longitudinal weld used as brittle fracture specimens have been examined in respect of longitudinal residual stresses (Figs. 8 and 179). The rectangular plate with intermittent slot weld used as a hot cracking specimen has also been analysed (Fig. 151). The weld longitudinal

Table 3. Maximum value of longitudinal residual stress, $\sigma_{l\,max}$, in welds compared with yield limit, σ_Y, for different base metals; after Vinokurov [8]

Base metal	σ_Y [N/mm^2]	$\sigma_{l\,max}$ [N/mm^2]
Mild and low-alloy steels	210–240	210–240
Austenitic steels, non-hardened	280–300	280–350
Aluminium alloy AlMg6	160	80–120
Titanium alloy	500–700	300–400

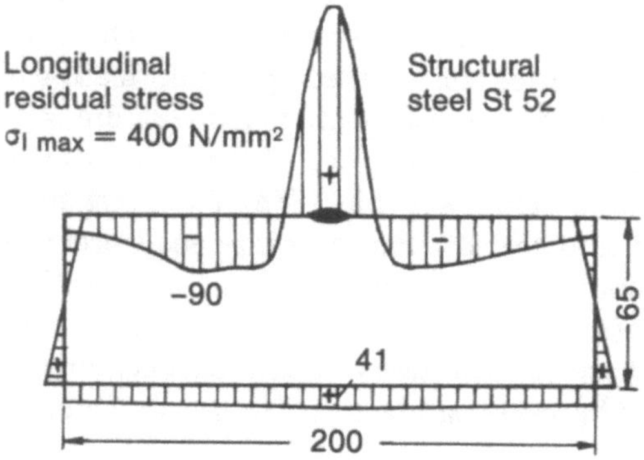

Fig. 179. Weld-longitudinal residual stresses in bead bend test specimen of low-alloy structural steel St52; experimental results after Klöppel [275]

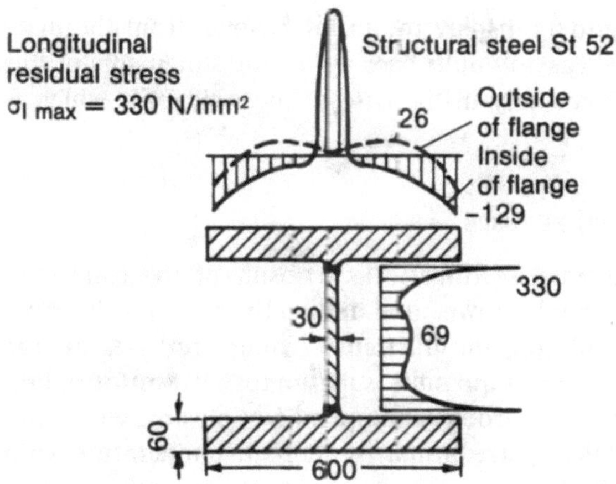

Fig. 180. Weld-longitudinal residual stresses in I section girder with web-to-flange welds, low-alloy structural steel St52; experimental results after Klöppel [275]

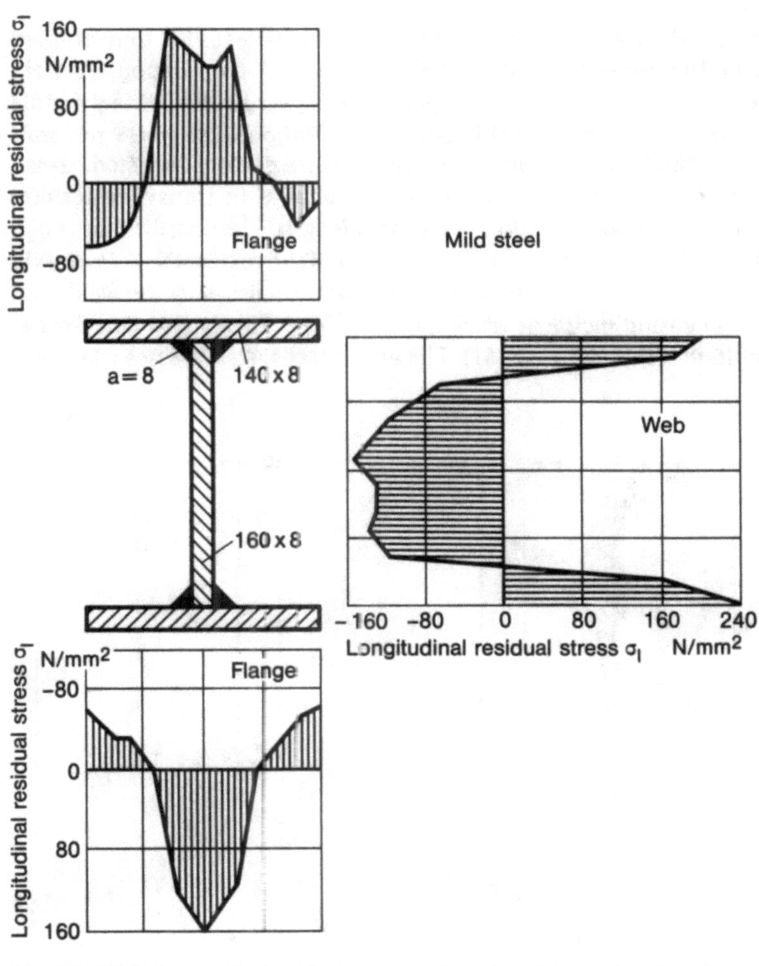

Fig. 181. Weld-longitudinal residual stresses in I section girder with web-to-flange fillet welds, mild steel; experimental results after Vinokurov [8]

residual stresses in I section girders display the profile known from the plate strip (Figs. 180 and 181). In the case of multi-pass welds, the statements of this section are approximately correct only in the case of the weld pass which is deposited last.

3.4.3 Weld-transverse residual stresses

Weld-transverse residual stresses arise directly as a result of the transverse contraction of the weld as it cools down, and indirectly as a result of its longitudinal contraction. In addition, the different cooling processes on the surface and in the interior as well as possible superimposed transformation processes are influencing factors. The groove gap opening or closing, which can be observed experimentally, takes place primarily at high temperature with a correspondingly low yield limit, so that it has scarcely any influence on residual stresses.

Transverse residual stresses at the butt weld in plates with unrestrained edges are primarily caused by restrained longitudinal contraction. A weld deposited instantaneously between two plates produces gaping in the middle section of the weld length if the weld cools down without transverse restraint (Fig. 182a). Accordingly, if the weld cools down transversely connected, transverse compression arises at the weld ends with a change to transverse tension when approaching the middle section of the weld length. The distribution which results for short plates is shown in Fig. 182b. A weld produced with slowly moving source, by contrast, results in high transverse tension at the weld end, thereby possibly causing incipient cracking (Fig. 205c). This is also true for the ends of intermittent welds (see Fig. 151). The above transverse stresses which are

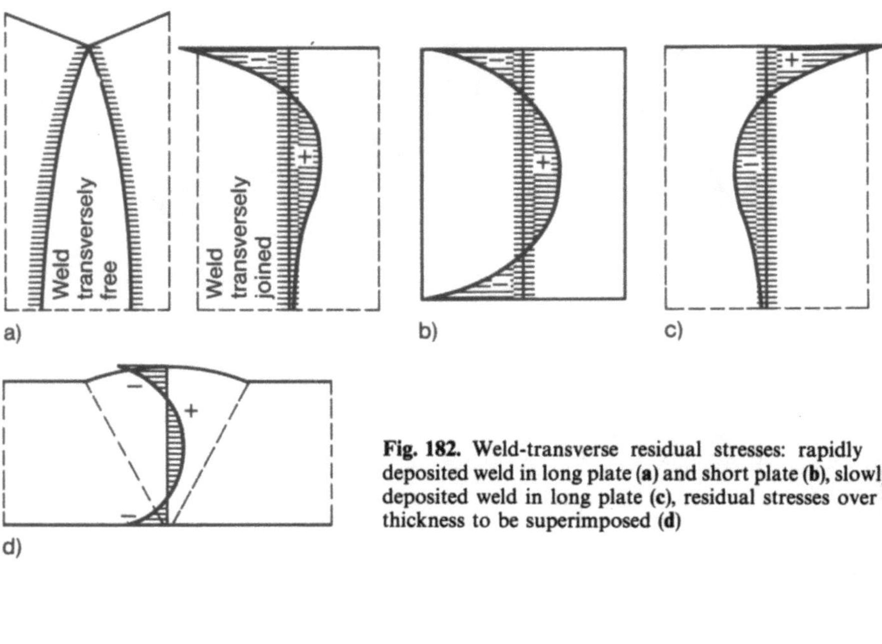

Fig. 182. Weld-transverse residual stresses: rapidly deposited weld in long plate (**a**) and short plate (**b**), slowly deposited weld in long plate (**c**), residual stresses over thickness to be superimposed (**d**)

constant over the plate thickness are superimposed by transverse stresses varying over the plate thickness (compression on the surface, tension in the interior) which result from different cooling conditions on the surface and in the interior (Fig. 182d). The three-dimensional stress state in a welded joint with V groove butt weld has been measured by Gunnert [4] (Fig. 183). The lack of the triaxial tensile stress state in the interior of the weld cannot be generalized. The applied method of measurement does not permit any all too accurate statement in the case considered.

In plates with restrained edges, the restrained transverse contraction of the weld and surrounding area determines the transverse residual stresses. Stresses resulting from restrained angular shrinkage may be superimposed.

The conditions are more complicated with metals subjected to microstructural transformation, and a reversal of the stress signs may occur. Figure 184 shows by way of example the transverse residual stresses in the surface of an electron beam welded plate made of an alloyed, martensitic-hardening steel, measured by the X-ray method. The stresses vary greatly within a confined area as a consequence of the concentrated high power heat input and large welding speed. They are consistent in qualitative terms with the calculated profile according to Fig. 140. The variety and complexity of possible distributions shown in Fig. 178 in respect to longitudinal residual stresses is also true for transverse residual stresses.

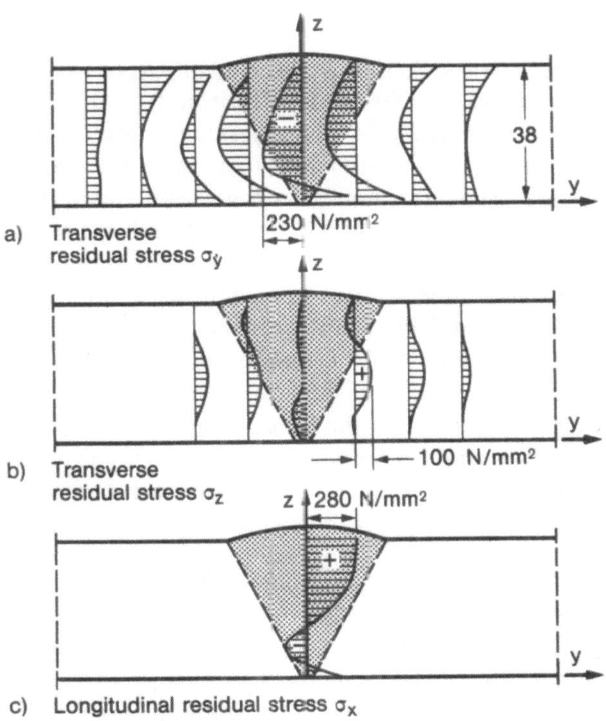

Fig. 183. Transverse residual stresses (**a**, **b**) and longitudinal residual stresses (**c**) in plate with V groove butt weld; experimental results after Gunnert [4]

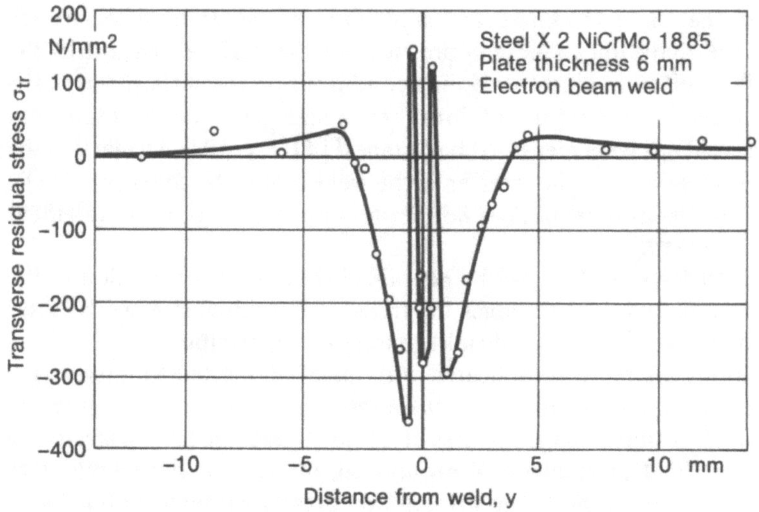

Fig. 184. Weld-transverse residual stresses in surface of electron beam butt welded plate of steel X2NiCrMo1885, thickness 6 mm, welding speed $v = 65$ mm/s, experimental results based on X-ray stress measurement; after Wohlfahrt [277]

The transverse residual stresses in the circumferential weld of cylindrical or spherical shells differ essentially from those in plane plates as a result of the superimposed bending stresses caused by the constriction of the shell (see Sections 3.2.3 and 3.3.3). A characteristic feature is the tendency to transverse tension at the weld root. Experimental results are detailed in Ref. [290].

In multi-pass butt welds in accordingly thicker plates, residual stresses which are irregularly distributed over the plate thickness occur to an increased extent. The last passes to be deposited encounter high shrinkage resistance in the longitudinal and transverse direction (in the plane of the plate), with the result that not only high longitudinal residual stresses occur but also high transverse residual stresses reduced somewhat by preheating as a consequence of the passes deposited earlier. The passes deposited first are subjected initially to tension, and then later to compression, superimposed by strong bending effects in the transverse direction. Transverse tension occurs at the root of the multi-pass butt weld deposited from one side where angular shrinkage is unrestrained, and transverse compression occurs where angular shrinkage is restrained. The residual stresses in the direction of the plate thickness, by contrast, are negligibly small. The triaxial tensile residual stress state is largely avoided by multi-pass welding.

The rises and drops of the transverse residual stress during multi-pass welding at the specimen of the rigid restraint hot cracking test is shown in Fig. 185. The reaction stress in the weld results from relating the measured reaction force to the weld cross-section which is enlarged pass-by-pass [7, 278]. The stress is highest at the third and fourth pass and exceeds the yield limit.

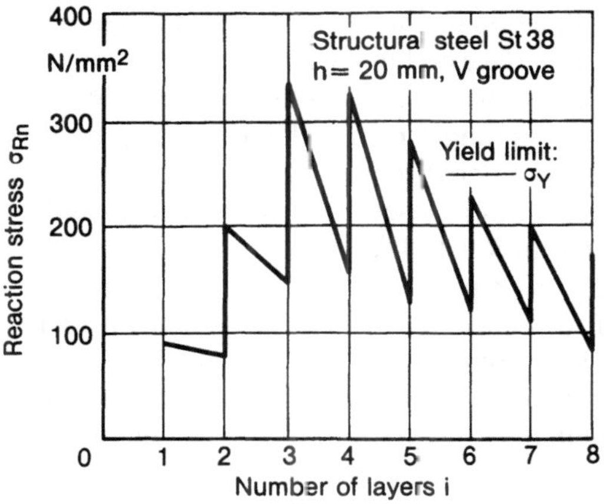

Fig. 185. Reaction stress transverse to multi-layer butt weld in V groove between rigidly restrained plate strips (RRC-test), experimental results; after Onoue [7, 278]

Consequently, enlarged straining lengths are necessary only up to the fourth pass in the components which are simulated in the test.

The multi-pass longitudinal weld with double-V groove and groove-symmetrical weld passes in the cylindrical shell displays a soft transverse restraint behaviour. With V groove and thus single-sided asymmetrical welding, a bending moment occurs as a result of restrained angular shrinkage, which reduces the tensile stresses at the weld root; particularly, however, the high plastic tensile strains in this area. In the multi-pass circumferential weld with V groove in the cylindrical shell, the shrinkage bending moment increases as a result of the suppression of angular shrinkage.

Residual stress measurements on multi-pass butt welds (e.g. in Ref. [8, 228, 229, 235, 280, 289]) generally confirm the above qualitative statements. They may also, however, show considerable deviations, which are caused by transformation processes, by preheating and post-weld heat-treatment measures as well as by altered welding sequence and restraint conditions. The residual stresses of multi-pass submerged-arc welds are compared in Ref. [279] with those of single-pass electron beam welds considering the transforming steel 20MnMoNi55. With the same plate thickness, the residual stresses in submerged-arc welds are higher and more extended than in electron beam welds.

The transverse (and longitudinal) residual stresses in the top and bottom surface of a MnNi-alloyed steel plate (1250 × 820 × 35 mm) with submerged arc multilayer welded butt joint have been determined using the cutting method with surface layer removal [289]. The filler metal in the single-bevel groove had a yield strength 50% higher than that of the base metal. The tensile longitudinal stresses reach the higher yield stress; the transverse stresses do so only on the bottom side at the weld root. The transverse stresses on the top side remain below or even change into compression. Fracture mechanics specimens transversely cut from the welded plate have the above longitudinal stresses mainly released.

The capability of the neutron diffraction method for measuring longitudinal and transverse residual strains on the surface and in the interior of welded joints is demonstrated by Ref. [287, 288]. The measuring results are compared with results from finite element analysis. A gas tungsten arc welded surfacing bead on a 8 mm thick stainless steel plate is considered in Ref. [287]. A hollow section made of a cold-formed channel section bar closed by a plate strip joined by longitudinal edge welds is considered in Ref. [288].

In electroslag welding, relatively thick plates ($h > 100$ mm) are joined in a single-pass operation by means of resistance fusion welding. The groove has a square shape with a gap width of approximately 30 mm (independent of plate thickness). The welding speed is small in accordance with the large cross-section of the weld. This results in the qualitative distribution of the residual stresses described for single-pass welding, the level of which depends, among other parameters, on the ratio of plate thickness to gap width. Compression prevails on the surface transverse to the weld, and tension in the interior. With a large plate thickness, the restraint of shrinkage in the plate thickness direction is no longer negligibly small. Tensile stresses build up in this direction. As a result, the triaxial tensile stress state, which causes part of the low notch ductility of the electroslag welded metal, arises in the interior of the weld. Further details of the residual stresses of electroslag welds in plane plates and cylindrical shells are reviewed in Ref. [8].

3.4.4 Residual stresses after spot-welding, cladding, and flame cutting

The maximum welding residual stresses in spot-welded joints manufactured from metals which undergo no major microstructural transformation occur on the inside surface of the plates at the weld spot edge in the form of triaxial tension, whereby the tangential stress may be increased beyond the (uniaxial) yield limit (see Fig. 137). These high tensile levels are partially reduced in spot-welded joints manufactured from metals with microstructural transformation (these include martensitic-hardening steels).

As regards the residual stress distribution in cladded steel, which is used in boiler and vessel construction in respect of corrosive media, path-breaking investigations have been performed by Schimmöller [272, 281] (see also Ref. [282, 283]). The aim of the investigations was a statement regarding the corrosion risk of the cladding and the proneness of the bounding interface to incipient cracking in the as-delivered state (roll-clad or explosive-clad) and following annealing (920 °C/0.5 h) or cold stretching (10%). Tensile residual stresses on the cladding surface or in the bounding interface should be avoided. The plane-hydrostatic residual stresses have been calculated from the measured strains which result from the removal of surface layers (see Section 3.6.3.2). The material investigated was boiler steel (plate thickness 9 mm) with austenitic CrNi steel cladding (cladding thickness 1 mm). In the as-delivered state, high compressive stresses were found on the cladding surface with a sharp drop to

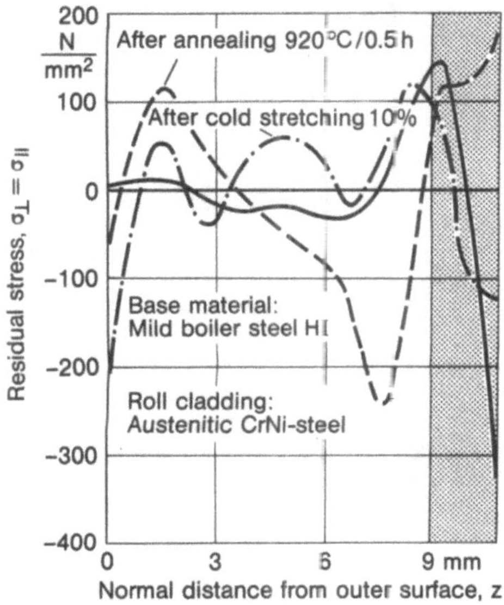

Fig. 186. Plane-hydrostatic residual stresses, $\sigma_{\perp} = \sigma_{||}$, in ferritic mild boiler steel with austenitic roll cladding, as-delivered, after annealing and after cold-stretching, experimental results; after Schimmöller [281] (ibid. mean values of σ_1 and σ_2 of specimen types 1.1 to 1.3)

a maximum tensile value in the bonding face of the cladding (Fig. 12). Subsequent annealing produces an unfavourable reversal of stress distributions, subsequent cold stretching a severe stress change on the outer surface of the cladding (Fig. 186).

Knowledge also exists regarding the residual stresses in flame cut edges and its change as a result of subsequent thermal or mechanical treatment [284, 285]. The longitudinal residual stresses in the flame cut edge have been calculated from the elastic resilience of related plate strips resulting from the removal of edge layers (see Section 3.6.3.1). Compressive residual stresses occur in the flame cut surface of carbon steels (St60-1 in Ref. [284]) as a result of martensitic-hardening (severe drop in compressive stress and in hardness over approximately 0.5 mm normal to the edge); further inside tensile residual stresses prevail. The level of the compressive residual stresses in the flame cut edge increases with the cutting speed and decreases with the heat output of the flame. No hardening occurs in the case of soft iron (α-iron) and thus also no compressive residual stresses in the flame cut edge, but only tensile residual stresses originating from thermal strains.

The effect of subsequent thermal or mechanical treatments has been investigated using structural steel StE36 [285]. The residual stress profile perpendicular to the flame cut edge is shown in Fig. 11. It changes little as a result of preheating (150 °C) during flame cutting, but hardening is reduced. Stresses and hardening are diminished as a result of subsequent annealing (650 °C/0.5 h); as a consequence of cold stretching (5 or 10%) high tensile residual stresses occur in the flame cut edge (Fig. 187). The stress reduction to be expected after cold stretching of homogeneous material does not occur here because the yield stress is heavily increased in the flame cut edge as a result of hardening. The machining

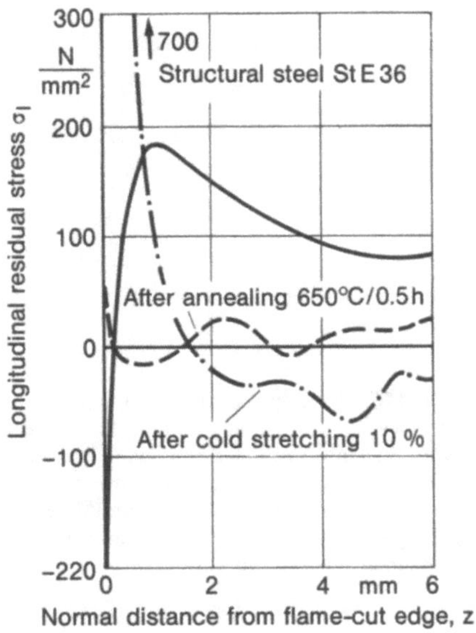

Fig. 187. Longitudinal residual stress σ_l in flame-cut edge of structural steel StE36 as-delivered, after annealing and after cold-stretching, experimental results; after Schimmöller [285]

of the non-post-treated flame cut edge, which is undertaken in practice for various reasons, results in a stress reversal into the tensile range already at a low machining depth (approx. 0.5 mm).

3.5 Welding distortion

3.5.1 Model simplifications

Depending on the type of deformation and component, welding distortion is assigned to the following classes:

- distortion in the plane of the welded component, e.g. longitudinal and transverse shrinkage of welded plates;
- distortion perpendicular to the plane of the welded component, e.g. plane or axisymmetrical angular shrinkage of welded plates;
- bending distortion of girders with longitudinal and transverse welds;
- twisting distortion of box girders with longitudinal welds;
- distortion of axisymmetric shells with meridional and circumferential welds.

Welding warpage, the buckling of thin-walled components as a consequence of shrinkage forces, represent a special class.

The primary consideration in welding residual stress analysis is the individual welded joint. All the details of the complex thermomechanical process should, if possible, be simulated. Shrinkage forces and shrinkage displacements (translation and rotation) following completion of the weld are the result of this analysis. By contrast, the primary consideration in welding deformation analysis is the entire structure connected by welded joints, proceeding from the shrinkage forces or shrinkage displacements of the individual welded joints, introduced as initial forces or initial displacements. To the extent that the latter can be determined, the stresses and deformations of the structure resulting therefrom can be calculated quasi-elastically according to the engineering formulae of structural analysis or, in more complicated cases, according to the finite element method. The warpages can also be determined approximately in this manner. The individual welds are instantaneously deposited in the model (i.e. there is no steady weld deposit). If the quasi-elastic analysis reveals that the yield limit is exceeded in the weld area, corrective reductions require to be made on the shrinkage parameters.

The considered shrinkage parameters of the welded joint are:

- the longitudinal shrinkage force F_s of straight and curved, butt and fillet welds;
- the transverse shrinkage Δ_{tr} of butt and fillet welds, determined on non-restrained plates;
- the angular shrinkage $\Delta\beta$ for butt and fillet welds, determined on non-restrained plates;
- the groove misalignment Δz perpendicular to weld and plate as a consequence of groove off-set during welding, e.g. in the case of vessel circumferential welds,
- the groove misalignment Δx in the direction of the weld as a consequence of groove longitudinal off-set during welding, e.g. in the case of twisting distortion of longitudinally welded box girders;
- the groove gap closing or opening Δy as a consequence of groove transverse off-set during welding.

The different types of welding deformation are dealt with in Sections 3.5.2 to 3.5.4 and the shrinkage parameters, which form the basis for the approximative analysis of the total structure, are stated there. Further data, formulae and application examples can be found in Ref. [10, 299–301]. Finally, warpage is dealt with in Section 3.5.5.

3.5.2 Transverse shrinkage and groove transverse off-set

The key for the quantitative description of transverse shrinkage lies in the non-restrained transverse off-set of the groove edges as a consequence of the heat source moving along the edges. This transverse off-set can be approximated by an elastic solution [8]. The elastic field of the infinite plate with stationary or moving heat source is known, see Section 3.1.2. The relevant equations are enlarged by a term for the particularly strong heat transfer to the surrounding air in the vicinity of the heat source, to improve modelling accuracy. The elastic field in the semi-infinite plate with stationary or moving source in the plate edge is gained by cutting the infinite plate in the centre line and cancelling the (transverse) stresses in this line through opposed, equally large edge forces.

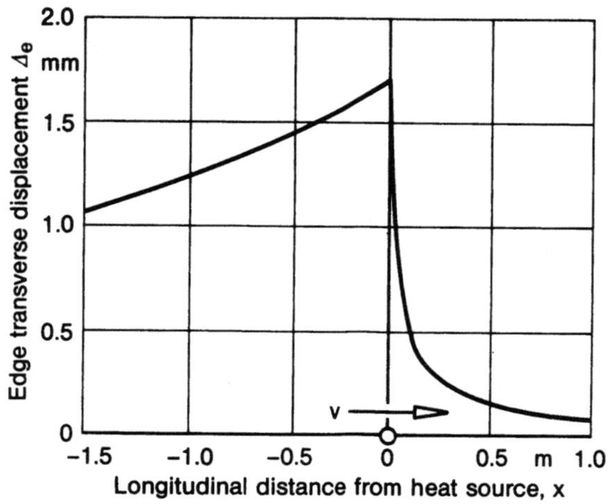

Fig. 188. Edge-transverse displacement of semi-infinite plate with heat source moving on edge for parameters selected close to practical application, analytical results; after Vinokurov [8]

Functional-analysis solutions are available for the semi-infinite plate subjected to edge forces of the considered type.

A numerical result for the semi-infinite plate with a heat source moving uniformly along the edge is presented in Fig. 188 for a specific practice-oriented parameter selection. The (groove) edge transverse displacement Δ_e rises rapidly ahead of the heat source, achieves its maximum value at the source and drops slowly after the source. The drop, however, only occurs with heat transfer to the surrounding air. In the absence of heat transfer, the transverse displacement remains constant despite the temperature equalization as a result of heat diffusion. The maximum value of the transverse displacement gained by the purely elastic solution is stated in Ref. [8] (with heat input q of the infinite plate):

$$\Delta_{\mathrm{emax}} = \frac{\alpha q}{c\varrho h v}. \tag{165}$$

The welding process is now considered with and without groove gap. With groove gap, the transverse closing displacement of the groove $2\Delta_e$ can take place unrestrained. At the maximum value, $2\Delta_{\mathrm{emax}}$, joining occurs through the weld seam. If adequate strength existed directly after the welding heat source, the transverse shrinkage would start immediately as a result of cooling. In reality, transverse plastic yield prevails at high temperature, so that the transverse shrinkage does not start within a small distance after the source (the distance can be estimated on the basis of the temperature field and the temperature dependence of the yield limit). The transverse shrinkage is thus reduced by a plastic portion, which depends on the material properties and the intensity of the heat transfer to the surrounding air. The transverse shrinkage Δ_{tr} after welding with groove gap is therefore given with $2\Delta_{\mathrm{emax}}$ according to equation (165), $q_w = q/v$ and transverse stiffness factor $\mu_t = 0.75$–0.85 (see Ref. [8]):

$$\Delta_{\mathrm{tr}} = \mu_t \frac{2\alpha q_w}{c\varrho h}. \tag{166}$$

In welding without gap or with fixed gap, the two groove edges press against one another ahead of the source, at first elastically, then elastic-plastically, so that $2\Delta_{emax}$ is additionally reduced. At the same time, however, the transverse plastic yield after the source is also reduced. According to Ref. [8], the transverse shrinkage during welding without gap is somewhat less than the transverse shrinkage during welding with gap. The lack of gap movability, however, may also exist as a result of tack spots, tack welds or groove wedges. The transverse shrinkage Δ_{tr} after welding without groove gap or with fixed groove gap is approximated by equation (166) with $\mu_t = 0.5$–0.7 (see Ref. [8]).

An expression for the transverse shrinkage Δ_{tr} in the form of equation (166) is obtained in a more simple manner from the unrestrained cooling shrinkage of a plate strip (length l, thickness h) with transverse weld. The heat input per unit length of weld q_w produces the mean temperature increase

$$\Delta T_0 = \frac{q_w}{c\varrho h l}. \tag{167}$$

With $\Delta_{tr} = \alpha \Delta T_0 l$, it follows that:

$$\Delta_{tr} = \frac{\alpha q_w}{c\varrho h}. \tag{168}$$

The older approximation formulae used in practice (see Ref. [280, 291–295]) are based partially on this formulation. After inserting the material characteristic values for mild structural steel and substituting the heat input per unit length of weld by the weld cross-section area (for example in accordance with equation (143) or on the basis of the groove cross-section), the result for the butt weld with a mean groove width w_g is for example (see Ref. [280]):

$$\Delta_{tr} = 0.17 w_g. \tag{169}$$

An alternative consideration, leading to equation (168), proceeds from the maximum temperature strain $\varepsilon_{T\max} = \alpha T_{\max}$ ($T_{\max}$ according to equation (49)) and integrates this strain (understood as transverse strain) transverse to the weld between weld and plate edge, while neglecting the temperature strains at vanishing yield stress at high temperature (e.g. $T_{\max} > 600\,°\mathrm{C}$). In addition, a correction for the transverse restraint caused by the cooling weld after the source is inserted as is also, as the case may be, an allowance for shrinkage of the material in the groove The relevant stiffness factor μ_t is, according to Ref. [2] or [7] (ibid., e.g. equation (101)), dependent on the length of the plastic zone in weld direction, l_{pl}, relative to the weld length, l_w (this is not the case for the longitudinal shrinkage strain in equation (149)). This results in equation (166) with

$$2\mu_t = \frac{1}{1 + \frac{1}{2}\frac{l_w}{l_{pl}}} + 0.335. \tag{170}$$

In intermittent welding without a groove gap, the transverse shrinkage, Δ_{tr}, is reduced in the ratio of the string weld length, l_{st}, to the sum of string length, l_{st}, and string interspace length, l_i, resulting in

$$\Delta_{ti} = \Delta_{tr} \frac{l_{st}}{l_{st} + l_i}. \tag{171}$$

In intermittent welding with a groove gap, the transverse shrinkage, Δ_{tr}, occurs without reduction.

The T joint with single-sided intermittent weld behaves in accordance with equation (171) in respect of transverse shrinkage. The double-sided staggered intermittent weld displays the transverse shrinkage of a continuous weld. The double-sided non-staggered intermittent weld doubles the value of the transverse shrinkage according to equation (171). The transverse shrinkage at the beginning of the weld is significantly smaller than indicated by equation (166). Welds placed alongside to each other or one above the other under unchanged thermal and mechanical conditions result in a summation of the individual shrinkages.

The transverse shrinkage of the first weld pass between thick plates has been examined in more detail in respect of cold cracking problems in multi-pass welding by Satoh [297, 266, 267] and Graville [298]. The relatively thin weld is surrounded by an axisymmetrical temperature field as a result of heat diffusion with temperature T_m^* in the weld centre plane at start of shrinkage. The integration of the relevant thermal strains between $r = 0$ and $r = \infty$ results in the transverse shrinkage

$$\Delta_{tr} = \alpha \sqrt{\frac{q_w T_m^*}{c\varrho}}. \tag{172}$$

Originally, the melting temperature was inserted for T_m^*. The transverse shrinkage becomes effective, however, only after cooling to a lower temperature. For this reason, $T_m^* = 800\,°C$ should be inserted for structural steel instead of $T_m^* = 1500\,°C$. Equation (172) applies in place of equation (168) for $h > h_t$ with

$$h_t = \sqrt{\frac{q_w}{c\varrho T_m^*}}. \tag{173}$$

In the range $h > h_t$ the transverse shrinkage Δ_{tr}, in accordance with equation (172) independent of h, can be considered to be superimposed from Δ_{t2} according to equation (168) and $\Delta_{t1} = \Delta_{tr} - \Delta_{t2}$. It consists of an early, short-duration phase 1 with Δ_{t1} resulting from weld-close heat diffusion and a later long-duration phase 2 with Δ_{t2} resulting from weld-distant heat dissipation (Fig. 189). The transverse shrinkage, Δ_{tr}, according to equations (168), (172) and (173) is presented in Fig. 190 for structural steel dependent on plate thickness h and heat input per unit length of weld, q_w. As regards the forces which are built up with restrained transverse shrinkage, reference is made to Section 3.3.2.

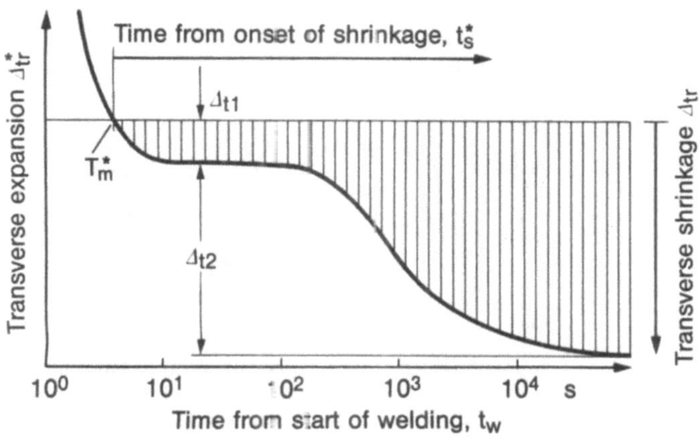

Fig. 189. Transverse shrinkage following transverse expansion, phases 1 and 2, for butt weld (first layer) between thicker plates, schematic representation; after Satoh et al. [266, 267, 297]

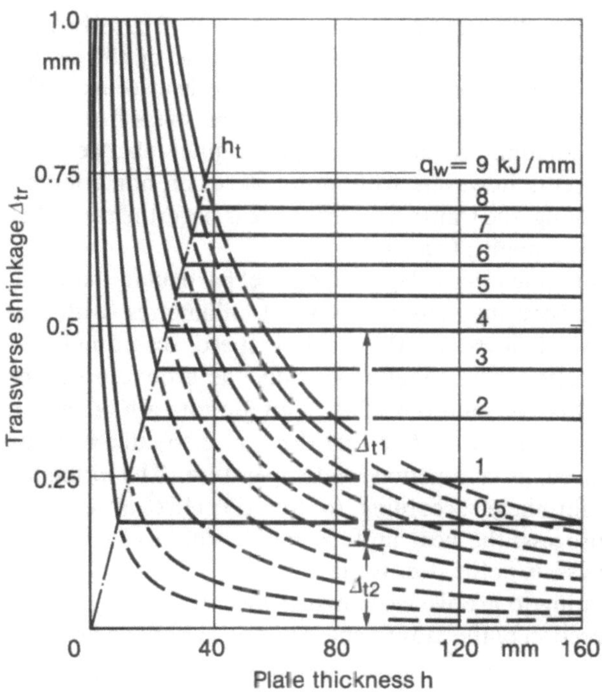

Fig. 190. Transverse shrinkage Δ_{tr} dependent on plate thickness h and heat input per unit length of weld, q_w, for structural steel: $\Delta_{tr} = \Delta_{t2}$ for $h < h_t$, $\Delta_{tr} = \Delta_{t1} + \Delta_{t2}$ for $h > h_t$, results from calculation; after Satoh et al. [266, 277, 297]

In multi-pass welding the transverse shrinkage Δ_{tr} is all the smaller the more passes are deposited in a given groove cross-section and the more intensive is the cooling process between the passes [268].

The groove face transverse offset during welding has been determined on large plates made from structural steel [234]. Groove opening occurred increasingly, intensified by transformation strains. The degree of edge restraint of the

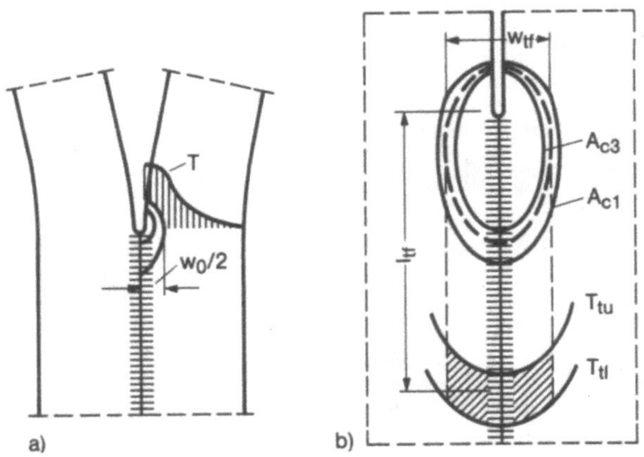

Fig. 191. Groove face movement during seam welding; groove opening (**a**) caused by inhomogeneous longitudinal thermal strains, high-temperature width w_0 nearly free of stresses, groove closing (**b**) caused by transformation strains in distance l_{tf} over width w_{tf}; after Vinokurov [8]

plates connected by welding had only a minor influence on the groove transverse movement.

In long narrow plate strips, the groove faces, as shown in Fig. 191a may temporarily bend apart ahead of the heat source. The cause of this is the longitudinal thermal strain in the groove faces ahead of the heat source. The high-temperature zone of width w_0 should be introduced as stress-free in the approximation calculation relating to the size of the angular change. This bending apart can only occur, however, with non-tacked or non-counter-restrained grooves. After cooling, the transverse shrinkage is nevertheless rather uniform, except at the starting point of the weld.

Groove face transverse offset in electroslag welding is of particular interest in practice. In this case, the groove may close by several millimetres (with gap widths of 30 to 40 mm), which increases the welding speed accordingly. In addition, the groove edges alter their angular position; the groove opens or closes. More accurate quantitative solutions of Russian authors are offered in Ref. [8]. The accuracy is enhanced by the fact that the heat dissipated into the surrounding air is introduced as a function of temperature. The principal dependence of transverse shrinkage according to equation (168) is retained. Considering the welding process, the heat input per unit length of weld, q_w, referred to the plate thickness, h, is the decisive factor.

The data in Ref. [8] relating to groove face angular movement are partially unclear in respect of the direction of the movement. Allowance is made not only for the groove longitudinal strain ahead of the source but also for the non-uniform transverse shrinkage after the source as an angle-changing element. The groove angular change at the start of welding may occur opposite to that at the end of welding. This can be presented in a particularly illustrative manner on the basis of the transformation strain after the source (Fig. 191b). Transformation strain as a consequence of $\gamma\alpha$-transformation during cooling occurs at the distance l_{tf} between the temperatures T_{tu} and T_{tl} from initiation and end of the transformation over the width w_{tr}, in which heating over A_{cm} (mean value of A_{c1}

and A_{c3}) occurs, i.e. into the γ mixed crystal range. The expansion of the above transformation area results in groove gap closing, as long as this area is moving within the first half of the weld length, and groove gap opening after this. The former state exists at the beginning of the weld, the latter at its end (provided the weld is sufficiently long).

3.5.3 Longitudinal and bending shrinkage

The longitudinal shrinkage Δl of plates, bars or girders of length l with centric longitudinal weld follows from equation (149) (with cross-sectional area A instead of hw),

$$\Delta l = \mu_l \frac{\alpha q_w l}{c \varrho A} \tag{174}$$

or also proceeding from the (active) shrinkage force F_s

$$\Delta l = \frac{F_s l}{EA}. \tag{175}$$

The longitudinal shrinkage force F_s is, according to Section 3.3.1, dependent on the heat input per unit length of weld, q_w, or on the weld cross-section area A_w, on the elastic modulus E as well as on the material characteristic value $\alpha/c\varrho$ (already expressed with equation (174)). Multi-pass welding as well as intermittent welding reduce the value of F_s.

With eccentric longitudinal welds, bending shrinkage is superimposed on longitudinal shrinkage. End cross-section rotation φ and deflection w occur in beams when subjected to the longitudinal shrinkage moment, $M_{sl} = F_s e$ (Fig. 192) (formulae according to beam theory; weld eccentricity e, geometrical moment of inertia I):

$$\varphi = \frac{F_s e l}{EI}, \tag{176}$$

$$w = \frac{F_s e l^2}{8EI}. \tag{177}$$

Equations (175) to (177) apply to plate strips and girders. If the plate or girder length is smaller than the width or height, this results in values which are too

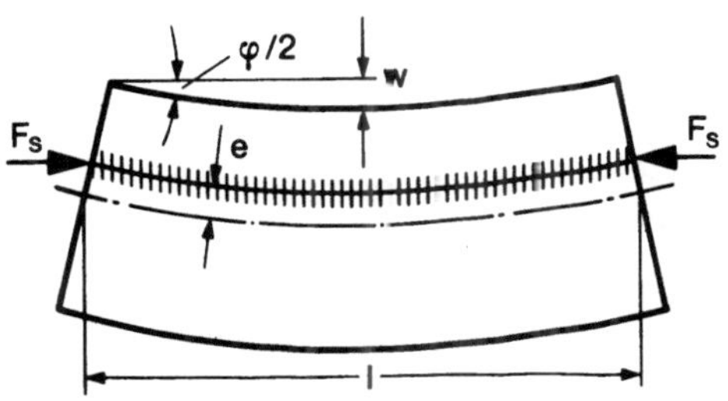

Fig. 192. Bending shrinkage of beam caused by eccentric longitudinal weld

high. On the other hand, the cross-sectional moment of inertia should be reduced if the cross-section is partially plastically deformed when subjected to the shrinkage force.

Considerable bending shrinkage occurs in girders as a result of eccentric transverse welds. The angular change $\Delta\varphi$ in the girder cross-section containing the transverse weld follows from the weld transverse shrinkage Δ_{tr}, the area moment S of the cross-section part with the weld and the geometrical moment of inertia I of the total cross-section:

$$\Delta\varphi = \Delta_{tr}\frac{S}{I}. \tag{178}$$

The shrinkage Δ_{tr} is reduced by plastic deformation if the girder has a large bending stiffness relative to the shrinkage force.

For the derivation of equation (178) an I section girder element of length l with a transverse weld in the flange is considered (Fig. 193). The transverse shrinkage force, $F_{st} = \Delta_{tr}EA_f/l$, which acts in the flange cross-section area A_f within the flange distance h_f from the girder centre line and thus produces the transverse shrinkage moment M_{st}, is derived proceeding from the transverse shrinkage Δ_{tr}. According to beam theory it follows that

$$\Delta\varphi = \frac{\Delta_{tr}EA_f h_f l}{lEI}. \tag{179}$$

Introducing $S = A_f h_f$, this leads to equation (178).

When welding components which are subjected to load (as for example during erection of structures, instant straightening or repairs) significantly increased bending shrinkage may occur as a consequence of the enlarged plastically deformed areas, while the residual stresses after load removal are reduced (see Ref. [7]).

Longitudinal and bending shrinkage are of special relevance for cylindrical, spherical and toroidal shells (Fig. 194) (see also Section 3.3.3). The longitudinal weld in the short cylindrical shell deflects the shell locally to the inside, the

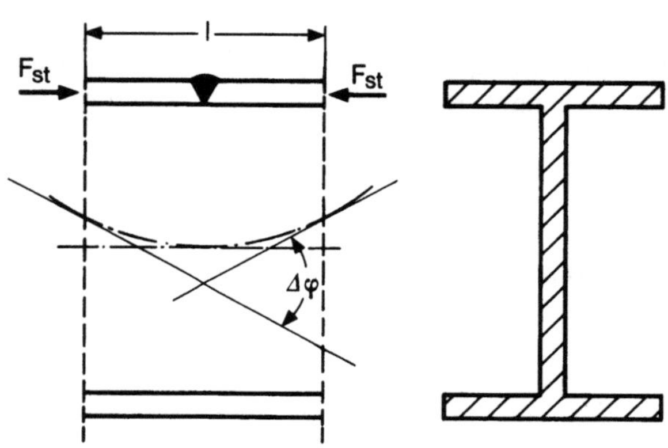

Fig. 193. Angular deformation $\Delta\varphi$ caused by transverse shrinkage of transverse flange weld

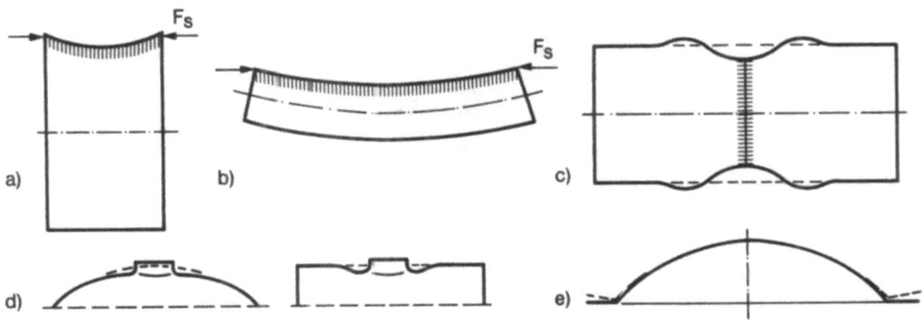

Fig. 194. Longitudinal and bending shrinkage at longitudinal weld (**a**, **b**), circumferential weld (**c**) and ring weld (**d**, **e**) of cylindrical and spherical shell

longitudinal weld on the long cylindrical shell may bend the entire cylinder and the circumferential weld constricts the cylinder (Figs. 194a to 194c). These shrinkage deformations are caused by the longitudinal shrinkage force. The "drawing-in" of block flanges or nozzles into spherical or cylindrical shells or also the flange tilting of a dished cover with circumferential flange weld, by contrast, are explained from the effect of the transverse shrinkage force (Figs. 194d and 194e). If axisymmetrical shells of different radial stiffness are joined by a circumferential weld (for example, a cylindrical shell joined with a shell of different thickness or a cylindrical shell joined with a dished end), edge offset may occur during welding because of a different, elastic-plastic deformation behaviour of the jointed parts.

3.5.4 Angular shrinkage and twisting distortion

Angular shrinkage occurs at butt, lap, T, cruciform and corner joints as a result of single-sided or also asymmetrical double-sided welding. The extent of angular shrinkage depends on the width and depth of the fusion zone relative to plate thickness, on the type of joint, on the weld pass sequence, on the thermo-mechanical material properties and on the characteristic parameters of the welding process (heat input per unit length of weld, distribution of heat source density).

The variety of influencing parameters explains the greatly differing or inadequately specified data in the literature relating to angular shrinkage. The basis for the sections which follow are the data derived on the basis of thermo-mechanical models by Kuzminov (in Ref. [8]).

Unrestrained angular shrinkage $\Delta\beta$ is considered first. The dependence on welding speed v, heat input per unit length of weld q_w and weld pass or plate thickness h which results for the single weld pass in mild and low-alloy structural steel, is shown in Fig. 195 (obviously compiled for conventional fusion zone profiles). For multi-pass welding, the angular shrinkages of the passes on the two

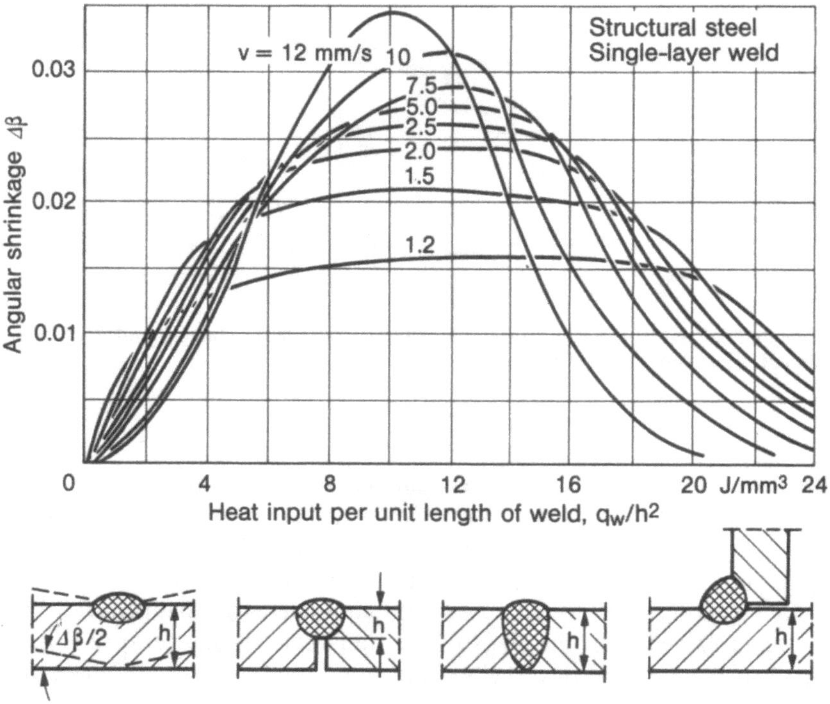

Fig. 195. Angular shrinkage $\Delta\beta$ dependent on heat input per unit length of weld, q_w, and weld or plate thickness h for different welding speeds v; after Vinokurov [8]

sides of the plate (indices i and j) are subtracted in accordance with:

$$\Delta\beta = \sum \Delta\beta_i m_i - \sum \Delta\beta_j m_j . \tag{180}$$

The correction factors m_i and m_j for the pass number i and j, respectively, should be inserted in accordance with Fig. 196a (m_i and $m_j \approx 1.0$ for the first passes on each side). The thicknesses h_i and h_j for $\Delta\beta_i$ and $\Delta\beta_j$, respectively, according to Fig. 195, should be selected in accordance with the respective weld pass total thickness of deposited weld passes. This is illustrated in Fig. 196b for a four-pass butt weld with X groove.

A distinction requires to be made in the case of fillet welds between the angular shrinkage $\Delta\beta$ of the through-plate and the tilting shrinkage $\Delta\beta^*$ between through-plate and butting plate. This is illustrated in Fig. 197 for the flange of an I section girder with single-sided web-to-flange fillet weld. The angular shrinkage $\Delta\beta$ is determined according to Fig. 193, whereby the heat portion attributable to the through-plate under study should be estimated in accordance with equation (150). The tilting shrinkage angle $\Delta\beta^*$, which is caused by the shrinkage of the approximately triangular weld cross-section in the direction of its hypotenuse, is largely independent of the dimensional parameters according to Ref. [8] when tilting is unrestrained, $\Delta\beta^* \approx 1.25°$. Where tilting is

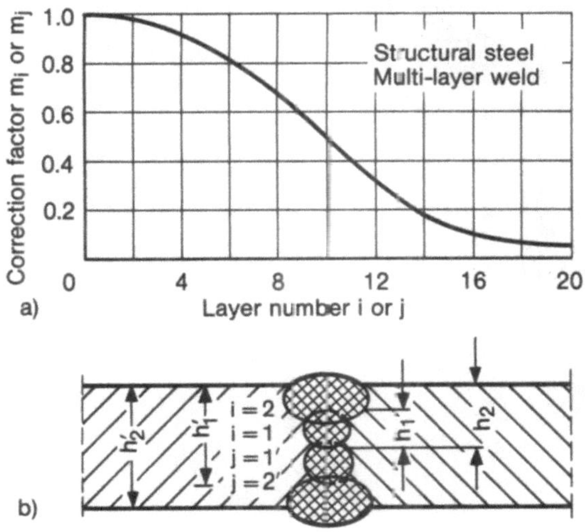

Fig. 196. Correction factors m_i or m_j dependent on layer number i or j (**a**); layer total thickness h_i or h_j of four-layer butt weld in double-V groove and layer sequence i = 1, i = 2, j = 1, j = 2 (**b**); after Vinokurov [8]

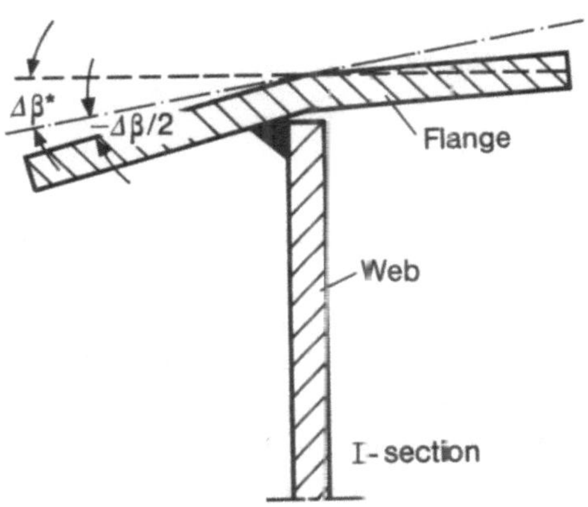

Fig. 197. Angular shrinkage $\Delta\beta$ and tilting shrinkage $\Delta\beta^*$ at I section girder with single-side web-to-flange fillet weld

restrained (with double-sided welds, restraint occurs by the fillet weld on the opposite side of the plate) the back tilting shrinkage angle $\Delta\beta_b^*$ (according to Ref. [8]) is

$$\Delta\beta_b^* = k_b \varepsilon_Y. \tag{181}$$

The parameter ε_Y is the yield limit strain, $\varepsilon_Y = \sigma_Y/E$. The factor k_b depends, according to Fig. 198, on the flange thickness h_{fl}, the web thickness h_{wb} and the throat thickness a ($\Delta\beta_b^* < 1.15°$).

With intermittent welds the (mean) angular shrinkage $\Delta\beta$ (or $\Delta\beta^*$) is reduced in the ratio of the string length l_{st} to the sum of string length and string interspace length l_i to the mean value

$$\Delta\beta_i^{(*)} = \Delta\beta^{(*)} \frac{l_{st}}{l_{st} + l_i}. \tag{182}$$

Allowance should be made for the restrained back tilting shrinkage in the case of

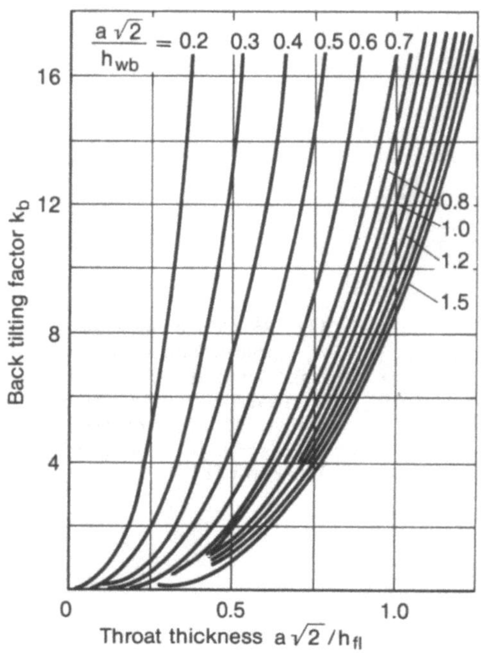

Fig. 198. Back tilting factor k_b dependent on throat thickness a in relation to flange thickness h_{fl} and web thickness h_{wb}; after Vinokurov [8]

double-sided (non-staggered and staggered) intermittent welds (not deposited simultaneously).

Twisting distortion occurs on longitudinally welded girders with closed cross-sectional contour as a result of shear deformation or shear stresses in the weld direction in the welding zone which are caused by unequal longitudinal deformation of the groove edges (e.g. of flange and web of a box girder). The twisting angle after welding (welds pre-tacked) is likely to be all the larger the longer the girder and the smaller the area enclosed by the cross-sectional contour (in accordance with the Bredt formula for twisted thin-walled bars with closed cross-section). In a box girder with four corner welds, the twisting distortion is particularly large if the corner welds are welded alternately in opposite direction (after pre-tacking). Twisting-like distortion also occurs in girders with open cross-sectional contour if the individual parts of the cross-section rotate relative to each other during welding.

3.5.5 Warpage of thin-walled welded components

Thin-walled welded structures ($h \lessapprox 6$ mm) may lose their stability under compressive welding residual stresses. This is referred to as welding warpage. Superimposed angular shrinkage may intensify the process, whereby, in terms of theory of structures, a limit load problem has to be solved instead of a stability problem. Examples are the buckling of plane vessel ends with circumferential weld, the web or flange buckling of thin-walled girders with longitudinal weld, the warpage of plate strips with longitudinal weld and the buckling in the outside area of patches inserted with ring weld into a plate.

Stability problems of this type concerning welded components can be treated on the basis of conventional stability theory by proceeding from the shrinkage forces. However, the degree of difficulty is increased by the following peculiarities of these problems. Thus it is not sufficient to solely determine the critical warpage or buckling stress. The acceptable compressive stress may be considerably above (or below) the critical stress if the post-critical (e.g. post-buckling) behaviour of the component is part of the phenomenon. What is of interest, therefore, is the entire post-critical stress and deformation behaviour, in particular also the magnitude of the warpage or buckling deflections which ultimately occur. This problem may be solved in simpler cases proceeding from shrinkage forces which do not vary in direction. In many cases, though, allowance must be made for the directional change of the shrinkage forces, which complicates the solution of the problem. Four examples of a solution are now discussed from Ref. [8]. Apparently, only Russian theoreticians have so far tackled the residual stress-related stability problem of welded structures.

The first component which is considered is the *plane end* (without the shell) of a vessel subjected to the longitudinal shrinkage force of the circumferential weld (Fig. 199). The buckling process is presented in a counter-loading diagram, in which, on the one hand, the drop of the radial shrinkage pressure σ_r in the shrinkage-tensioned ring weld (circumferential stress σ_{ci}, width w_{pl}), and on the other hand the rise of pressure resistance in the end plate (radius R), is plotted over the radial displacement u_r of the ring weld (positive-directed to the inside). The equilibrium condition for the semicircular ring section subjected to the internal pressure σ_r provides the relation between σ_r and σ_{ci}:

$$\sigma_r = \sigma_{ci} \frac{w_{pl}}{R}. \tag{183}$$

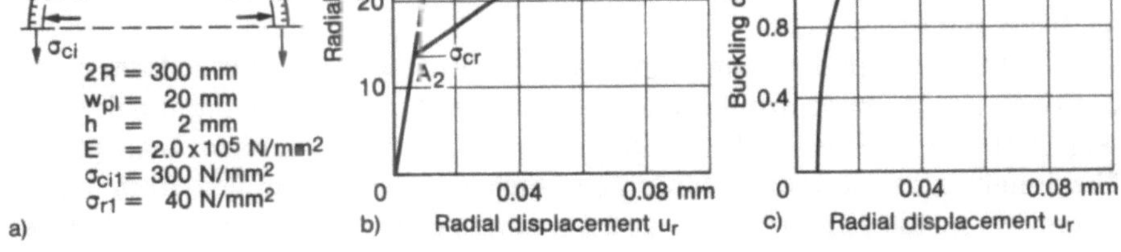

Fig. 199. Buckling of vessel end with circumferential weld (**a**), counter-loading diagram (**b**) and buckling deflection Δw (**c**) dependent on radial shrinkage displacement u_r of inner edge of plastic zone (width w_{pl}), analytical results; after Vinokurov [8]

The reduction of σ_r as a result of radial displacement u_r causes a simultaneous reduction of σ_{ci} ("stress relieving"). With a rigid end plate ($u_r = 0$) the full radial pressure occurs. With an elastic end plate (without instability) point A_1 identifies the state of equilibrium. If, however, the end plate buckles at σ_{cr} (point A_2) and if shape-hardening occurs in the post-buckling range, point A_3 is then reached. The shrinkage displacement u_r is relatively large (with correspondingly large buckling deflection Δw) and the circumferential weld is relatively strongly stress-relieved.

The critical buckling stress of the circular plate with bending free edge support is determined from

$$\sigma_{cr} = 0.385 \frac{Eh^2}{R^2}. \tag{184}$$

The buckling deflection Δw in the post-buckling range results, according to Ref. [8], in:

$$\frac{\Delta w}{h} = 1.96 \sqrt{\frac{\sigma_r}{\sigma_{cr}} - 1}. \tag{185}$$

Buckling deflections in the post-buckling range can also be determined by a geometrical transformation of the radial displacement u_r into a circular arc deflection (subtracting the compressive strains).

The second component considered is a *circular patch* welded into a ring plate (e.g. vessel end), which exerts the corresponding radial tension σ_r on its (outside) surrounding area as a consequence of shrinkage with tangential tension σ_t (flange, nozzle or heat spot present a similar problem). For the infinite plate with a circular opening (radius R) subjected to tension σ_r, this results (see Ref. [8]), independent of the support condition at the edge of the opening, in:

$$\sigma_{cr} = \frac{Eh^2}{4(1-\nu^2)R^2}. \tag{186}$$

Two buckling waves (m = 2) form at the edge of the opening. If the infinite plate with circular opening is substituted by the corresponding finite plate, i.e. a ring plate, σ_{cr} then rises as the radius ratio R_o/R_i becomes smaller (Fig. 200). This rise is accompanied by the formation of three, four and more buckling waves. The influence of the support condition at the edge of the opening on the result is relatively minor.

The third component considered is a *thin-walled bar* with cruciform cross-sectional shape made from plate strips joined through longitudinal welds (Fig. 201). The I girder with longitudinal welds may be analysed analogously. The longitudinal shrinkage force F_s (which can be determined from $w_{pl} =$ 50 mm, $\sigma_Y = 400$ N/mm^2, $h = 2$ and 6 mm respectively), divided by the cross-section area A, results in the longitudinal compressive stress σ_c in the plate strips. The instability study can be conducted for the plate strip transversely rigidly restrained at one longitudinal edge (but longitudinally unrestrained) and

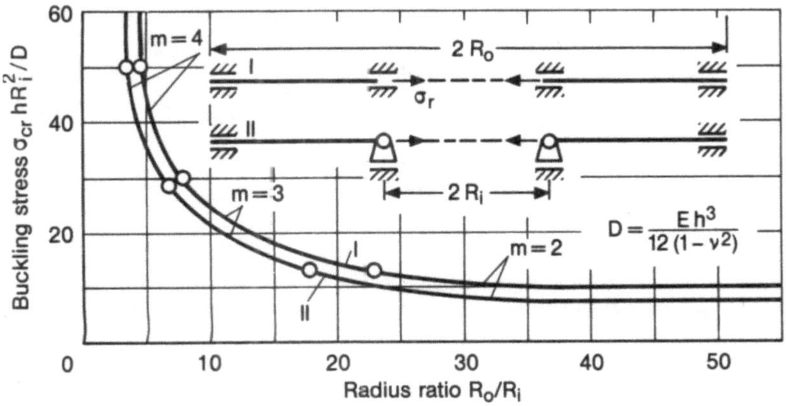

Fig. 200. Buckling stress σ_{cr} of ring plate subjected to radial tension at inner edge, two support conditions, number of buckling waves m, analytical results; after Vinokurov [8]

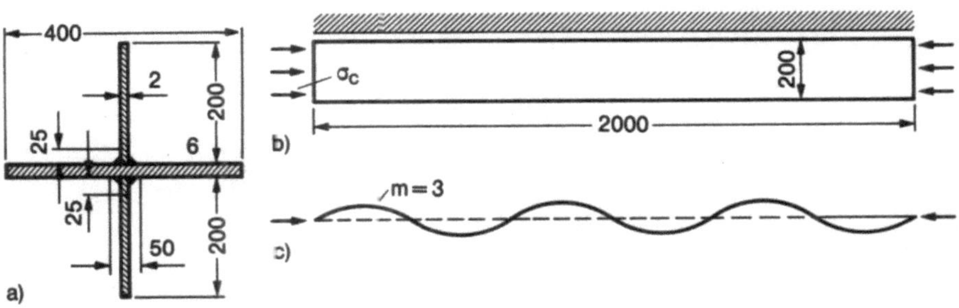

Fig. 201. Buckling of thin-walled bar with cruciform cross-sectional shape made from plate strips joined by longitudinal fillet welds; dimensions (**a**), buckling model (**b**), buckling waves (**c**), analytical results; after Vinokurov [8]

subjected to longitudinal pressure. The result for this model is

$$\sigma_{cr} = k \frac{\pi^2 E h^2}{12(1-\nu^2)w^2} \tag{187}$$

with $k = 1.328$ and $m = 3$ for $l/w = 10$ (dimensions according to Fig. 201). The buckling deflection Δw in the post-buckling range is calculated in accordance with the approximation method stated in Ref. [8] and mentioned in respect of equation (185).

The fourth component considered is a plane *plate strip* with centric longitudinal weld, subjected to warpage as a consequence of the longitudinal shrinkage force as shown in Fig. 202. This instability problem is solved in Ref. [8] (ignoring the angular shrinkage) by minimizing the potential energy. The solution shows that the plate strip deflects in a manner which reduces the shrinkage stress

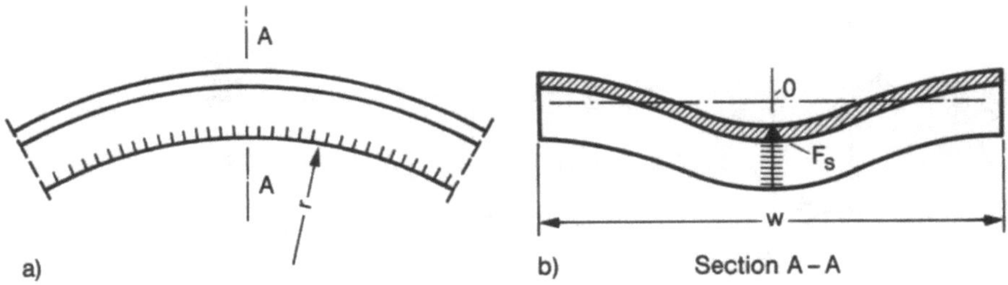

Fig. 202. Warping of plate strip with centric longitudinal weld; longitudinal curvature (**a**) and transverse curvature (**b**), centre of gravity 0 of cross section and shrinkage force F_s; after Vinokurov [8]

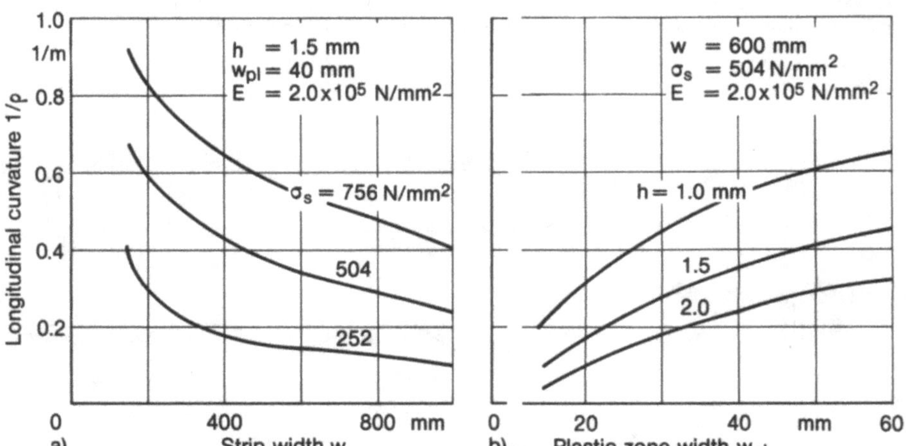

Fig. 203. Longitudinal curvature $1/\varrho$ of warping plate strip with centric longitudinal weld (without angular shrinkage transverse to weld) dependent on strip width w and shrinkage stress σ_s (**a**) as well as dependent on width of plastic zone, w_{pl}, and plate thickness h (**b**), analytical results; after Vinokurov [8]

in the weld area. The bending moment required for the longitudinal curvature is created as a result of the fact that the shrinkage force is offset relative to the centre of gravity of the deformed cross-section. Instability already occurs at an arbitrarily small but finite shrinkage force connected with correspondingly small warpage. The longitudinal curvature $1/\varrho$ is presented in Fig. 203 dependent on plate thickness h, plate width w, shrinkage stress σ_s and width of the plastic zone w_{pl}. The plate length l is of no influence. The (longitudinal) shrinkage after occurrence of instability can be presented in a similar manner (see Ref. [8]). The analysis results have been confirmed by measurements.

3.6 Measuring methods for residual stress and distortion

3.6.1 Significance of test and measurement

The presentation hitherto has been oriented to theoretical models with reference to numerical solutions and experimental findings. Whereas the numerical methods were dealt with in more detail in this connection, this is not the case for the measurement methods. For this reason, methods of residual stress and distortion measurement are presented separately below. They form a part-area of the general methods of stress and deformation measurement [305, 322]. Finally, elementary similarity relations are stated for test welding in the reduced or enlarged geometrical scale.

From what has been stated previously, it is clear that theoretical models and numerical solutions include strong simplifications to limit the calculation effort. The processes are approximated only in their principal characteristics. It is, therefore, essential to examine the possible simplifications by testing and to check to what extent the mathematical solution reflects reality. In many cases, test and measurement are also preferred without an accompanying calculation for reasons of the shorter time required, although the information gained in this way can be transferred and generalized to a lesser extent only.

3.6.2 Strain and displacement measurement during welding

The need for measurement exists primarily in the high temperature zone of the welded joint as this is where the maximum strains occur during welding. Consequently, strain measurement methods are required which are executable at high temperature. The feet of mechanically attached strain gauges must be cooled. In the case of slow processes it is possible to repeatedly place the uncooled detachable strain gauge for short time intervals on the markings designating the measuring base. Also, high temperature resistance strain gauges are commonly used. The arrangement of the measuring feet transverse and longitudinal to the weld seam is shown in Figs. 204a and 204b; the marking of the measuring base by means of balls in Fig. 204c. The temperature at the point

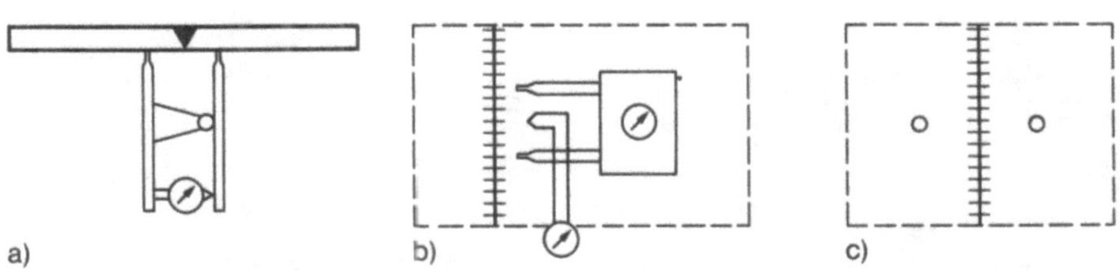

Fig. 204. Measurement of transverse and longitudinal strain at weld seam using measuring feet (**a**, **b**) or measuring balls (**c**) and combined with thermocouple (**b**); after Vinokurov [8]

of strain determination is also measured (thermocouple in Fig. 204b) in order to distinguish the strain components and to draw conclusions regarding the stresses. The thermal and transformation strains are then separated proceeding from the dilatometer curve. The elastic stress follows from the measured total strain ε_{tot} after subtracting the thermal and transformation strain ε_T in accordance with Hooke's Law with temperature-dependent elastic modulus, while observing the fact, however, that the temperature-dependent yield limit σ_Y cannot be exceeded. This process is illustrated in Fig. 205. The curve $\varepsilon_{tot} - \varepsilon_T$ follows from the cooling curve ε_{tot} and from the dilatometer curve ε_T. During the initial rise of $\varepsilon_{tot} - \varepsilon_T$ the (tensile) yield limit strain $\varepsilon_Y = \sigma_Y/E$ is not exceeded by

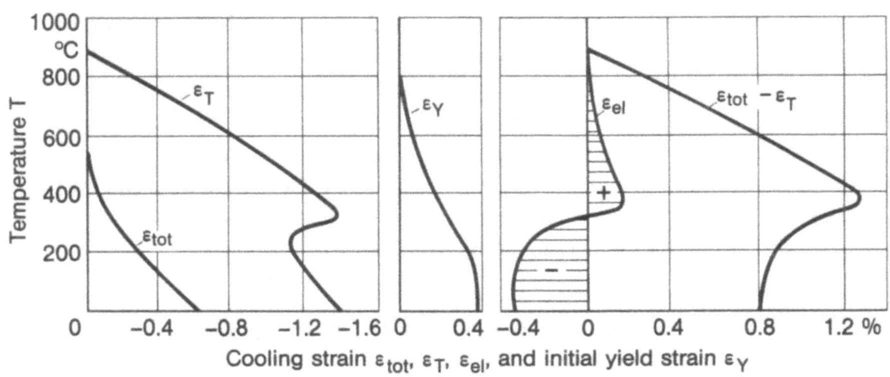

Fig. 205. Cooling strain profile at weld seam ε_{tot} and in dilatometer ε_T, initial yield strain ε_Y and elastic strain profile at weld seam ε_{el} over temperature; after Vinokurov [8]

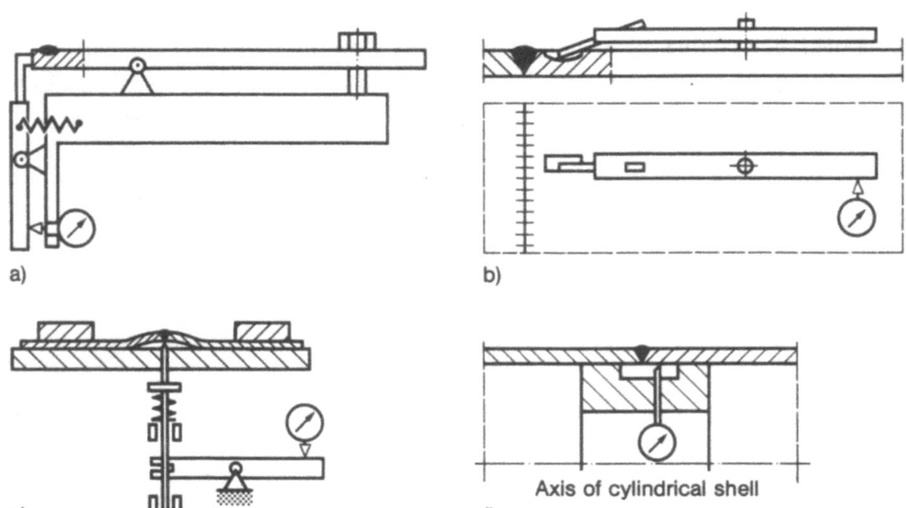

Fig. 206. Measurement of transverse and longitudinal displacements (**a**, **b**) as well as of deflections (**c**, **d**) at weld; after Vinokurov [8]

ε_{el}. During the subsequent drop of $\varepsilon_{tot} - \varepsilon_T$, caused by the onset of transformation, elastic unloading occurs in which, in the considered case, the compressive yield limit strain is not reached. The stress profile $\sigma = E\varepsilon_{el}$ over temperature follows from the corresponding elastic strain profile ε_{el}. This procedure for uniaxial stresses, originating from Okerblom [2, 3], may be extended to biaxial surface stresses using modern measurement and analysis techniques.

The displacement measurement in individual points of the high temperature zone relative to a reference point in the cold area of the component or even outside of the component is solved, adapted to the individual case. The transverse and longitudinal displacement measurement at the weld seam as well as the vertical displacement measurement perpendicular to plane plates and cylindrical shells (rotary inside ring with measuring instrument) are presented in Fig. 206. The displacement measuring instruments act mechanically, optically, inductively, capacitively or on the basis of electrical resistance.

3.6.3 Destructive residual stress measurement

A distinction is made between destructive and non-destructive residual stress measurement methods [302–304]. The former are also termed "mechanical", the latter "physical", which linguistically is unsatisfactory to the extent that mechanics is a sub-sector of physics. In the case of destructive methods, a distinction requires to be made between complete and partial destruction of the component. The latter include small holes or ring grooves in the surface of the component, which permit (restricted) further use of the part.

The principle of destructive (or partially destructive) residual stress measurement is now explained, proceeding from the better known load stress measurement [305]. Load stress measurement is relatively simple provided the material can be assumed as elastic and only the surface of the component is to be considered, which is generally adequate in practice. The elongation or shortening of a small measuring base on the surface of the component is determined while the component is subjected to loading or unloading. The measurement requires to be performed in at least three directions in order to completely determine the biaxial stress state. The strains result from the measured displacements, by relating them to the length of the measuring base; the stresses from the strains by means of Hooke's law. Resistance strain gauges, detachable strain gauges and photoelastic surface layers are used primarily for such measurements in practice. The measurement and analysis technique is simplified if the principal stress direction is known, this being determined, for example, by means of brittle lacquer.

Residual stresses, just like load stresses, can be determined by relieving of forces, not, however, by relieving the component of external forces (by definition residual stresses are internal forces without external forces), but by relieving the local material volume which comprises the measuring base of internal forces. What is measured is the relieved strain or the resilience of the completely or partially cut-out part or of the residual body. In the case of the cut-out part, an adequately large volume, as homogeneously stressed as possible, should be available. The residual stress results from the relieved strain in accordance with Hooke's law. In the case of the residual body, material is removed in layers from beams, plates or cylindrical bodies, also from the base of blind holes; and a back-calculation made from particular locally relieved strains or from global resiliences of the residual body to obtain the residual stress relieved during removal as well as the residual stress originally existing in the removed part. This possibility of back-calculation exists only for bodies of a relatively simple geometry, for the elastic behaviour of which theoretical solutions are available (e.g. beam, plate, cylindrical or spherical shell).

Temperature constancy is an absolutely essential requirement for precise measurements with all residual stress measurement methods. Particular measures for assuring temperature constancy are required during cutting, drilling or layer removal. When measuring the resilience of the residual body, the yield limit must not be exceeded if high accuracy demands are made. This limits the resilience methods to residual stresses markedly below the yield limit.

3.6.3.1 Measurement of uniaxial welding residual stresses

In many cases it is acceptable to determine welding residual stresses assuming a mainly uniaxial effect. In general, this uniaxial residual stress is distributed inhomogeneously over the cross-section of the component, e.g. the weld longitudinal residual stress in I girders with web-to-flange weld or the transformation stress at flame cut plate edges.

In the *strip cutting method* shown in Fig. 207, which became more widely known by the investigations of Thürlimann [306], the component is cut in direction x of the residual stress to be measured into a fairly large number of narrow strips, from the relieved strain of which follows the stress:

$$\sigma_x = -E\varepsilon_x . \tag{188}$$

Cutting is performed by a saw. The relieved strain is measured by means of detachable strain gauges or bonded resistance strain gauges. The robust detachable strain gauges act by means of measuring balls which are inserted into the surface of the component at the end points of the measuring base. The measur-

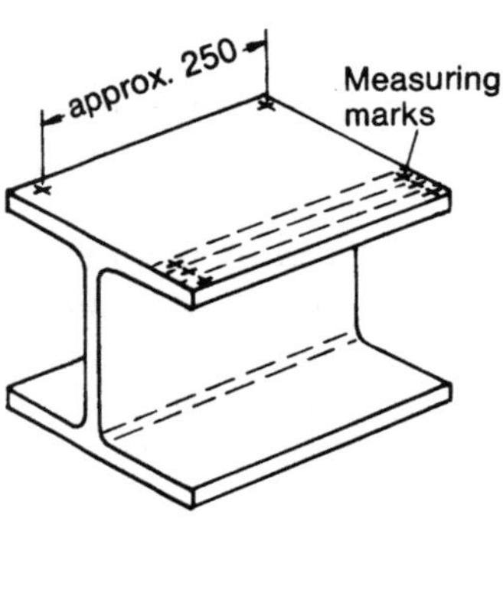

Fig. 207. Strip-cutting method applied on I section girder; after Thürlimann [306]

ing base length of 100 to 250 mm is comparatively large, which makes it appear desirable to have non-varying residual stresses within this length. The more sensitive resistance strain gauges (but with connecting wires which may impede the cutting process) permit a shorter measuring base length. One point to note is that strain relief is already partially activated when cutting the component transversely; all the more so, the narrower the component section which is to be further cut into strips. Occasionally, the strips are also left connected with one end in the cross-section (cut open in a comb-like fashion).

In the *resilience method* according to Stäblein [307] (Fig. 208), bars with an inhomogeneous distribution of the uniaxial longitudinal residual stresses are subjected to layer removal in the plane clamping condition (planing, grinding, etching, spark erosion, or transverse saw cuts). The bars are unclamped between the layer removal operations and the bar curvature which results from the resilience is determined by means of the back-spring deflection, curvature or strain measured on the side opposite to the removal side. The initial residual stresses σ_x in the bar-longitudinal direction over the bar height direction z are calculated according to the following equation, stated originally by Treuting and Read [308], on the basis of the beam bending theory:

$$\sigma_x = \frac{E}{3l^2}\left[h^2\frac{\mathrm{d}w}{\mathrm{d}h} + 4hw - 2\int_h^{h_0} w(z)\mathrm{d}z\right]. \tag{189}$$

In the above equation h_0 is the initial height of the bar and h its variable height following the removal of the layers, l the bar length over which the deflection w occurs, and E the elastic modulus. Accordingly, the initial residual stress in the distance h from the measurement side results from differentiation and integration of the deflections w measured following the removal of the layers. Alternatively, it is possible to proceed instead from the deflections w from the relieved strains ε on the measurement side. The relation between w and ε is expressed by elementary formulae of the beam bending theory.

The edge stress σ_{x0}, which is of particular significance in practice, follows from equation (189) with $h = h_0$ and $w = w_0 = 0$:

$$\sigma_{x0} = \frac{E}{3}\left(\frac{h_0}{l}\right)^2\left(\frac{\mathrm{d}w}{\mathrm{d}h}\right)_{h=h_0} \tag{190}$$

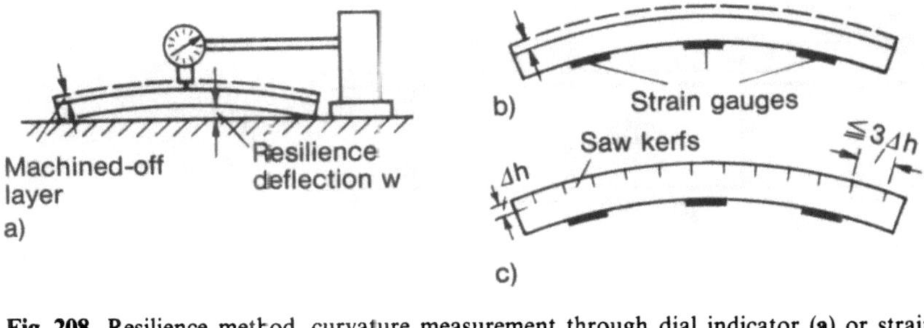

Fig. 208. Resilience method, curvature measurement through dial indicator (**a**) or strain gauges (**b, c**), stress relief achieved by machining (**a, b**) or saw cuts (**c**); after Stäblein [307]

The differentiation and integration in equation (189) were originally performed graphically [307, 308], and later numerically [284]. A significant increase in accuracy is achieved by means of balancing and interpolation polynoms (numerical smoothing according to Ref. [284]).

The above resilience method has been used with particular success for determining residual stresses in flame cut plate edges [284, 285].

A special variant of the resilience method is the longitudinal cutting-open of ring-shaped slices taken from thin-walled cylinder shells followed by measuring the difference in outside diameter and calculating the local circumferential membrane residual stresses on that basis [332].

3.6.3.2 Measurement of biaxial welding residual stresses

Determining biaxial welding residual stresses is a slightly more involved process. In the *block cutting method*, the orthogonal directions x and y of the normal stresses σ_x and σ_y to be measured are given (these may, but need not be the principal stresses) and the stresses are assumed as non-varying over the plate thickness which is an acceptable assumption in the case of thin sheet material. The measuring bases or strain gauges are usually applied to both sides of the plate. Following this, the plate is cut into rectangular blocks (approx. $30 \times 30\ \text{mm}^2$) [8, 309] (Fig. 209). The residual stresses σ_x and σ_y result from the relieved strains ε_x and ε_y:

$$\sigma_x = -\frac{E}{1-\nu^2}(\varepsilon_x + \nu\varepsilon_y), \tag{191}$$

$$\sigma_y = -\frac{E}{1-\nu^2}(\varepsilon_y + \nu\varepsilon_x). \tag{192}$$

At least three measuring directions, combined in a three-element strain gauge rosette, are required for determining the complete plane stress state [302]. The method has been used for verifying the effectiveness of stress relieving measures on large tanks.

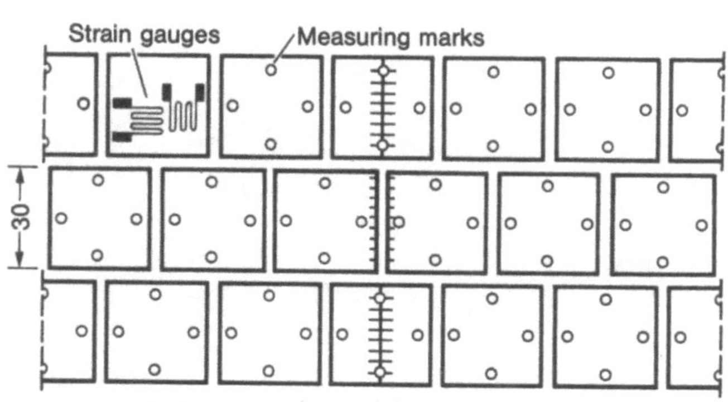

Fig. 209. Block-cutting method applied on plate with butt weld; after Vinokurov [8]

An extension of the block cutting method for thick plates and varying stress level over thickness consists in an additional cutting-step releasing the surface layer of the block with the strain gauge attached [289].

The method proposed by Peiter [328, 329] to separate edge strips from a square plate in order to determine the residual stresses in the plate evaluating resilience effects has no sound theoretical basis (see the commentary in Ref. [328]).

In accordance with another already standardized method [312], i.e. the *hole drilling method* according to Ref. [310–313], the relief of radial strain when drilling a small through-hole into a plate is evaluated (Fig. 210) (the extension in the form of the blind hole method is dealt with later). The measuring bases extend across the drilled hole or are located outside of the hole.

In the original version, the relative displacements of the drilled hole edges (drilled hole diameter d_0) are measured applying detachable strain gauges with measuring bases which bridge the drilled hole (end points on circle with diameter d). A back-calculation is made from the relative displacements Δx and Δy to the residual stresses σ_x and σ_y [321]:

$$\frac{E}{2(1+\nu)}\frac{2\Delta x}{d} = \left\{(1-\nu)\left(\frac{d_0}{d}\right)^2 + \frac{1}{4}\left[1-\left(\frac{d_0}{d}\right)^2\right]\left(\frac{d_0}{d}\right)^2\right\}\sigma_x - \\ - \left\{\frac{1-2\nu}{2}\left(\frac{d_0}{d}\right)^2 + \frac{1}{4}\left[1-\left(\frac{d_0}{d}\right)^2\right]\right\}\sigma_y, \tag{193}$$

$$\frac{E}{2(1+\nu)}\frac{2\Delta y}{d} = \left\{(1-\nu)\left(\frac{d_0}{d}\right)^2 + \frac{1}{4}\left[1-\left(\frac{d_0}{d}\right)^2\right]\left(\frac{d_0}{d}\right)^2\right\}\sigma_y - \\ - \left\{\frac{1-2\nu}{2}\left(\frac{d_0}{d}\right)^2 + \frac{1}{4}\left[1-\left(\frac{d_0}{d}\right)^2\right]\right\}\sigma_x. \tag{194}$$

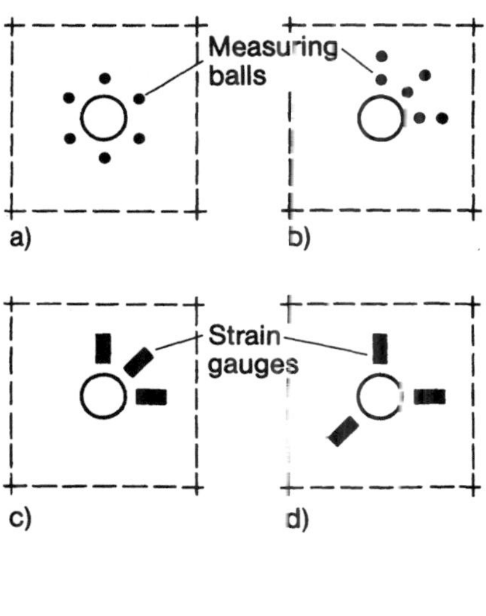

Fig. 210. Hole-drilling method with measuring balls (**a**, **b**) or strain gauges (**c**, **d**)

In the case of $d_0 = 12$ mm, $d = 16$ mm and $\nu = 0.3$, the stresses σ_x and σ_y follow from:

$$\sigma_x = E\left(0.989\frac{2\Delta x}{d} + 0.378\frac{2\Delta y}{d}\right), \tag{195}$$

$$\sigma_y = E\left(0.989\frac{2\Delta y}{d} + 0.378\frac{2\Delta x}{d}\right). \tag{196}$$

At least three measuring directions are needed for completely determining an unknown biaxial residual stress state.

In the current version, the measuring bases are located radially outside the drilled hole, and the detachable strain gauge is replaced by bonded resistance strain gauges. The relation between the relieved radial strain ε_r under the angle α relative to the x-axis at measuring points on a circle with diameter d outside the drilled hole with diameter d_0 and the residual stresses relieved within the drilled hole follows from the solution according to the theory of elasticity for the infinite plate with a hole subjected to biaxial membrane stress σ_x and σ_y (understood as principal stresses):

$$\varepsilon_r = (A + B\cos\alpha)\sigma_x + (A - B\cos\alpha)\sigma_y, \tag{197}$$

$$A = -\frac{1+\nu}{2E}\left(\frac{d_0}{d}\right)^2, \tag{198}$$

$$B = -\frac{1+\nu}{2E}\left[\frac{4}{1+\nu}\left(\frac{d_0}{d}\right)^2 - 3\left(\frac{d_0}{d}\right)^4\right]. \tag{199}$$

Identical equations are incorrectly stated in Ref. [304] for the measuring base bridging the hole.

To completely determine an unknown biaxial residual stress state (two principal stresses σ_1 and σ_2 and the principal stress direction β) it is necessary to evaluate the relieved radial strain ε_r in at least three measuring bases with different measuring directions (e.g. by applying a three-element strain gauge rosette).

It holds for the outside measuring base arrangement [312] with $\alpha = 0°$, $\alpha = 45°$ and $\alpha = 90°$ (appertaining ε_{00}, ε_{45} and ε_{90}) according to Fig. 211 (the diagonal measuring element at bottom left can also be arranged at the top right) that:

$$\sigma_{1/2} = \frac{\varepsilon_{90} + \varepsilon_{00}}{4A^*} \pm \frac{\sqrt{2}}{4B^*}\sqrt{(\varepsilon_{90} - \varepsilon_{45})^2 + (\varepsilon_{45} - \varepsilon_{00})^2}, \tag{200}$$

$$\tan 2\beta = \frac{\varepsilon_{00} - 2\varepsilon_{45} + \varepsilon_{90}}{\varepsilon_{00} - \varepsilon_{90}}. \tag{201}$$

An alternative form of equation (200) is:

$$\sigma_{1/2} = \frac{\varepsilon_{90} + \varepsilon_{00}}{4A^*} \pm \frac{1}{4B^*}\sqrt{(\varepsilon_{90} - \varepsilon_{00})^2 + (2\varepsilon_{45} - \varepsilon_{90} - \varepsilon_{00})^2}. \tag{202}$$

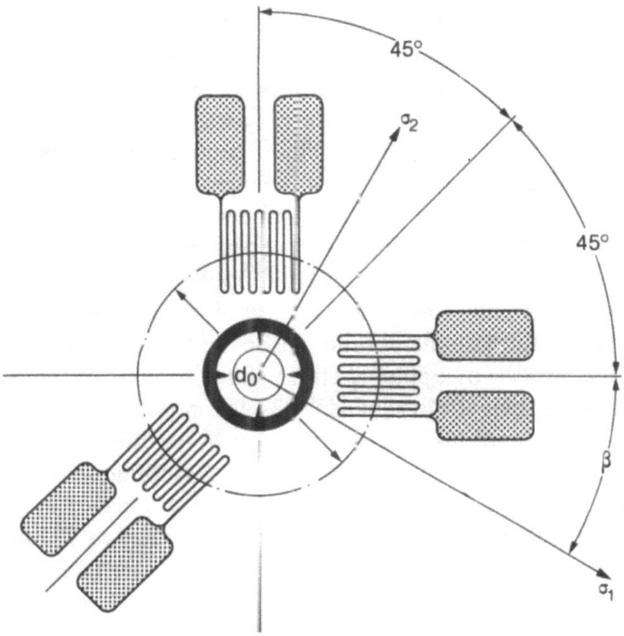

Fig. 211. Three-element strain gauge rosette for use in hole-drilling method, angular positioning and diameters; according to ASTM-Standard [312]

The constants A^* and B^* are identical to A and B according to equations (198) and (199) in the (fictitious) case of point-like acting measuring elements. In fact, length and width of the measuring elements must be allowed for in A^* and B^*; corresponding formulae are stated for example in Ref. [312, 313]. The constants A^* and B^* can also be gained from an experimental calibration of the measuring setup in the uniaxial tensile test.

The measuring sensitivity of the hole drilling method is all the greater the closer the measuring elements can be moved to the edge of the hole. Measuring balls can be positioned at a distance of 1 mm from the edge. Strain gauges can be applied with $2.5 < d/d_0 < 3.4$.

The size of the drilled hole depends on the size of the measuring elements (diameters d_0 of 1.5 to 3.0 mm and measuring base lengths of 1.5 mm are used). The strain gauges can be positioned most accurately around the drilled hole if available as three element rosettes.

The accuracy of the hole drilling method is impaired if the hole is drilled close to component edges or other drilled holes, for the method is calibrated to the conditions in the infinite or at least largely extended plate. Plastic deformations can likewise falsify the results of measurement. These occur in the case of residual stresses close to the yield limit or if the drilling technique is improperly applied.

A variant of the hole drilling method (see Ref. [8]) determines the axisymmetrical inhomogeneous residual stress state (e.g. in a spot-welded joint) assumed as constant over the plate thickness, by reboring the hole to gradually increasing diameter within the measuring base (arrangement according to Fig. 210a). In a similar way as for the resilience method, a complicated back-calculation is required from the changes in length of the measuring base to the residual stresses in the initial state. This procedure is identical to the hole drilling method for cylindrical bodies if, in place of the measuring base bridging the hole, a tangentially arranged bonded resistance strain gauge is applied further to the outside, i.e. attached to the outer edge surface in the case of circular plates.

The hole drilling method with plate membrane stresses can be extended to superimposed plate bending stresses by applying strain gauge rosettes on both sides of the plate and evaluating their strain signals [302]. It can also be applied to larger plate thicknesses and body-like components with residual stresses which may be constant or variable over the depth coordinate. Assuming that the residual stress state below the surface of the component is not changing, or only slightly, over the depth coordinate z (in respect of level, multiaxiality and direction), hole drilling can be performed with minor modification of the constants A^* and B^* in the analysis equations, also with blind holes (*blind hole method*). The blind hole depth should be slightly larger than the blind hole diameter, $h_0 = 1.2d_0$. This results in residual stresses averaged over the blind hole depth.

However, the variation also in the residual stress level over the depth coordinate z can be determined in cases of a marked stress gradient normal to the surface [314, 315, 334, 340]. For this purpose, the profile of relieved strain $\varepsilon = \mathrm{f}(h_0)$, at the component with depth-variable stress, is compared with that at the tensile model or tensile specimen with depth-constant stress, calibration curve $\varepsilon_0 = \mathrm{f}_0(h_0)$ in accordance with Fig. 212. The ratio of the gradients $(\mathrm{d}\varepsilon/\mathrm{d}h_0)/(\mathrm{d}\varepsilon_0/\mathrm{d}h_0)$ with identical h_0/d_0 and d/d_0 results directly in the stress just relieved at the bottom of the hole in the considered component relative to the stress in the calibrating specimen. It is only approximately correct to equate this stress at the bottom of the blind hole to the original residual stress. Despite these positive statements, the method is not well suited for stress fields with severe gradients normal to the surface or in the plane of the surface [330, 334].

An alternative to hole drilling procedure is the *cutting-out method*, according to Gunnert [4] and Kunz [309], in which the relieved strain is determined from cylindrical blocks cut out of thin plates using detachable strain gauges or bonded resistance strain gauges. The *ring groove method* is used as an alternative to the blind hole method for depth-independent and depth-dependent residual stresses in body-like components (Fig. 213). The ring groove of adequate depth (recommended values shown in Fig. 213) relieves the stress state in the surface within the ring groove completely. Further enlarged ring groove depth no longer alters the measuring signal. What applies to the measuring base arrangement within the ring groove according to Fig. 213a (indexing of ε according to the angle of inclination of the measuring base versus the horizontal direction)

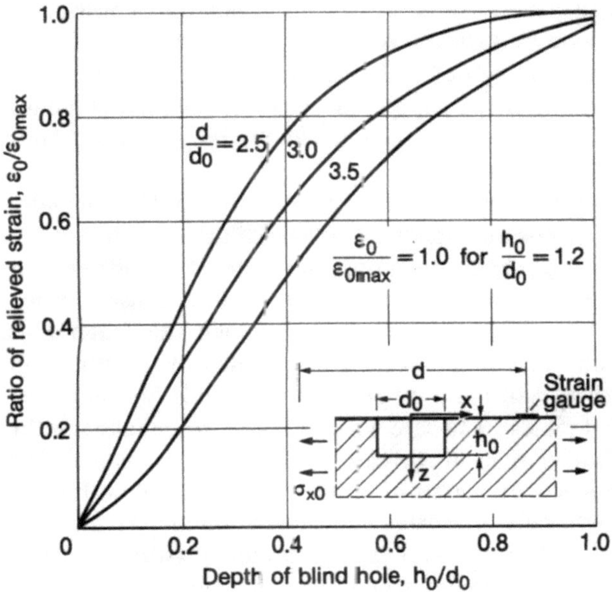

Fig. 212. Calibration curves for stress relief by blind hole in homogeneous tensile stress field: after Häusler, König and Kockelmann [315]

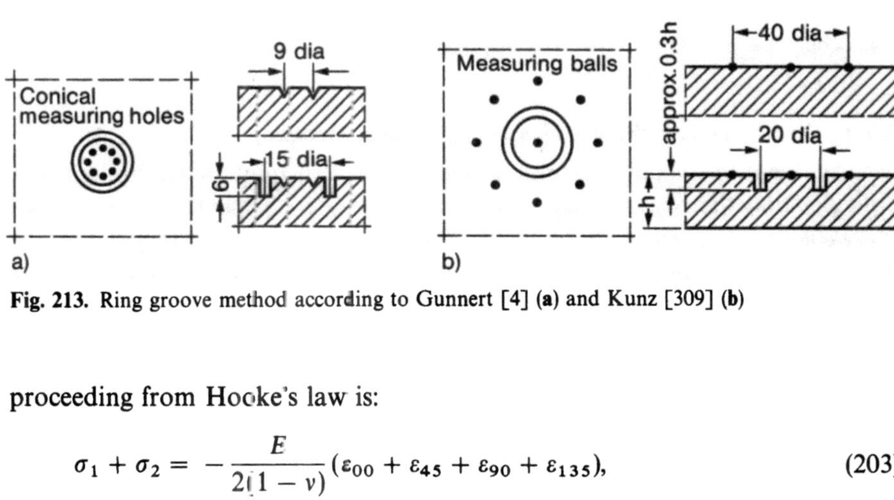

Fig. 213. Ring groove method according to Gunnert [4] (**a**) and Kunz [309] (**b**)

proceeding from Hooke's law is:

$$\sigma_1 + \sigma_2 = -\frac{E}{2(1-\nu)}(\varepsilon_{00} + \varepsilon_{45} + \varepsilon_{90} + \varepsilon_{135}), \tag{203}$$

$$\sigma_1 - \sigma_2 = \frac{E}{1+\nu}\sqrt{(\varepsilon_{00} - \varepsilon_{90})^2 + (\varepsilon_{45} - \varepsilon_{135})^2}, \tag{204}$$

$$\tan 2\beta = \frac{\varepsilon_{135} - \varepsilon_{45}}{\varepsilon_{00} - \varepsilon_{90}}. \tag{205}$$

The ring groove in combination with the measuring base arrangement bridging the groove according to Fig. 213b or outside the groove acts just as a blind hole. Consequently, relations similar to equations (200) to (202) apply, modified according to the changed number and position of the measuring bases, if outside strain gauges are used. Using identical equations for the evaluation of measuring bases bridging the ring groove as stated in Ref. [304] is incorrect.

Hole drilling, blind hole, cutting-out and ring groove methods are widespread in practice, hole drilling being supported by an application standard [302]. But scientific investigations also are conducted applying these methods [194, 286].

The *resilience method* extended from beam to plate problems is particularly suitable for biaxial, homogeneous and depth-variable welding residual stresses. Knowledge of the spring-back deflection alone, however, is insufficient for solving the biaxial problem. Strain gauge rosettes are therefore used, which indicate the complete biaxial state of relieved strain when the material is removed in layers. Equations for evaluation have been stated by Treuting and Read [308] for homogeneous plates and by Schimmöller [272] for laminated plates (in respect of cladded plates). The care which requires to be taken in the case of the plate is even higher than in the case of the beam. Layer removal without mechanical force effects by means of spark erosion and smoothing of the measurement results by means of a balancing calculation are essential. As the clamping forces increase as layer removal progresses and thus also the risk of the yield limit being exceeded, it is recommended to remove layers only up to approximately the centre plane of the plate, i.e. to perform the removal and measuring operations on two identical plates proceeding from the one and from the other surface. The method is applied in Ref. [281] on cladded plates.

3.6.3.3 Measurement of triaxial welding residual stresses

Measurement of triaxial (welding) residual stresses encounters the fundamental difficulty that the relevant material volume to be stress relieved is positioned relatively inaccessible in the interior of the component. Nevertheless, certain methods can be applied here also if the principal directions of the residual stress state are known beforehand, e.g. if they coincide with the main directions of the geometry of the body. These methods are presented below.

The simplest step for determining the triaxial residual stress state consists in additionally measuring the change in thickness when applying the block cutting method. It is then possible (see Ref. [8]) to conclude from this what the averaged stress perpendicular to the plate plane is.

Another method from Ref. [8] which avoids averaging over the total plate thickness consists in drilling narrow deep blind holes to the point of the residual stress to be measured, positioning a miniaturized longitudinal strain measuring element in them, then cutting out the relevant material volume together with holes and measuring element and, in so doing, stress and strain relieving it (*three-dimensional cutting method*) (Fig. 214).

The axial directions of the three holes coincide with the three principal directions of the residual stress state. If the two holes in the plate plane become impractically long, inclined holes are provided instead (Fig. 214a), from the inclined extension strains of which, ε_{yz} and ε_{xz}, the horizontal strains, ε_x and ε_y,

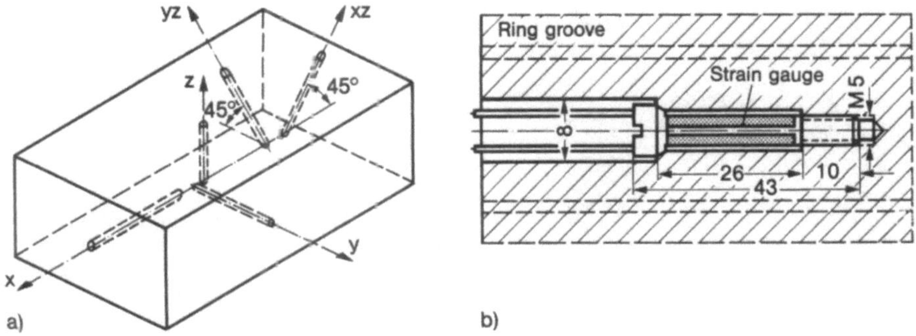

Fig. 214. Measurement of the triaxial residual stress state in the interior of the component using blind holes (**a**) fitted with measuring elements (**b**), strain relief by ring groove; after Vinokurov [8]

are determined with the vertical strain ε_z being known:

$$\varepsilon_{xz} = \frac{1}{2}(\varepsilon_x + \varepsilon_z), \tag{206}$$

$$\varepsilon_{yz} = \frac{1}{2}(\varepsilon_y + \varepsilon_z). \tag{207}$$

The ends of the holes with the measuring elements should be as close as possible to each other; on the one hand, in order to be able to assign the three relieved strains to a single stress state also in an inhomogeneous stress field. On the other hand, however, they should not be too close because they will otherwise cause mutual strain-relieving. The distance between the ends of the holes should be approximately ten times the diameter of the hole in order to limit this interaction effect. A slightly pretensioned bolt with measuring shank and measuring gauge is a suitable measuring element (Fig. 214b). Drilling the hole and tightening the bolt cause measuring errors which should be minimized by calibration or by comparative evaluations.

An attempt has also been made to use only a single hole instead of three and to apply strain gauge rosettes to the bottom and wall of the hole.

According to the *slice cutting method* proposed by Ueda et al. [316, 317], the residual stress sources are determined as initial strains by dissecting the component in order to elastically back-calculate from them the residual stresses by means of the finite element method. For the special case of the welding residual stress state in the centre part of a centric, double-symmetrical butt weld between plate strips (residual stresses constant in longitudinal direction of the weld, residual stresses symmetrical to the weld and to the plate centre plane, weld-longitudinal residual stresses subject to the plane strain condition), the initial strains are determined decoupled from each other in a slice cut-out perpendicular to the weld and in slices cut-out parallel to the plate surface. The thin slice separated out perpendicular to the weld is further dissected while measuring the relieved strains. The stresses corresponding to the plane strain condition result

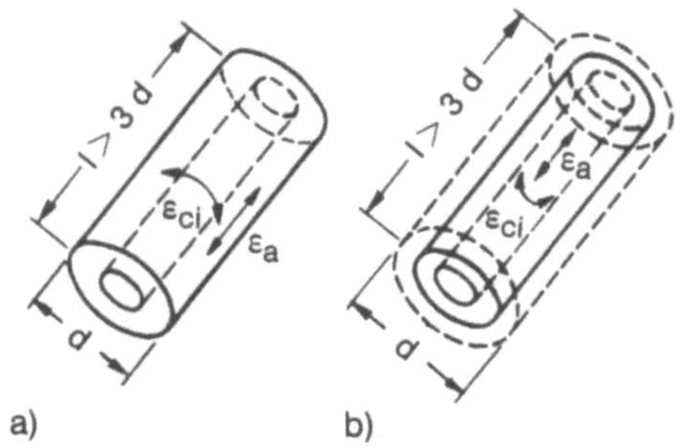

Fig. 215. Boring-out method (**a**) and turning-off method (**b**) with measurement of the circumferential and axial strains on the surface opposite to the removal side

from the plane stress state of the slice dividing these stresses by $(1 - \nu^2)$. In addition, a weld-longitudinal stress component is produced as a result of the strain suppression in this direction. The thin plates separated out longitudinal to the weld and parallel to the plate surface are dissected longitudinally in order to determine the basic component of the weld-longitudinal residual stresses from the relieved strains. The method, which has been verified in this form also by means of measurements, proceeds in the general formulation of Ref. [316] from the incorrect assumption that the "effective" initial strains determining the residual stress state (these being the initial strains effective after elastic releasing) remain unchanged when the transverse or longitudinal slices are cut.

The residual stresses in long solid and hollow cylindrical parts (rods, shafts, tubes) can be determined in accordance with the *boring-out or turning-off method* according to Heyn and Baur [318], Mesnager [319] and Sachs [320], assuming axisymmetry and axial independence of the stress state (Fig. 215). The material is removed in layers on the inside and outside, respectively, of the cylinder. On the measuring side opposite the removal side, the relieved strains are measured in the circumferential and axial direction with strain gauges. The initial residual stresses are calculated back from the measured values in accordance with the basic formulae of elastic cylinder behaviour. This relatively accurate and satisfactory measuring method for triaxial residual stresses has been used in the case of welded joints for the butt joint between cylindrical parts and for biaxial residual stresses in circular plates (Fig. 9). The method variant for plates is mentioned in Section 3.6.3.2 (enlarging a drilled hole to gradually increasing diameter).

3.6.4 Non-destructive residual stress measurement

Residual stresses or strains are measured non-destructively by means of the *X-ray method.* X-rays are diffracted crystal lattices and produce interference phenomena, from which it is possible to draw conclusions relating to the interplanar spacing of the lattice. The load stress or residual stress is determined from the change in the interplanar spacing compared to the stress-free state.

The reflection method is used in practice [302, 324–327] (Fig. 216). An X-ray striking the surface of the component produces, after reflection, interferences

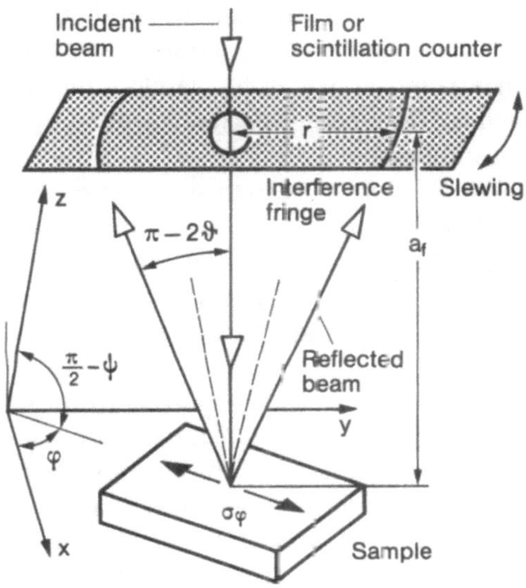

Fig. 216. Residual stress measurement by X-rays, reflection method, Bragg angle ϑ, azimuth angle φ, angle of inclination ψ; after Peiter [304]

which appear on a slewed film (according to Debye-Scherrer) as an interference ring. More modern equipment uses a scintillation counter instead of the film. The interference line, of the first order, satisfies Bragg's law (with n = 1), according to which the glancing angle depends on the interplanar spacing d_A of the atoms in the crystal lattice and on the wavelength λ of the X-radiation (practicable 0.002–2.0 Å, 1 Å = 10^{-7} mm):

$$2d_A \sin\vartheta = n\lambda. \tag{208}$$

The Bragg angle ϑ can be determined with high precision from the radius r of the interference ring and the spacing a_f between specimen and film:

$$\vartheta = \frac{1}{2}\arctan\left(-\frac{r}{a_f}\right). \tag{209}$$

The biaxial stress state on the surface is determined from the change in radius of the interference ring under three azimuth angles φ (for example φ, $\varphi + \pi/4$, $\varphi + \pi/2$) and the radiographic elastic constants. The X-ray beam is applied in each case at different angles of inclination ψ (relative to the surface normal) and the lattice strain, is averaged linearly over $\sin^2\psi$ (procedure according to Macherauch and Müller). Both the principal stress total and the individual stress components are determined according to the $\sin^2\psi$ method (basic equation of X-ray stress measurement [302]). At the same time, the first order residual stresses can be separated from those of higher order. It is also possible to distinguish between the stresses in the various components of the microstructure [333].

The measuring area covered is 0.1 to 1 mm^2, the measuring depth being approximately 10 μm. The stress distribution can also be determined approximately in a greater depth by locally removing surface layers. Modern diffractometers are transportable and centre free. They permit measurements on the component, at least under laboratory conditions.

The main advantage of X-ray residual stress measurement is that it is non-destructive. A focal point of applications on welded joints are surface measurements with the maximum possible local resolution, particularly measurements in the direct vicinity of weld seams (see Fig. 184). Measuring problems may occur in the case of materials with a strong texture (e.g. produced by cold forming).

Another more recently developed non-destructive stress or strain measuring technique is the *neutron diffraction method* [331, 287, 288]. Neutrons are scattered by the atomic nucleii, whereas X-rays are scattered by the electron shells, with the consequence that neutrons penetrate far deeper than X-rays, i.e. up to 3 cm in steel or 30 cm in aluminium alloys. Thus stresses or strains can be measured in the interior of the components. Typically, a neutron beam from the reactor core (approx. 50×50 mm^2) is transmitted by a monochromator crystal. Slits in absorbing cadmium masks are placed between the component and both the incident and the diffracted beam in order to reduce the sampled volume of residual stress (Fig. 217). Parallel scattered beams originate from the crystallographic planes which have the correct orientation for diffraction. By studying

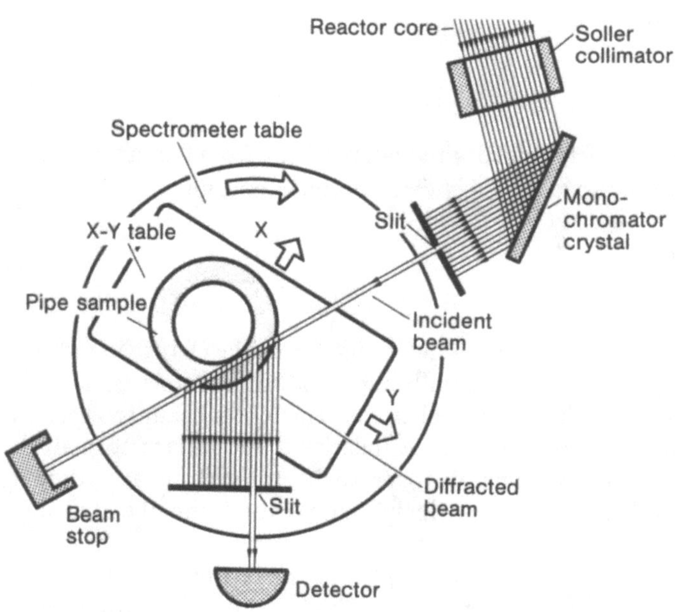

Fig. 217. Residual stress measurement based on neutron diffraction, applied on electron beam welded pipe sample; after Hosbons et al. [331]

both the peak location and intensity of the diffracted beam, data on stress or strain and texture can be obtained. The practical application of the method on welded components is described in Ref. [287, 288].

Further non-destructive residual stress measurements are the *ultrasonic method* and the *magnetostriction method.* Both are in the development stage and are not yet mature enough for widespread application. One version of the ultrasonic method [338, 339] is based on the (rather slight) speed dependence of ultrasound on the stress state. The measurements performed are echo time measurements with two transversal waves, which are orthogonally polarized. The required measuring volume is relatively large for reasons of accuracy. In most cases, the principal stress difference averaged over the component thickness is evaluated. Another version of the ultrasonic method [336, 337] uses Rayleigh waves and longitudinal waves in the surface of the component to determine the residual stress there evaluating differences in velocity caused by a slight nonlinearity of the strain-displacement relations. In the magnetostriction or Barkhausen noise method [335] the stress state is deduced from the value of the local magnetization restraint. Hardness measurements also contain information regarding residual stresses. Tensile stresses reduce hardness proportionally; compressive stresses leave it unchanged [302].

3.6.5 Distortion measurement after welding

Distortion in the post-weld cooled state is determined by applying length and angular measuring techniques common in practice without any need for

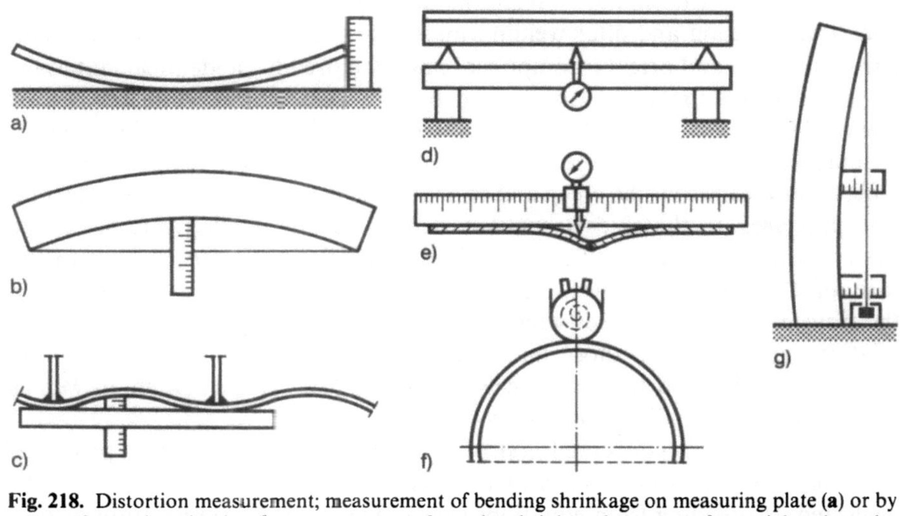

Fig. 218. Distortion measurement; measurement of bending shrinkage on measuring plate (**a**) or by means of a tensioned string (**b**), measurement of angular shrinkage by means of a straight-edge ruler (**c**), continuous measurement of deflection (**d**, **e**), measurement of constriction (**f**), measurement of inclination (**g**); after Vinokurov [8]

welding-specific adaptation. Application examples are shown in Fig. 218. Transverse and longitudinal shrinkage is most easily determined using a measuring tape. Bending or angular shrinkage can be measured on a measuring plate by means of a tensioned string (measuring should be performed in the horizontal plane because of the sag of the string) or by means of a straight-edge set against the component (Figs. 218a, 218b, 218c). The deflection can also be measured continuously in order to determine the bending or angular distortion profile (Figs. 218d, 218e). On vertically extended components such as pillars, supports and tank walls, inclination and deflection are measured by means of strings hanging exactly vertically. The tensioning weight of the strings is immersed into liquid in order to prevent oscillations (Fig. 218g). Circumferential measurements on spherical and cylindrical shells are performed by means of a string wrapped around the structure, the string being applied tightly to all sides by means of two tensioning rollers which are twisted against each other by a constant spring force (Fig. 218f). The change in circumferential length from which the deflections can be calculated follows from the twisting angle of the tensioning rollers applied to the component with weld compared with the component without weld.

3.6.6 Similarity relations

In the case of very large and thick-walled components, it is cost-saving to conduct the welding tests on a model reduced in size. The opposite procedure may also occur, namely that the welding tests are carried out on an enlarged model when considering very small and thin-walled components. This is done in order to be able to better handle the tests. In both cases, the question which is posed relates to the similarity relations for forces and displacements, stresses and strains during and after welding under the condition of geometrical similarity of model and original component manufactured from identical materials.

The similarity relations are derived below in accordance with Ref. [8] for identical temperature fields (i.e. identical temperatures in corresponding points). It must be accepted that the time-related gradients of the temperature field differ. This means that the creep and relaxation effects as well as the transformation processes, to the extent that they depend on the cooling rate or on the austenitizing time, occur differently. Creep and relaxation can, at least in the case of the single-pass weld, be ignored. The situation is different with the transformation processes which depend on the cooling rate. The similarity relations apply only to those materials for which transformation can be excluded. These include austenitic steels, steels with low carbon content as well as many aluminium and titanium alloys.

Consideration is concerned first of all with (high-power) welding processes on solid bodies in which the heat transferred to the surrounding air can be neglected. Proceeding from equations (31) and (32) for the temperature field of a moving point source on the semi-infinite solid, the following relations must be satisfied in order to obtain identical temperature fields in the original component (without index) and in the model (index m):

$$\left(\frac{q}{2\pi\lambda R}\right)_{\mathrm{m}} = \frac{q}{2\pi\lambda R}, \tag{210}$$

$$\left(\frac{vR}{2a}\right)_{\mathrm{m}} = \frac{vR}{2a}. \tag{211}$$

The scale factor γ of the geometrically similar model is introduced as

$$\gamma = \frac{R_{\mathrm{m}}}{R} = \frac{x_{\mathrm{m}}}{x} = \frac{y_{\mathrm{m}}}{y} = \frac{z_{\mathrm{m}}}{z} = \frac{h_{\mathrm{m}}}{h} \tag{212}$$

Hence, there follow from this and from equations (210) and (211), respectively, the similarity relations:

$$\frac{q_{\mathrm{m}}}{q} = \gamma, \tag{213}$$

$$\frac{v_{\mathrm{m}}}{v} = \frac{1}{\gamma}. \tag{214}$$

The dwell time Δt is proportional to the heat input per unit length of weld, $q_{\mathrm{w}} = q/v$, e.g. in accordance with equation (102). Hence, it follows that

$$\frac{\Delta t_{\mathrm{m}}}{\Delta t} = \gamma^2. \tag{215}$$

For the weld cross-section A_{w} (groove cross-section in the case of single-pass welding, layer cross-section in the case of multi-pass welding), it holds that:

$$\frac{A_{\mathrm{wm}}}{A_{\mathrm{w}}} = \gamma^2. \tag{216}$$

For the electrode feed rate v_{e} the relation in the case of a similarly changed weld cross-section is:

$$\frac{v_{\mathrm{em}}}{v_{\mathrm{e}}} = \gamma. \tag{217}$$

The stresses σ result identically; the displacements u by contrast are modified (however, with identical angles of rotation),

$$\sigma_{\mathrm{m}} = \sigma, \tag{218}$$

$$u_{\mathrm{m}} = \gamma u. \tag{219}$$

With low-power welding processes on plates, particularly also with electroslag welding, the heat transferred to the surrounding air plays a more important role and should be allowed for in the similarity relation. Proceeding from equation (33) for the moving line source in the plate, what applies additionally to equations (210) and (211) is:

$$\left(\frac{br^2}{a}\right)_{\mathrm{m}} = \frac{br^2}{a}. \tag{220}$$

It follows with $b = 2(\alpha_c + \alpha_r)/c\varrho h$ assuming identical heat transfer coefficients $(\alpha_c + \alpha_r)$ in the model and in the original component and with $r_{\mathrm{m}}/r = \gamma$:

$$\frac{h_{\mathrm{m}}}{h} = \gamma^2. \tag{221}$$

If the component dimensions change with γ, the plate thickness must therefore be changed with γ^2. The heat input q_{m} is proportional to the thickness:

$$\frac{q_{\mathrm{m}}}{q} = \gamma^2. \tag{222}$$

The other equations, equations (214) to (219), remain valid.

Application of the above similarity relations to the stress and deformation analysis in the case of electroslag welding is presented in Ref. [8] (ibid., p. 215). The heat input which is altered in accordance with equations (213) and (222), respectively, generally necessitates different welding methods on model and original component. In the considered case, the model was gas tungsten arc welded with the filler metal wire placed on the square groove.

4 Reduction of welding residual stresses and distortion

4.1 Necessities and kinds of measures

Welding residual stresses have a negative or positive effect on the strength of the component depending on their type, sign, direction and distribution. Triaxial tensile residual stresses in combination with cracklike defects promote brittle fracture. Uniaxial or biaxial tensile residual stresses diminish corrosion resistance and enhance the stability limit; compressive residual stresses improve the fatigue strength. Details of this are provided in Chapter 5 and in Ref. [437] (ibid., Sections 2.3.2 and 3.4). Components with welding residual stresses may distort during subsequent machining, storage and service loading. A particularly disturbing effect is the back-spring deformation during metal cutting. Welding distortion reduces the fatigue strength and limit load of the components. Specified manufacturing tolerances may be exceeded as a result of welding distortion. It is therefore necessary to minimize welding residual stresses and welding distortion or, as far as possible, to control them according to the respective requirements.

A basic problem in this connection consists in the fact that welding residual stresses and welding distortion to a large extent behave in a contrary way. A component rigidly fixed when welded displays high residual stresses. If, by contrast, it is welded without any restraint, the distortion is large and the residual stresses are relatively smaller. It is not, therefore, readily possible to manufacture a component which is at one and the same time low in residual stresses and in distortion, as is often desired in practice. Consequently, it is necessary to distinguish measures which keep residual stresses low (for example, stress relief methods) and those which keep distortion low (for example, straightening methods). Many of these measures relate simultaneously to residual stresses and distortion by way of a compromise.

The basic problem may indeed extend even further. Longitudinal and transverse residual stresses of a weld also behave in a contrary manner if, namely, the widely held view (e.g. in Ref. [5]) is correct that steeper temperature gradients during welding cause higher residual stresses. This view cannot indeed relate to the (global) transverse residual stresses of a weld, for these are certainly reduced as a consequence of the lower, but more concentrated heat input. On the other hand, the maximum level of weld longitudinal residual stresses, with the exclusion of transformation stresses, always reach the yield limit or exceed it as a consequence of strain hardening, no matter the temperature concentration with which welding is performed. The higher temperature concentration does, however, produce larger plastic strains in the weld zone as it cools down, so that higher demands are made in this case in respect of ductility and resistance to aging. The higher temperature concentration is also likely to cause greater variations in residual stresses in the weld thickness direction and, coupled to this, in the transverse direction of the weld, so that at least locally increased stresses and degrees of multiaxiality are conceivable (not confirmed in Ref. [279] in respect of the transforming steel 20MnMoNi55). An excessively high temperature concentration should be avoided, however, not because of possibly local increased residual stresses but because of undesired microstructural changes which are connected to the temperature gradient over time.

A further problem in reducing welding residual stresses and welding distortion consists in the fact that this involves exhausting the ductility of the material. As a result, the stress-relieved or straightened component may be closer to the failure limit than the as-welded component showing residual stresses and distortion. Such critical conditions occur mainly at the root of sharp notches and cracklike defects. Stresses and strains concentrate here locally, both the stresses and strains produced by the welding process as well as by the external load. Only recrystallization annealing completely undoes such local damage.

The measures for reducing welding residual stresses or welding distortion may be further subdivided into those prior to, during and after welding. Prior to welding, a decision is made regarding design and material measures. This can involve preshaping, bracing or fixing. The welding sequence and the process parameters are specified. Measures during welding may involve preheating or cooling. Post-weld measures are related to stress-relieving or straightening. The sections which follow below are arranged according to design, material and manufacturing measures. The latter are subdivided according to measures prior to and during welding and to post-weld measures.

Reducing welding residual stresses comprises the following possibilities:

- decreasing the level of the residual stresses, in particular the maximum tensile levels,
- decreasing the zones or volumes with high residual stresses,
- decreasing high degrees of multiaxiality of tensile stress.

Reducing welding distortion refers to decreasing the most striking permanent deformations.

4.2 Design measures

Welded structures should be designed in such a way that welding residual stresses and welding distortion remain controllable during manufacture, thus assuring the welding feasibility. Design measures are those specifications which are the sole responsibility of the design engineer. These include configuring and dimensioning the structure and subdividing it by welded joints, selecting the type of joint (butt, overlap, cruciform and T joint, corner, inclined-T and edge joint) and specifying the throat thickness to the extent that this can be selected irrespective of the plate thickness. Deciding on the material of the welded structure is treated in a separate section. All wider-ranging decisions, particularly also those relating the groove shape and welding sequence as well as the choice between continuous and intermittent welds, are assigned to manufacturing. This delineation of responsibilities does not always completely conform to organizational commitments in companies.

The principal design measures for limiting welding residual stresses and welding distortion are:

- smallest possible weld length,
- smallest possible plate thickness,
- smallest possible throat thickness,
- intermittent weld more favourable than continuous weld,
- fillet weld more favourable than butt weld,
- large straining length of the parts connecting to the butt weld,
- favourable subdivision into assemblies.

A statement in favour of the smallest possible weld length is the well-known slogan: "The best welded structure is the one which does not even have a weld".

A practical application of the principle of the smallest possible weld length is shown in Fig. 219. The least distortion is displayed by the girder (a) assembled from two channel sections joined by butt welds, whereas the girder (b) assembled from chord and web plates joined by four fillet welds distorts to a greater extent, but is characterized by the possibility of increasing the chord thickness compared to web thickness. The girder (c) assembled from two angle section bars is less suitable because of asymmetrical distortion.

Weld accumulations and weld crossings are regarded as particularly detrimental. Where weld seams are accumulated in parallel, welding residual stresses acting in the same direction and plastically deformed areas superimpose to some extent. Where weld seams cross, high residual stresses are built up in two directions. In both cases the twice applied thermomechanical cycle may exceed the ductility of the material locally, for instance at notches and defects.

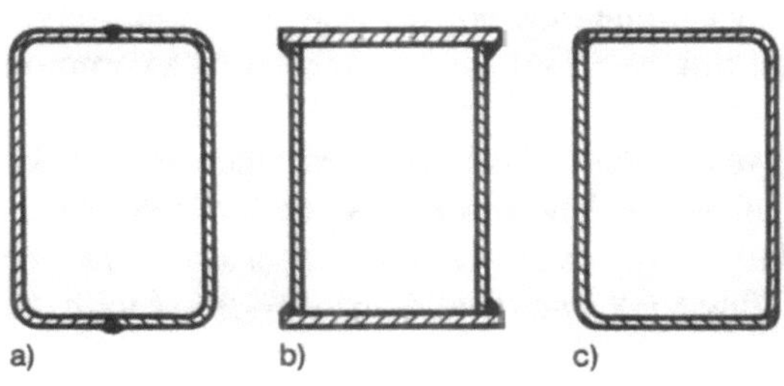

Fig. 219. Smallest possible weld length demonstrated for box girders: little distortion with two symmetrically positioned butt welds (**a**), medium distortion with four symmetrically positioned fillet welds (**b**), heavy distortion with two asymmetrically positioned corner welds (**c**); after Malisius [5]

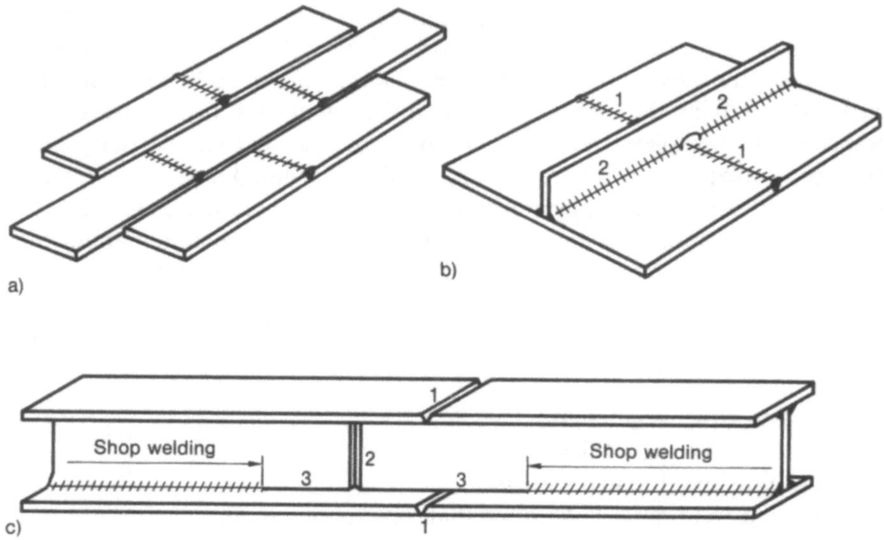

Fig. 220. Avoidance of weld crossings and favourable welding sequence: plate panel with staggered pattern of transverse welds (to be manufactured first) and continuous longitudinal welds (to be manufactured last) (**a**); plate stiffened by rib, transverse welds to be manufactured first and longitudinal welds thereafter, weld crossings to be avoided by cutouts (**b**); I section girder joint with staggering of chord and web welds, the staggered transverse welds to be manufactured first and the longitudinal welds thereafter (**c**); after Malisius [5]

A counter-measure is to arrange the transverse welds in a staggered pattern between continuous longitudinal welds, for instance in plate panels, (Fig. 220a), or in I section girder joints (staggering of chord and web weld) (Fig. 220c). The amount of staggering should be at least twenty times the plate thickness. Another counter-measure is cutting-free the weld seam intersection points so as to avoid intersecting overwelding (Fig. 220b). Cutouts of this type, however, cannot generally be recommended without reservations in respect of fatigue strength.

The smallest possible plate thickness is called for in order to limit the level and extent of the triaxial tensile stresses which promote brittle fracture. Only with a larger plate thickness can the third principal stress in the plate thickness direction acquire major proportion. That is why specifications often prescribe stress relief annealing for welded components exceeding a certain plate thickness. As another counter-measure, however, it is also possible to replace thick chords by lammellar packs and thick-walled cylindrical vessels by multi-layer designs. In the latter case, the metal ring packs are joined together by a common circumferential weld.

The throat thickness of fillet welds can be freely selected whereas it is the same as the plate thickness for butt welds. The throat thickness of fillet welds should not be greater than necessary on the basis of static dimensioning, for the weld heat input and thus the shrinkage force and the distortion increase with the throat thickness.

Fastening welds (these being welds which do not serve for direct transmission of forces according to design) may be executed as intermittent welds in order to reduce the total heat input. The weld strings of the double fillet intermittent weld can be arranged in a staggered pattern. Where there is a risk of corrosion, the weld strings should be closed through cutouts. Reservations exist in connection with intermittent welds and cutouts in respect of fatigue strength and they are therefore permissible only in the case of fastening welds. A further low-distortion type of fastening, e.g. in the case of stiffening ribs, is (oblong) plug welding (which can also be replaced by spot welding). The plug welded rib mentioned is compared in Fig. 221 to the version with intermittent fillet welds.

The fillet weld in the cruciform, T, corner and lap joint is superior to the butt weld in respect of residual stresses and distortion (not so in respect of fatigue strength). The gap and force flow diversion in the case of the fillet weld reduce the stiffness of the joint with the result that the transverse residual stresses build up to a lesser extent. Particularly in older literature, butt welded structures are termed "monolithic" and therefore more prone to brittle fracture. It is preferable in view of the statements above to overlap subsequently attached patches and not to weld them flush. The fillet weld is also superior to the butt weld in respect of the required dimensional tolerances of the components to be joined. A higher degree of transverse and angular misalignment is acceptable than is the case with butt welds, the groove of which must be particularly accurately aligned (which, in practice, leads to prestressing by pushing out or drawing in). The stiffness drawback of the butt weld can be countered in design terms by an adequately large component length transverse to the weld ("straining length"). In respect of the angular shrinkage of the T joint, the weld between three members is superior to the more common double fillet weld (Fig. 222), although this is achieved at the expense of higher (i.e. butt weld like) transverse residual stresses.

Finally, the subdivision of the structure, particularly well known from shipbuilding and bridge construction, into assemblies (or sections) prefabricated to specified dimensional accuracy, is a particularly effective measure for controlling welding residual stresses and welding distortion in the total structure (in addition, the construction time is reduced as a result of parallel fabrication). The dimensional accuracy of the assemblies is a guarantee of dimensional accuracy

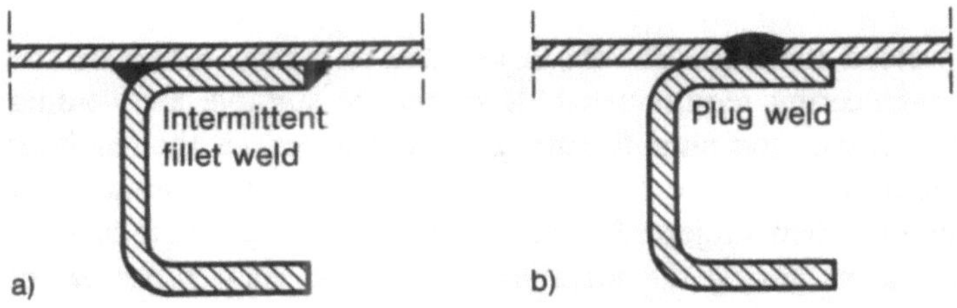

Fig. 221. Reduction of distortion with fastening welds: panel plate on top of channel section bar, heavy distortion with intermittent fillet welds (**a**), little distortion with plug welds (**b**); after Malisius [5]

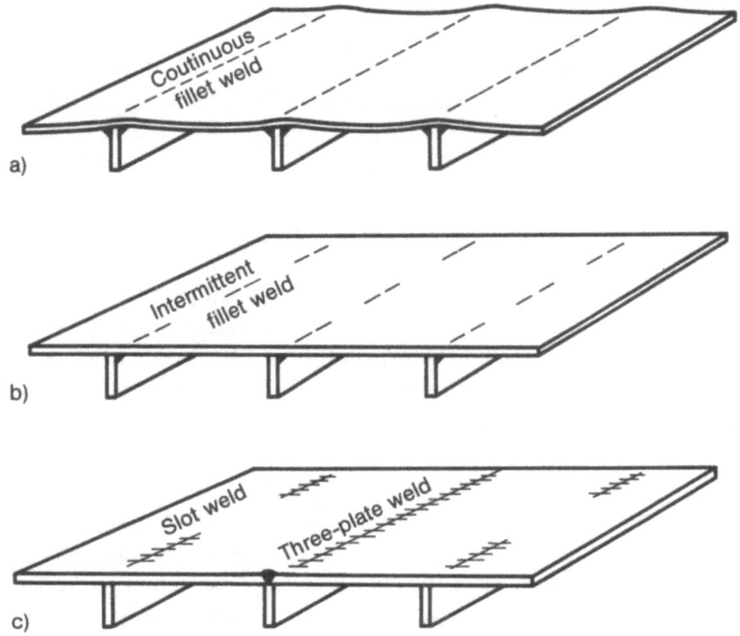

Fig. 222. Reduction of distortion in plate with rib: heavy distortion (sharp bends) with continuous fillet weld (**a**), medium distortion with intermittent fillet weld (**b**), little distortion with slot weld or three-plate weld (**c**); after Malisius [5]

for the total structure. In addition, the assemblies can be manufactured in the shop, and only the total structure is welded in the open. In bridge building and structural engineering, welding of joints on-site is avoided to the maximum possible extent. On-site joints are joined with high-strength bolts. In any case, the aim should be to achieve the highest possible degree of symmetrization of the assemblies and thus of the weld positions in the total structure.

4.3 Material measures

4.3.1 Starting points

The material for welded structures should be selected to a large extent on the basis of welding engineering requirements. It should be suitable for welding under the respective design and manufacturing circumstances, i.e. the resulting welded joints should prove to be free of cracks, resistant to fracture and adequately deformable when subjected to service loads. Strength is reduced if hot or cold cracks form during the metallurgical and thermal processes of welding or if the crack formation potential when subjected to subsequent loading is drastically increased. Welding residual stresses play an important role in crack formation and crack formation potential. Compressive residual stresses impede crack formation, whereas tensile residual stresses, particularly those of a multiaxial nature, promote it

In view of the special theoretical and experimental problem of sufficiently reliably determining welding residual stresses in a concrete application, welding residual stresses scarcely emerge in appropriate professional discussions and testing techniques in an explicit and quantified form. They are contained only implicitly and unquantified in the specifications relating to specimen size, specimen geometry and weld heat input in hot and cold crack test procedures. This state of affairs is the cause of the deficient transferability of test results obtained from hot or cold crack specimens to the welded component. Only the occasionally encountered verification by measurement of compressive residual stresses in the weld of the test specimen seems to be convincing for a favourable strength forecast, although, in this case also, the question of transferability remains open. It is likely, however, that a fundamental improvement in the situation will occur with the progressive improvement in numerical welding residual stress analysis. Nevertheless, welding residual stress fields, notably in the case of microstructural transformation, are so complex that general statements will be difficult. It will only be possible to conduct a detailed analysis of individual cases.

It was not hitherto possible to state in single-valued quantified terms the principal material parameters for welding suitability from the aspect of welding residual stresses and welding distortion, quite apart from the even more difficult question relating to crack initiation and crack propagation. This results from the fact that longitudinal and transverse residual stresses and longitudinal and transverse distortion are generated according to different mechanisms, that the process proceeds in a strongly non-linear manner, that the thermomechanical material characteristic values are greatly temperature-dependent and that, in addition to thermal expansion, the transformation processes may play a significant role.

4.3.2 Material characteristic values in the field equations

The field equations for the linear-elastic basic models of transient thermal stress during welding are known, from which the decisive material characteristic values are evident to the extent that the linear-elastic solution determines the, in fact, greatly non-linear elastic-plastic behaviour. The latter is the case because the welding operation proceeds in a highly localized manner and is elastically supported. In the sections which follow, known plate solutions are evaluated and regarded as representative also for compact section rods (with butt joint) and solid bodies (with surfacing weld), which is not strictly the case. The thermal diffusivity a, decisive for transient heat conduction processes, is also in these cases introduced, in order to achieve formal uniformity, where more briefly the coefficient of thermal conductivity λ could be used ($a = \lambda/c\varrho$).

For the transient temperature field in the infinite plate with momentary line source perpendicular to the plate plane (model for spot welding) according to equation (25) (with $T_0 = 0$, $b = 0$), the radial stress is known from equation (110). For the quasi-stationary temperature field around the line source perpendicular to the plate plane moving uniformly and linearly in the infinite plate (model for seam welding) according to equation (33) (with $T_0 = 0$, $b = 0$), equation (112) applies to the longitudinal stress and equation (174) to the (mean) longitudinal shrinkage. The maximum transverse displacement of the edge of the semi-infinite plate with line source perpendicular to the plate plane moving uniformly upon the edge (model for welding with open groove) is given by equation (168). The transverse stress follows from this according to equation (151) introducing the elastic "straining length".

Although these equations do reveal material characteristic values which are decisive for welding residual stresses and welding distortion, the situation is, nevertheless, unclear because the process characteristic values Q and $q_w = q/v$ (heat input per unit length of weld) are determined to a not inconsiderable extent by material characteristic values.

The influence of the material on Q or q is first clarified. The welding heat serves to fuse base and filler material. During this process, heat also passes into the surrounding base material, is transferred to the ambient air and radiated to the surroundings. If Q and q are understood as weld-effective heat, the ambient loss (3 to 80%, depending on welding method, see Table 1) remains disregarded. The weld-effective heat is diffused for the greater part and for the lesser part (up to 48%, see Section 2.3.1.3.2) contributes to fusion. The fusion-effective part of Q or q may be set proportional to $c\varrho T_m$ (T_m melting temperature, $c\varrho$ averaged between 0 °C and T_m) if the latent melting heat and also workpiece temperatures below or above 0 °C remain disregarded. The thermal diffusion part of Q or q may, on the other hand, in a first approximation be set proportional to $ac\varrho T_m$ (equations (25) and (33) with $T = T_m$ and $T_0 = 0$). For a general assessment of welding suitability, it may suffice to set the entire weld-effective heat proportional to $ac\varrho T_m$. In equations (110), (112), (174), (168) and (151) the expression aT_m then replaces $Q/c\varrho$ or $q/c\varrho$, respectively (apart from the proportionality factor). From the thus transformed equations, it is clear that α, E, a and T_m together are the decisive material characteristic values and not α, E, λ and c, as seemed to be the case on the basis of the original form of the equations.

In view of the fact that the derivations to the stress and deformation equations referred to above are based on elastic material behaviour, the mechanical material characteristic value characterizing the maximum residual stress level, the yield limit σ_Y, does not yet occur. It should be added to the above characteristic values.

4.3.3 Traditional consideration of the influence of the material

The influence of the material on welding residual stresses and on welding deformations can be comparatively assessed essentially on the basis of the familiar material characteristic values T_m, a, α, E, σ_Y (see Table 4), the temperature dependence of a, α, E, σ_Y being initially disregarded.

Table 4. Material characteristic values (at 0 °C) determining the welding suitability in respect of welding residual stresses (relative to brittle fracture) and post-weld deformations, welding suitability indices calculated from the mean material characteristic values; after Radaj [343].

Base metal	Melting temperature T_m [°C]	Thermal diffusivity a [mm²/s]	Thermal expansion α[1/°C]	Elastic modulus E [kN/mm²]	Yield limit σ_Y [N/mm²]	Fracture toughness K_{1c} [N/mm$^{3/2}$]	Welding suitability indices λ_σ [−]	λ_ε [−]
Low-alloy steel	1520	7.5–9.5	11×10^{-6}	210	200–700	> 800	1.0	1.0
High-alloy steel	1400	5.0–7.5	16×10^{-6}	200	250–550	> 800	1.07	0.86
Aluminium alloy	600	75–100	24×10^{-6}	65	80–280	> 600	0.28	0.01
Titanium alloy	1800	6	8.5×10^{-6}	110	500–700	> 2200	8.13	1.08
Copper alloy	1080	120	18×10^{-6}	130	30–420	> 800	0.10	0.02
Nickel alloy	1435	15	13×10^{-6}	215	120–630	> 800	0.49	0.43

The melting temperature, T_m, acts in the same direction on welding residual stresses and welding deformations. A higher melting temperature results in higher stresses and deformations. The same holds true for the volume-specific fusion heat. From the aspect of T_m, aluminium would be relatively suitable for welding and titanium relatively unsuitable.

The thermal diffusivity a, which characterizes the rate of temperature equalization with transient heat conduction, acts in the same direction on stresses and deformations, the effect of λ, c and ϱ being combined in a. Small λ with high $c\varrho$ would be particularly favourable from the aspect of a, which speaks in favour of steel or titanium and against aluminium or copper. A high temperature concentration should be aimed for, provided undesired microstructural transformation does not preclude this.

The coefficient of thermal expansion α acts in the same direction on stresses and deformations in a particularly pronounced manner, although this may be confined by the occurrence of transformation strains acting in the opposite direction. From the aspect of α, titanium would be relatively suitable for welding and aluminium relatively unsuitable.

The elastic modulus E, including the less varying Poisson's ratio ν, acts in the same direction on welding residual stresses and in the opposite direction on welding deformations. Instability phenomena (warpage), in particular, are restrained by a high elastic modulus. From this aspect, aluminium would be low in residual stresses and high in deformations. Steel, titanium and copper would be higher in residual stresses and lower in deformations.

The yield limit σ_Y, including the hardening modulus, likewise acts in the same direction on welding residual stresses and in the opposite direction on welding deformations. A higher yield limit makes higher residual stresses possible, both in respect of peak values as well as in respect of the average stress level. The deformation energy stored in the welded structure may thus be increased, promoting the risk of brittle fracture. On the other hand, the deformation, including offset groove edges ahead of the weld pool, is diminished by the smaller and less extensive plastic strains. These statements also apply in respect of increasing the yield limit at elevated temperature (for example, for high-temperature steels). The latter increases the risk of cracking at elevated temperatures. This must be countered by high-temperature ductility (i.e. forgeability). Cast iron is less suitable for welding because of the lack of high-temperature ductility.

4.3.4 Derivation of novel welding suitability indices

From the preceding discussion of the principal material characteristic values determining welding residual stresses and post-weld deformations according to Table 4, the tendencies which are revealed are in some cases contradictory. For example, aluminium from the aspect of T_m is relatively suitable for welding, from that of a and α relatively unsuitable and, finally, from that of E and σ_Y, requires to be assessed differently regarding the requirements of "low residual stresses" and "low welding deformations". An overall evaluation can be made using the welding suitability indices λ_σ (for welding residual stresses relative to brittle fracture) and λ_ε (for welding deformations), which include a multiplicative superimposition of the individual characteristic values relative to a reference material, indicated by an asterisk [343]:

$$\lambda_\sigma = \frac{T_m^* a^* \alpha^* E^* K_{Ic}}{T_m a \alpha E K_{Ic}^*}, \tag{223}$$

$$\lambda_\varepsilon = \frac{T_m^* a^* \alpha^* E \sigma_Y}{T_m a \alpha E^* \sigma_Y^*}. \tag{224}$$

In λ_σ, the determining factor on the materials side for the level of residual stresses is introduced with $T_m a\alpha E$, which must be set in proportion to σ_Y because higher residual stresses should be sustainable at a higher yield limit. The higher yield limit, however, can only be exploited without brittle fracture if the fracture toughness K_{Ic} (or for less brittle materials the critical crack opening displacment δ_c) rises in the same ratio. Hence there follows in equation (223) the factor $\sigma_Y/\sigma_Y^*[(K_{Ic}/K_{Ic}^*)/(\sigma_Y/\sigma_Y^*)] = K_{Ic}/K_{Ic}^*$. In λ_ε, the determining factor on the material side for the level of welding deformation is stated with $T_m a\alpha$, which is modified in the opposite direction by the material elasticity E and the yield limit σ_Y in respect of components subjected to residual stresses. In the case of the values of λ_σ and λ_ε in Table 4, low-alloy steel with $\sigma_Y = 450$ N/mm^2 is selected as reference material. The average values of the material characteristic values, which are stated in ranges, are used in λ_σ and λ_ε. The melting temperatures T_m and T_m^* are introduced relative to a working temperature of 0 °C.

It should be noted that the indices λ_σ and λ_ε identify welding suitability only from the aspects of welding residual stresses (relative to brittle fracture) and post-weld deformations. The processes at elevated temperatures during welding which lead to hot or cold cracks or to offset groove edges, which are likewise important for welding suitability, are not yet covered with these welding suitability indices. It is conceivable, however, that the temperature-related plotting of λ_σ and λ_ε (with T_m relative to the considered temperature in each case) permits more extensive statements. Transformation influences are not, yet, covered in λ_σ and λ_ε. Hardening and ageing, by contrast, can be allowed for in the strength values introduced.

From Table 4 it is clear that aluminium and copper alloys appear on average to be less suitable for welding, whereas titanium alloys come off well (the possible embrittlement of titanium as a result of gas absorption is not covered). High-alloy steel is on average as suitable for welding as low-alloy steel; the possible embrittlement of steel as a result of hydrogen, ageing or hardening is not covered. Nickel alloys are on average considerably less suitable for welding than steel. The individual alloy may, however, vary greatly from the average values of λ_σ and λ_ε included in Table 4.

The welding suitability indices can also be used to quantify the influence of preheating on residual stresses and distortion; see Section 4.4.2.4.

The welding suitability indices λ_σ and λ_ε, which characterize the welding suitability of the material group regarding welding residual stresses and distortion, supplement the weldability indices or hardening equivalents which are used for steel in respect of undesired hardening in the heat-affected zone [344].

4.4 Manufacturing measures

4.4.1 Starting points

The range of possible manufacturing measures for reducing welding residual stresses and welding distortion is large, but yet limited in its applicability and effectiveness. Incorrect design or material decisions cannot be remedied, or only partially, by manufacturing measures. However, on the other hand, the design and material possibilities are considerably enlarged by sophisticated manufacturing techniques.

Just as welding residual stresses play a determining role regarding weldability even during the welding process, this applies equally to welding distortion. If the edges of the welding groove are offset too strongly as a result of distortion (opening, closing, off-set), manufacturability may be restricted or rendered impossible. Consequently the measures presented below relate not only to residual stresses and distortion in the completed welded structure, but equally to residual stresses and distortion during the manufacturing process.

Organizational formalism of manufacturing measures is first of all considered. For welded structures, the residual stresses and distortion of which are controlled by well established production measures, it suffices to have a welding schedule with, by way of example, the following tabular data (after Ref. [5]):

- welding method,
- weld types,
- shop or assembly welds,
- base materials,
- filler materials,
- test methods,
- heat treatment.

For structures more complicated to manufacture with numerous mutually influencing welds, this should be supplemented by a welding sequence schedule, which, by way of example, contains the following additional data (after Ref. [5]):

- prefabrication of the parts,
- dimensional inspection of the parts,
- straightening of the parts,
- tacking or fixing of the parts,
- layer sequence in the weld cross-section,

- welding sequence over the weld length,
- welding direction in the weld sections,
- back-step welding,
- sequence of welds,
- application of jigs and fixtures,
- simultaneous assignment of several welders.

Welding schedule and welding sequence schedule are jointly compiled by the design and welding engineer.

The multitude of possible manufacturing measures are explained in detail below. In this connection, a distinction is made between measures prior to and during welding and post-weld measures. In manufacturing practice, such a distinction is only conditionally correct because welding manufacture proceeds as a multiply staggered process. Semi-finished parts are joined to form components, components to form assemblies, assemblies to form structures. The indication "prior to and during welding" or "post-weld" can therefore only relate to the particular manufacturing stage in question.

4.4.2 Measures prior to and during welding

4.4.2.1 Overview

The principal manufacturing measures prior to and during welding are first of all listed, then combined to form subsections, explained there in detail and finally demonstrated by taking the examples of three typical applications – a girder joint, a plate panel and patch welding:

- prefabrication,
- dimensional accuracy,
- groove cross-section and groove shape,
- multi-layer welding and layer sequence,
- tandem-welding,
- weld sequence,
- back-step welding,
- symmetrically simultaneous welding,
- tacking,
- preheating,
- simultaneous cooling,
- counter-aligning,
- preshaping,
- bracing,
- fixing.

4.4.2.2 General measures

The *prefabrication* of structural components and assemblies which are then joined to form the total structure (see Section 4.2) is a precondition for limiting the residual stresses and distortion of the total structure on the basis of effectively controlling residual stress and dimensional accuracy after each fabrication step. Construction time is reduced as a result of parallel manufacture. Components and assemblies can be manufactured in the shop.

The *dimensional accuracy* of the parts to be joined by welding is a prerequisite for limiting residual stresses and distortion. Lack of dimensional accuracy of the parts usually necessitates pressing and fixing prior to welding until the weld can be laid in the desired manner. This causes prestresses and predeformations which are superimposed on the welding residual stresses and welding distortions. The German Standard DIN 8570 [345], which specifies different classes of dimensional accuracy for manufacture, can be used to specify the permissible dimensional deviations for welded structures.

4.4.2.3 Weld-specific measures

The *groove cross-section* should be dimensioned as narrow as possibe to permit the smallest possible heat input combined with the strongest possible heat concentration of the welding method (by analogy to the tight dimensioning of the fillet weld according to Section 4.2). As a result, the fusion zone and the plastically deformed zones remain small, which reduces residual stresses and distortion. The selected groove shape has a considerable influence on the groove cross-section. The square groove permits the smallest possible groove cross-section. The U shape is more favourable than the V shape. Symmetrical weld grooves reduce angular shrinkage while residual stresses are increased. For this reason the butt weld double-V groove may be used in place of the butt weld V groove (or Y groove). By analogy, the double-sided fillet weld may be selected in place of the single-sided fillet weld. The reduction in angular shrinkage is only fully achieved, however, if the welds are laid simultaneously on both sides or if alternate multi-layer welding is possible.

Multi-layer welding is applied on thick-walled components primarily because of the limited heat output of the welding processes, and has the advantage, compared to single-layer welding, from the point of view of residual stresses that no appreciable tensile residual stresses are built up in the direction of plate thickness, which enhances the brittle fracture resistance. Transverse residual stresses and transverse and angular distortion, by contrast, are generally increased by multi-layer welding. The longitudinal residual stresses of the layers deposited first are relieved, however, by the layers placed over them. In multi-layer welding in the asymmetrical V or Y groove, the transverse residual stresses depend considerably on whether angular shrinkage is permitted or not. If it is permitted, high transverse tensile residual stresses occur at the weld root. In two-layer welding in the symmetrical double-V groove, angular shrinkage can be held at a low level by simultaneous single-layer welding on both sides. Moreover, the better heat efficiency achieved as a result of this is advantageous (an important aspect especially for gas welding of aluminium and copper). Alternate multi-layer sequences on the two sides are also used for reducing angular shrinkage, although these require easy-to-use turnover jigs for the

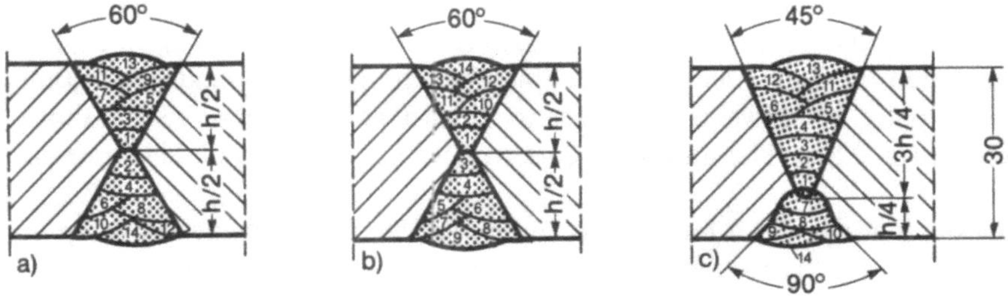

Fig. 223. Reduction of angular distortion by alternating weld pass deposition in double-V groove: turn-over after each pass (**a**), turn-over twice with symmetrical double-V groove (**b**), turn-over twice with asymmetrical double-V groove and grooved-out root (**c**); after Malisius [5]

component. Layer sequences for double-V groove welds which have proven satisfactory in practice are shown in Fig. 223. The asymmetrical double-V groove is used where the weld root is grooved out. For the T joint with double-sided weld, the connected parts are inclining in the direction of the first weld. The weld on the opposite side can only partially compensate for the inclination caused by the first weld, which has already cooled down. Aligning in opposite direction prior to welding is common practice. In the case of the corner joint with internal and external fillet weld, the inclination as a result of the internal weld is greater than that resulting from the external weld. High shape accuracy is, therefore, achieved by laying the external weld first. Longitudinal shrinkage is reduced in multi-layer welding compared to single-layer welding; transverse and angular shrinkage, on the other hand, are increased.

Tandem welding designates welding processes in which two or even more heat sources are arranged in series and move together one after the other producing the weld seam. Such a procedure is well established in submerged arc welding. Up to three electrode wires may be used, melting into the same longitudinally extended weld pool or connected with separate weld pools. In laser beam welding the tandem effect is achieved by splitting the laser beam [395]. The effect of tandem welding into a single weld pool is, besides higher welding speed, a lower cooling rate connected with better evaporation. The thermal control effects of multilayer welding are available if separate weld pools are used. Aspects of preheating and post-weld heat treatment are put into effect to some extent in tandem welding.

The *welding sequence* may have a considerable impact on residual stresses and distortion. It is, therefore, precisely specified in the welding sequence schedule for high-quality welded structures. The supreme principle is to assure "unrestrained transverse shrinkage" or "large straining length" for butt welds. In the case of plate panels with transverse welds arranged in a staggered pattern to avoid weld intersections (Fig. 220a), the discontinuous transverse welds are welded first, followed by the continuous longitudinal welds. A similar procedure is adopted for cylindrical vessels where the longitudinal welds are executed first, followed by the circumferential welds. In plate assemblies with stiffeners, the butt welds between the plates and between the stiffeners should be welded first, and only then the fillet welds between plate and stiffener (Fig. 220b). In girders prefabricated for field assembly, a section of the longitudinal weld in the joint area is initially left unwelded in order to at least partially observe the above principle (Fig. 220c). The unwelded girder plates act as the available straining length.

Back-step welding is a measure to counteract the wedge-shaped opening or closing of the weld groove ahead of the weld pool, which may especially occur with long welds and low welding speed. If the groove displacement is suppressed by pretacking or by rigid lateral restraint, this may result in a permanent roof-shaped arching of the weld joint. A layer sequence which has proven satisfactory in practice for back-step welding is shown in Fig. 224. First of all, short tack welds of length $l \approx 2.5h$ are executed, spaced $e^* \geq 25h$ apart. Spacing should be all the closer, the slower the welding speed in order to reliably prevent groove edge displacement. The first weld layer is subsequently deposited in back-step from tacking point to tacking point opposite to the global weld direction. Finally, the remaining layers are applied continuously, alternating the direction. Back-step welding also reduces transverse and longitudinal shrinkage of the weld joint as a whole. Welding residual stresses remain limited if the welding operations are performed in succession without the part cooling down completely in between. This is of significant benefit compared to the section-wise joining of the structure with the (unavoidable) complete cooling down of the welded sections before the remaining groove openings are closed. Back-step welding is a wide-spread method for large structures such as ships or tanks.

Symmetrically-simultaneous welding reduces distortion. Welds arranged symmetrically in the component reduce distortion only if they are laid for the most part simultaneously. Consequently, the circumferential welds of large vessels or tubes are executed symmetrically-simultaneously by several welders (the first layer in the back-step method). As cylindrical parts of the type considered do not react locally soft, but only stiff as a whole, welding, if not performed simultaneously, would cause the first layer to tear and result in unacceptable misalign-

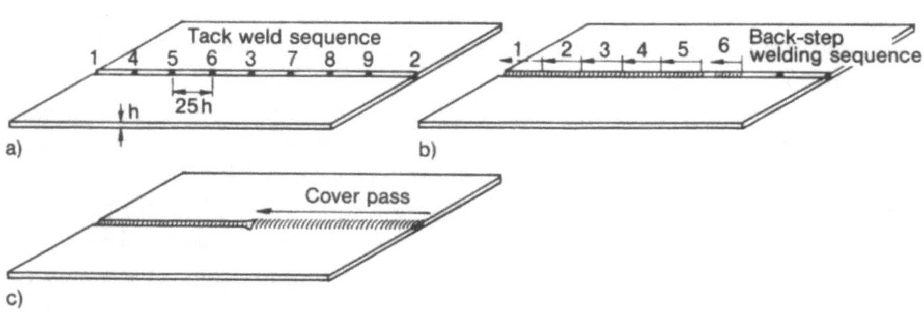

Fig. 224. Reduction of transverse shrinkage as well as groove gap distortion by back-step welding, practicable only with arc welding: tack weld sequence (**a**), back-step welding sequence in first layer (**b**) and cover pass (**c**); after Malisius [5]

ment. Two welders are assigned to the top and bottom of large I section girders (especially of girders without a straining length provided by unwelded longitudinal welds) to execute the chord joints symmetrically-simultaneously. If the bottom chord is welded overhead, it is then necessary for two welders to work at the bottom, compared to one welder at the top, because of the slower overhead working speed. The symmetrically opposite longitudinal welds of slender girders are welded simultaneoulsy to reduce distortion. The longitudinal curvature caused by a longitudinal weld on one side of the girder is only partially compensated for by a weld seam which is later on welded on the opposite side. In the case of groove cross-sections of appropriate symmetry (e.g. double-V groove weld), layers arranged symmetrically to the centre plane of the plate should likewise be welded simultaneously. This rules out angular shrinkage and improves heat efficiency.

The *tacking* of the aligned parts of a structure, mentioned above in respect of back-step welding, before actual seam welding, has the purpose of achieving particularly accurate manufacture in terms of shape and dimensions. Tacking, however, must be employed with care. First of all, tacking must not be performed at those grooves where high transverse residual stresses can be avoided by unrestrained shrinkage. The tack welds must then be sufficiently long in order to reliably transmit the shrinkage forces which will in other cases occur. A weld length equal to two to three times the plate thickness is recommended [5]. As relatively small-volume tack welds are applied to relatively large-volume components, the cooling rates which occur with tacking are high and, as a consequence, hardening and cracking are frequently accompanying phenomena. These undesired phenomena are countered by preheating when tacking. Finally, the formation of cracks and defects is possible when overwelding the tack welds. The counter measures used in multi-layer welding, such as preheating, slag removal, removal of defects, are also common practice in the case of tacking.

4.4.2.4 Thermal measures

Welding residual stresses (in level and extent) and welding distortion can be reduced by *preheating*. That this is so (ignoring transformation influences and influences by the temperature gradients over time, see Section 4.1) is revealed most clearly from the welding suitability indices according to equations (223) and (224), in which the temperature difference $T_m - T_p$ with preheating temperature T_p should be introduced in place of the melting temperature T_m. The remaining material characteristic values (without asterisk) should also be introduced for the preheating temperature, but they vary less strongly relative to $T_m - T_p$. In this way, λ_σ and λ_z increase with higher T_p and lower T_m. The heat input of welding and the width of the fusion zone are diminished. Preheating from this aspect is important for materials with a high thermal diffusivity such as aluminium and copper alloys (preheat temperature $\leq 200\,°C$ and $\leq 700\,°C$, respectively). Preheating may also be necessary for reducing residual stresses at elevated temperatures during welding of materials with a low ductility at high temperature (cast iron, cast light metal). This consideration is correct, however, only for global preheating with unrestrained cooling down. Local preheating may involve additional residual stresses and distortion (shown, by way of example, for the preheated circumferential weld of a tube by Köppel and Reuschling [390]). It is recommended to preheat largest possible surface areas, i.e. wide areas on both sides of the weld. For the present, the conditions involved in the additional occurrence of influences from microstructural transformation are unexplored. Local preheating cannot, therefore, be unrestrictedly termed as beneficial from the aspect of welding residual stresses and distortion.

By contrast, preheating in the case of steel may, for metallurgical reasons, be necessary [357]. In this case, it is a matter both of the reduction in cooling rate in the heat-affected zone which improves the microstructure as well as of the decrease in hydrogen diffusion which prevents cold cracking (hydrogen for example from a moist electrode covering). The preheating temperature T_p necessary to obtain a definite cooling time $\Delta t_{8/5}$ in single-pass welding is determined from the relations in Section 2.4.3.3, e.g. on the basis of equations (98) and (99) introducing $T_0 = T_p$ [391].

The heat-affected zone of low-alloy ferritic higher-strength structural steels (they contain strength-enhancing alloying elements including CMn and are partially supplied in a heat-treated condition) is jeopardized by cold cracking which occurs in martensite hardened zones in the presence of diffusible hydrogen and may lead to brittle fracture. Frequent types of cold cracks are under-bead cracks. Hardening and hydrogen diffusion can be decreased by preheating. Preheating slows down the cooling process. A lower cooling rate means a more ductile microstructure. Degrees of hardness ≤ 350 HV are aimed at. The preheating temperature for welding steels (up to 300 °C) is selected dependent on carbon equivalent and ratio of (single-layer) weld thickness to plate thickness (or a more precise cooling parameter). Preheating to at least 20 °C is common practice to reduce the risk of brittle fracture if the working temperature is below 5 °C. Preheating is not necessary for austenitic steels because martensitic hardening does not occur in these steels.

Preheating of the weld zone is performed locally with gas torches, heating mats or induction coils. Asbestos mats are used for reducing the heat losses and as a protection for the welders. The aimed-at preheating temperaure is checked by means of temperature indicating crayons or adhesive thermometers. Global preheating in a furnace is only occasionally performed. The regulations of the codes in respect of preheating are summarized in Ref. [351]; the terms and processes in Ref. [346].

Gas welding, compared to arc or beam welding, may be considered as a process with a kind of local preheating in view of the heat input over a larger area. As a consequence, residual stresses occur in a less pronounced manner.

Cooling simultaneously with heating by welding is employed in order to limit the width of the fusion zone. The weld surface in electroslag welding is shaped by water-cooled copper hoops. In flash butt welding, the workpiece is grasped by water-cooled copper rings bridging a short working length. Water-cooled copper electrodes are used for resistance spot welding. Welding on a root-shaping copper bar is also performed from the aspect of better heat diffusion (i.e. cooling). Reducing the size of the weld pool has a reducing effect on residual stresses and distortion. Nevertheless, the triaxial tensile stress state occurs to an increased extent in thick-walled components. Unacceptable hardening may also be an argument against cooling.

The circumferential weld of pipes made of austenitic stainless steel are multipass-welded applying cooling with flowing or sprayed water inside the pipe ("heat sink welding") in order to prevent the axial tensile stresses at the weld root and convert them to compressive stresses. Cooling is applied during all weld passes following the root pass or only during the last pass. The objective of the procedure is to prevent stress corrosion cracking in cases of sensitive materials and high oxygen content of the medium in the pipe.

The residual stresses after heat sink welding have been analysed using an axisymmetrical ring element model [426], the object being to optimize the process. The heat input per pass should be kept low and axial restraint during welding is recommended. The temperature at the cooled inside of the pipe should not exceed 200 °C. Further results based on residual stress measurements [427] refer to heat sink welding, last pass heat sink welding, induction heating stress improvement (see Section 4.4.3.3.4) and other stress improving measures.

4.4.2.5 Mechanical measures

Distortion-compensating aligning (*counter-aligning*) prior to welding is possible to a certain extent in respect of angular shrinkage. It is possible, in this way, to counteract the tilting of the chord with successively welded double-sided longitudinal fillet welds by inclining the chord relative to the web prior to welding. Likewise, it is possible to proceed with single-sided or double-sided fillet welds in the case of corner joints. Plate strips joined with a longitudinal weld in V groove can be aligned in a roof shape to compensate for angular shrinkage. The distortion-reducing measures also include wedge-shaped aligning of the gap of the weld groove and control of this gap by a progressively withdrawn wedge at the end of the groove, which ensures that the welding groove is held open. This procedure can be applied on not excessively long welds with transversely movable groove edges. It is widespread practice in the case of gas welding of copper and aluminium, in this case replacing back-step welding, which is not practical here because of the high heat losses. The various counter-aligning measures are presented in Fig. 225.

Preshaping is employed as roof shaping of girder chords in order to compensate for their angular shrinkage. The heart-shaped kink, which is produced by angular shrinkage and directed inwards at the longitudinal welds (V groove open to the outside) of cylindrical tanks, is most simply countered by leaving the ends of the sheet plates plane when rounding the rest. It is possible to counter the "drawing-in" of patches, block flanges or nozzles inserted with ring weld into vessels or pipes by reverse predeformation (i.e. reverse bulging) of the cylindrical or spherical shell. Local preheating can act in the same direction. Pipe ends may be slightly bulged at the circumferential weld joint in order to obtain after

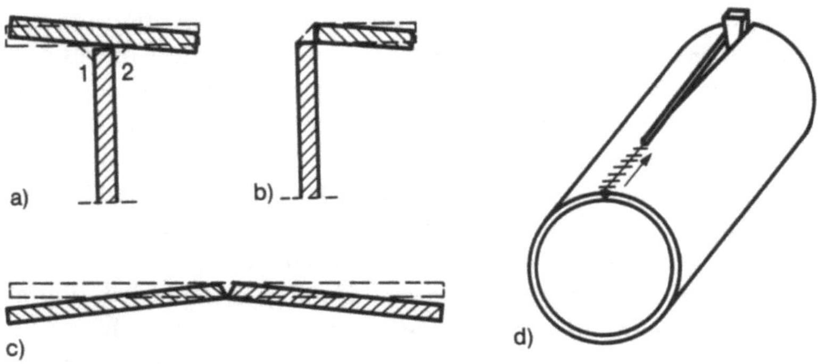

Fig. 225. Reduction of distortion by counter-aligning measures with girders (**a**, **b**) and plates (**c**), pre-welding position traced in solid lines, post-welding position in broken lines; fixing of groove gap by wedge in single-pass gas welding (**d**); after Malisius [5]

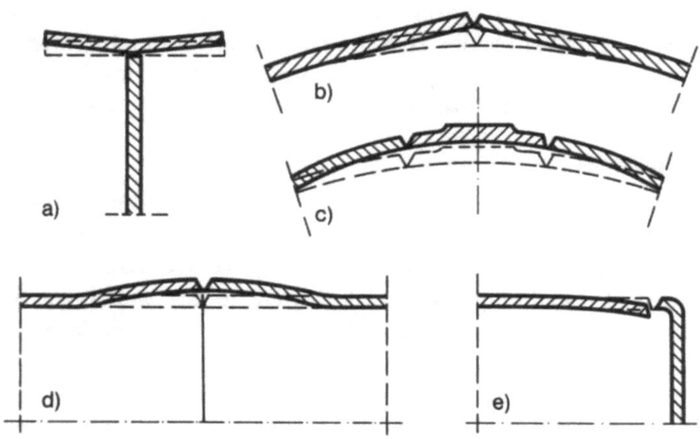

Fig. 226. Reduction of distortion by preshaping, pre-welding shape traced in soild lines, post-welding shape in broken lines: roof shaping of girder chords (**a**), plane end sections of cylindrical shells (**b**), outward bulging of spherical shell with block flange (**c**), outward bulging of pipe with circumferential weld (**d**), inward drawing of pipe at plane end (**e**); after Malisius [5] and Vinokurov [8]

welding the (internally streamlined) smooth shape without any kink to the inside. The weld misalignment to the outside in the case of circumferential welds between cylindrical shell and end plate can be countered by drawing-in the end of the shell. The various preshaping measures are presented in Fig. 226. They are employed for relatively thin-walled structures. In such cases, shrinkage occurs to a larger extent. On the other hand, the required forces for cold preshaping can still be produced with common workshop presses.

Elastic positive-contact *bracing* is employed for hollow girders with single-sided longitudinal welds in order to compensate for bending shrinkage (Fig. 227). The magnitude and direction of the bending deflection are selected so

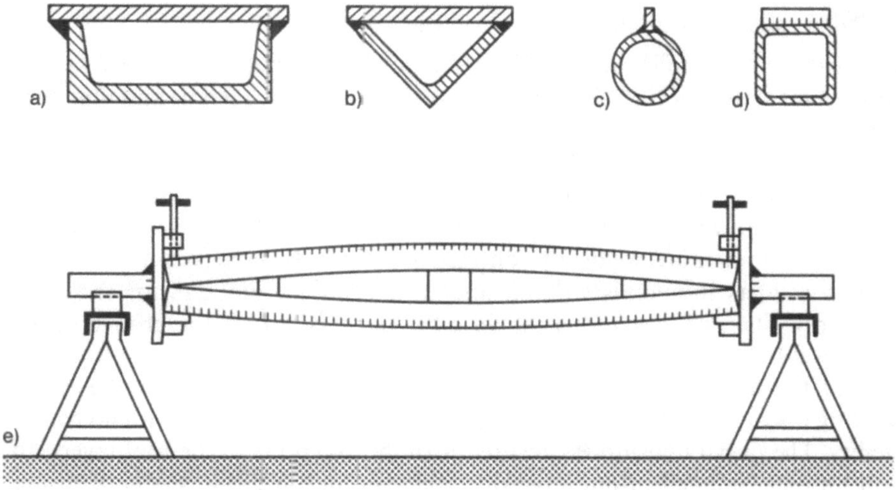

Fig. 227. Elastic bracing of hollow girders with single-side longitudinal welds (**a**, **b**, **c**) or transverse welds (**d**) during welding in rotary device (**e**); after Malisius [5]

that the girder springs back into the straight shape after welding. The weld is placed on the convex girder side. Bending with positive contact is necessary because the girder cross-section is "softened" during welding. The bending must be selected slightly larger than the likely bending shrinkage when the girder is not braced. Bracing can be achieved in a particularly economic manner by counter-bracing two girders in a rotary device for applying a suitable sequence of layers in the gravity position. Bracing on a curved surface is also employed for compensating angular shrinkage, e.g. in sheet metal plates with welded stiffening ribs (in shipbuilding). The heart-shaped kink to the inside at the longitudinal welds (V groove open to the outside) of cylindrical tanks, which is caused by angular shrinkage, is countered among other things by fixing measures taken at the groove. With overlapping or butt welded thin sheet metal plates, the purpose of bracing on a slightly curved surface is to avoid corrugation in the direction of the weld and a corresponding misalignment of the groove edges during welding.

Fixing the parts to be joined by welding in a frame as rigidly as possible is an effective countermeasure against angular and bending shrinkage. In many cases, the load from the weight of the connected parts is sufficient without any special fixing elements. However, it is only possible to reduce relatively slightly the back-spring shrinkage when the component is released (approximative analysis in Ref. [8]). The arching of thin sheet metal plates at the weld pool is minimized by fixing measures close to the groove. Fixing is a largely unusable practice in respect of transverse and longitudinal shrinkage. The very large forces which occur in this connection rapidly exceed the frictional locking limits of the fixing elements. The transverse displacement of the groove edges is minimized better by groove gap inserts than by fixing.

4.4.2.6 Typical applications

The first typical application considered according to Ref. [5] is the *I section girder* joint shown in Fig. 228. As the cross-section acts as a whole and thus very stiffly, even when welding short weld sections, the aim must be to lay the welds simultaneously as far as possible. First the web is joined, and then the chords. As the web cross-section is significantly smaller than the chord cross-section, the restraint on shrinkage exerted by the web on the chords is not very large. The top and bottom chord welds, by contrast, are deposited largely simultaneously by two welders (numerals without and with prime in Fig. 228 for the layer sequence). The two executions shown differ in the shape of the groove of the chord weld. Shape (a) with double-V groove requires a rotary device (rotation around the axis of the girder) to permit the gravity position for the entire welding operation. Shape (b) with U groove is intended for joints with no rotary facility. The gravity position is ensured with the exception of the root layers. In both cases, a rounded cutout is provided between web and chord weld to ensure defect-free welding over the entire width of the chord. The cutouts, however, have a negative effect on fatigue strength so that it may be advisable to groove out and weld the transition between web and chord. In compact I cross-sections, the chord weld may also be welded first (Fig. 220c). The compressive residual stresses which occur transverse to the chord weld as a result of this

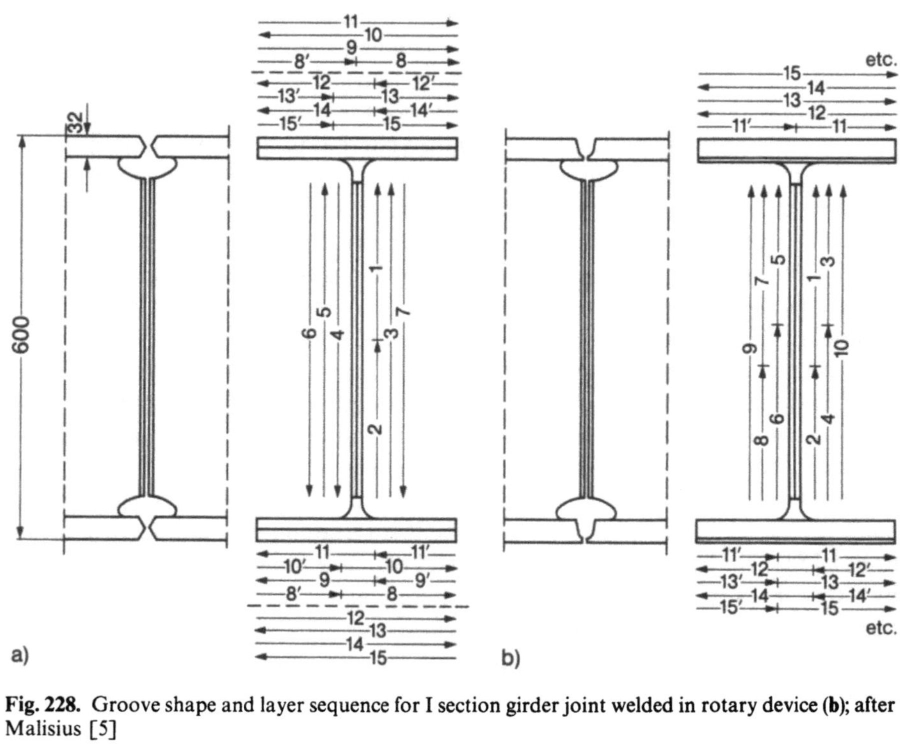

Fig. 228. Groove shape and layer sequence for I section girder joint welded in rotary device (**b**); after Malisius [5]

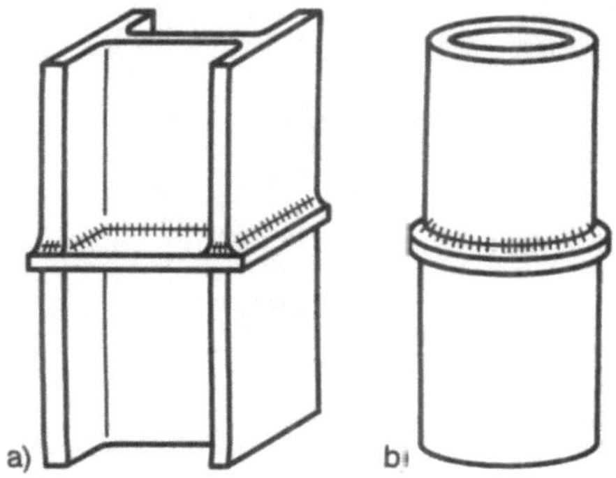

Fig. 229. Fillet welded joint with cross plate used in compression loaded I section girder (**a**) or tube (**b**); after Malisius [5]

increase its fatigue strength. Rolled I section girder joints subjected to static compressive load are better manufactured with a cross plate and fillet welds (Fig. 229). The fillet weld is easier to manufacture free of defects, and shrinkage and transverse residual stresses are less pronounced. It is not necessary to observe a particular welding sequence.

The second typical application considered according to Ref. [5] is the manufacture of large *plate panels* of tank bottoms, ship decks or ship bulkheads. The sheet metal plates including the weld grooves are cut to shape with a high degree of dimensional accuracy. The plates should be free of corrugations or dents. In rectangular bottoms or walls the plates are arranged so that the longitudinal welds run continuously parallel to the long edges whereas the transverse welds are arranged in a staggered pattern (see Fig. 220a) in order to avoid the crossing of the welds with the particularly unfavourable superimposition of tensile residual stresses. The shape and arrangement of the metal plates and the welding sequence for a tank bottom are shown in Fig. 230. The transverse seams are welded first. The outside plates are provided with radial weld grooves, which are initially half-filled and are not fully filled until the shell-to-bottom fillet welds have been executed. They are thicker than the interior plates in order to diminish warpage and to reinforce the lower edge of the shell. The flat bars placed under the butt welds also serve for reinforcement besides backing the melting bath. The longitudinal welds of slightly tapered bottoms are exclusively radial. They tend to form circumferential corrugations as a consequence of the radial residual stresses. This is countered with the flat bars placed underneath. The use of high-performance welding is recommended to keep the transverse and angular shrinkage to a relatively low level. In the case of the octagonal panel with the welding sequence shown in Fig. 231, each radial weld is completed before the next joint is tacked and welded in back-step. Unrestrained transverse shrinkage is thus possible. Where warpage may occur as a result of angular and longitudinal shrinkage, this can be avoided by hammering the weld before the next plate is joined by welding. Thin plates ($h \leq 3$ mm) tend to warp more easily than thick plates. Edge welds may produce dents in the interior, interior welds may produce corrugations on the edge side. It is widespread practice to brace thin plates on a slightly curved surface for

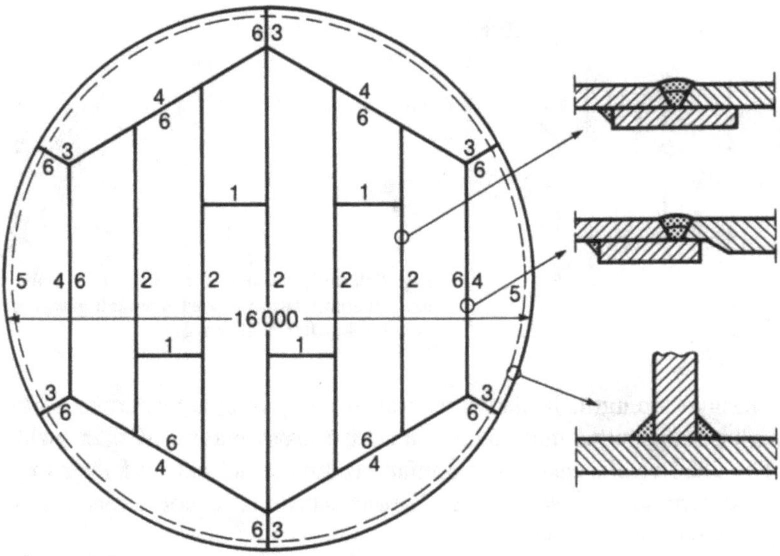

Fig. 230. Joint design and welding sequence for plate panel of tank bottom; after Malisius [5]

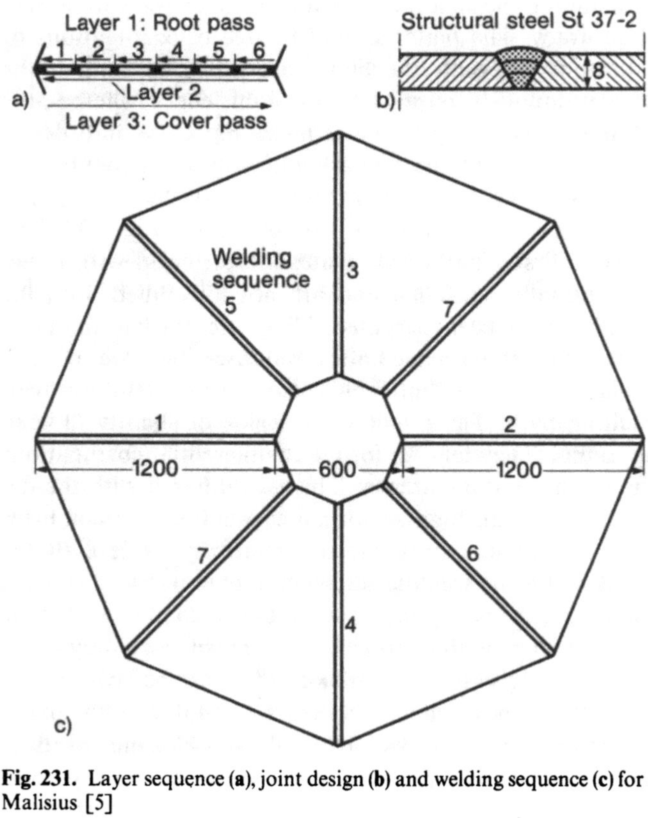

Fig. 231. Layer sequence (a), joint design (b) and welding sequence (c) for octagonal plate panel; after Malisius [5]

welding. Hammering after welding is advisable. Welding on a root-forming copper backing has proven satisfactory in practice.

The third typical application considered according to Ref. [5] is *patch welding* in which dimensionally precise cutting to shape and favourable welding sequence play an important role. Patches are inserted or placed overlapping in order to close cutouts which have arisen as assembly openings, as a consequence of design modifications or by removing defective material. Patches placed overlapping, which represent no problem in respect of distortion and residual stresses, have the drawback of a low fatigue strength, are prone to crevice corrosion and interrupt the smooth wall surface. Patches inserted flush, by contrast, require a well reflected welding sequence if residual stresses and distortion are to remain under control. The basic principle is to subdivide the circumference of the patch into two halves, which are finish-welded in turn, each

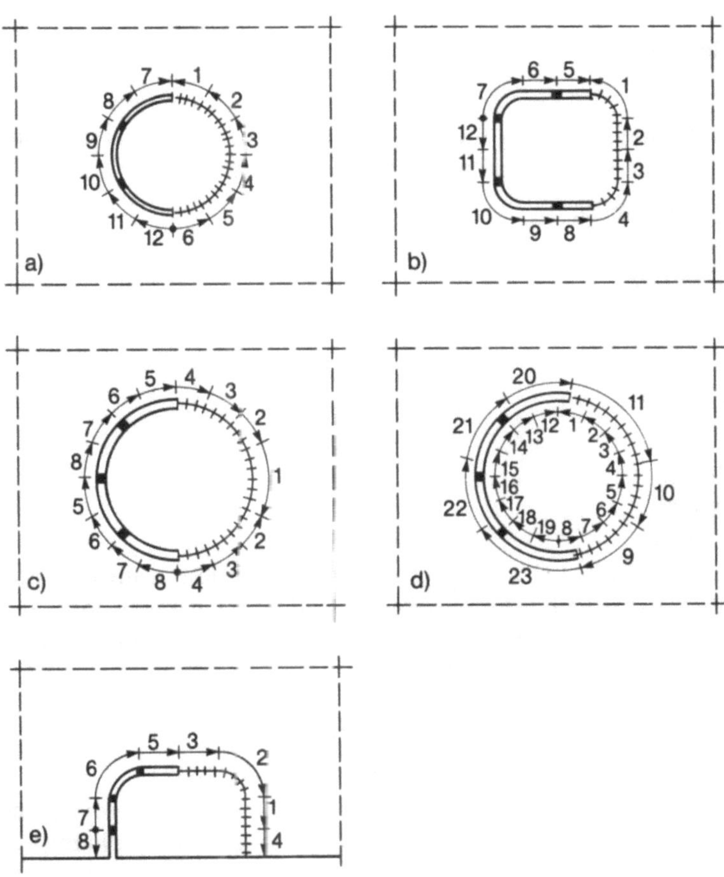

Fig. 232. Welding sequence for inserted patches: circular and square patch (**a, b**), circular patch with two welders simultaneously acting (**c**), circular patch with multi-layer weld (**d**) and rectangular edge patch (**e**), after Malisius [5]

in back-step. This assures relatively unrestrained transverse shrinkage of the first circumferential half without interfering twisting of the patch. The tack points on the opposite side should indicate the transverse shrinkage by tearing. Suitable welding sequences for circular and rectangular patches are presented in Fig. 232. Larger patches are best welded simultaneously by two welders (Fig. 232c). Thicker patches are welded in the first layer with a small back-step, in the subsequent layers with a large back-step in alternating direction (Fig. 232d). For double-V groove welds, one side of the circumferential half is completed first. Rectangular patches are well rounded at corners to enable continuous welding around. The slight bulging of the patch, which is occasionally employed, does not offer the aimed-at gain in compliance. Stretching the weld zone by hammering is a useful means of reducing residual stresses in a ductile material. Local annealing, by contrast, does not offer any major improvement.

4.4.3 Post-weld measures

4.4.3.1 Overview

Possible post-weld manufacturing measures aimed at reducing (macroscopic) tensile welding residual stresses and welding distortion are:

- annealing for stress relief,
- cold stretching,
- flame stress relieving,
- vibration stress relieving,
- hammering,
- rolling,
- spot compression,
- spot heating,
- hot straightening,
- cold straightening,
- flame straightening.

Reference is made to Ref. [15] in respect of the presentation of (macro- and micro-) residual stress reduction from a more metallographical viewpoint and not restricted to welding residual stresses. A stimulating overview of stress relieving measures is offered by Ref. [349].

4.4.3.2 Hot stress relieving (annealing for stress relief)

4.4.3.2.1 Hot stress relieving in practice and relevant codes

Welding residual stresses can be reduced while simultaneously improving the microstructure, and the risk of distortion and cracking can be diminished (stabilization) by heating the component to above recrystallization temperature [347–358, 392]. The elastic modulus and the yield limit are substantially reduced at this temperature and creep processes occur to a greater extent. The recrystallization temperature is approximately half the melting temperature, measured in degrees [K]. The tensile stress areas are stretched as a result of yielding during heating and the compressive stress areas upset accordingly. During subsequent annealing, stress relaxation occurs as a result of creep, with then only minor deformation and stress redistribution. Of importance for stress relief as a result of yielding is the fact that the yield limit decreases to a greater extent during heating than the elastic modulus, which occurs when the temperature approaches recrystallization temperature, so that the thermal strains are not directly transformed into elastic stress relief, for this purely thermoelastic process is reversed during cooling. Subject to this condition, the residual stresses are reduced not only for the period of annealing but also after cooling down.

Heating and cooling must proceed slowly in order to avoid major temperature differences between the surface and the interior of the component. Related thermal stresses may otherwise cause cracks and prevent the desired reduction in residual stresses as a result of new residual stresses being generated. The related distortion would also be prohibitive. A prolonged warming-through and annealing or holding time, in turn, is necessary to allow stress relaxation to progress sufficiently uniformly and completely. Stress relief annealing reduces not only the macroscopic residual stresses considered here, but also the microscopic residual stresses. The residual stresses in markedly heterogeneous materials (e.g. flame cut edges or cladded plates) cannot be reduced easily by annealing, see Section 3.4.4.

Post-weld heat treatment with stress relieving and microstructural recovery or recrystallization is performed on mild and low-alloy steels in the temperature range of $450 \leq T \leq 700\,°C$ with annealing times of $1\,h \leq t_a \leq 3\,h$. The heating rate is selected depending on plate or shell thickness h, e.g. 5 °C/min for $h = 10$ mm and 1 °C/min for $h = 50$ mm. The cooling rate should only be half as great. Alloyed austenitic steels may require higher temperatures (up to 1050 °C) combined with a shorter annealing time and rapid heating and cooling in order to produce sufficient relaxation and a favourable microstructure without crack formation. This particularly applies to high-temperature steels, which do not relax until the elevated temperature is reached and may embrittle beforehand as a result of precipitations. Precipitation hardening aluminium alloys (e.g. AlMgSi-type alloys) are first solution annealed at 500 °C and then aged at 150 °C.

Post-weld heat treatment can be performed on not too large components globally in the annealing furnace or, in the case of a component size in excess of this, locally in the weld area, with gas torches, heating mats or induction coils in addition to asbestos mats for heat insulation [350]. Large vessels can be annealed globally by global heat input combined with external insulation. Local annealing is also applied on smaller components following repair and site welding. The desired temperature profile at the component for global or local annealing is presented schematically in Fig. 233. Thermocouples or pyrometers are necessary for reliably monitoring the temperature. In multi-pass welding, the deeper passes are post-weld heat-treated by overwelding. The related temperature patterns are presented in Section 2.4.4.

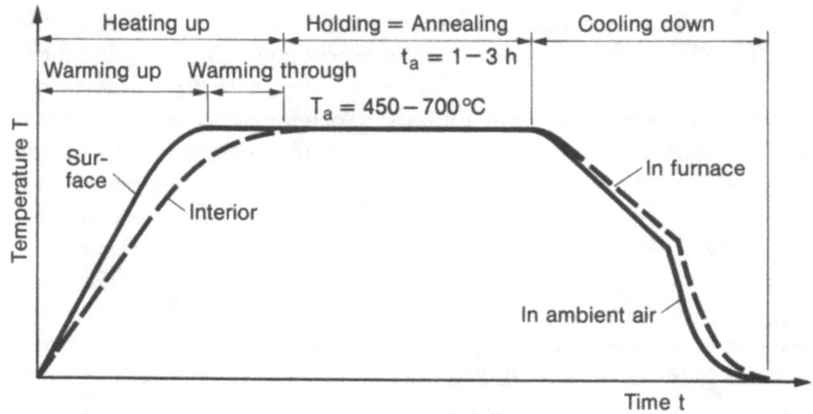

Fig. 233. Temperature profile over time of post-weld heat treatment, schematic representation

In the case of global post-weld heat treatment the residual stresses, the maximum values of which are at or above the yield limit, can be reduced under most favourable circumstances to approximately 15% of their initial value, i.e. approximately to half the hot yield limit at annealing temperature. This reduction is acceptable in practice especially as the permissible load stresses remain in some distance from the yield limit. The limited annealing time and annealing temperature, the triaxiality of the stress state, the inhomogeneous cooling and the elastic resilience are responsible for only limited stress relief. If a welded joint consists of different materials, this alone may result in considerable cooling stresses being again built up following annealing for stress relief (see Sections 3.4.4 and 4.4.3.2.6 concerning cladded components). If only the cold and hot yield limits in the weld and heat-affected zone of an otherwise homogeneous material differ (e.g. as a consequence of hardening), the cooling stresses, in contrast, are not increased.

With local post-weld heat treatment, it is by no means possible to reduce residual stresses to the favourable value of approximately 15%, stated above in respect of global annealing. What can be achieved in the individual case depends on the design, material and process parameters. Whereas large-area annealing is generally recommended, arguments are presented in Ref. [389] in favour of a locally concentrated heat input combined with simultaneous cooling of the surrounding areas.

To achieve large-area annealing, a minimum width $2w_a$ of the heated zone with adjoining gradual temperature drop should be obeserved, e.g. a multiple of the wall thickness h or, in the case of circumferential welds of pipes or vessels, a multiple of the square root on cylindrical shell radius R times shell thickness h [351, 356]:

$$2w_a \geqq 5\sqrt{Rh}. \qquad (225)$$

As is generally known, the circumferential weld is subjected to the post-weld risk

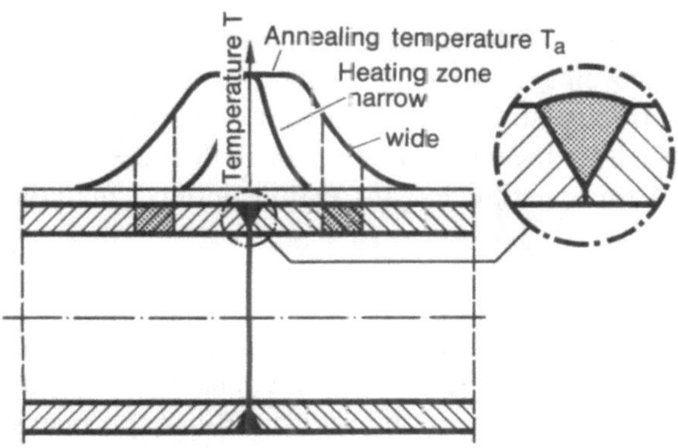

Fig. 234. Local post-weld annealing with narrow and wide heating zone, circumferential weld in pipe; after Vinokurov [8]

of bending stresses transverse to the root. Post-weld heat treatment of the narrow weld zone would only, after cooling, produce a similar stress state as welding itself. If, however, the post-weld heat treatment zone is selected wide, the high transverse bending stresses are shifted to an area in some distance from the weld where the temperature drop occurs (Fig. 234). In applying local post-weld heat treatment on multiply connected structures, heating and cooling should be performed as symmetrically as possible in order to avoid high and widespread cooling stresses.

The value of local post-weld heat treatment is thus based in a certain controllability of the residual stress state and in the improvement of the microstructure. Local post-weld heat treatment is performed to reduce the risk of brittle fracture in the weld zone. It is less suitable for improving dimensional stability because the residual stresses are, in most cases, only relocated.

In view of the practical significance of global and local hot stress relieving, mandatory relevant data are presented in codes. In the German codes, for example, they are found in the Steel-Iron Material Sheets of the VDEh and of the VdTÜV, in the DIN Steel Standards, in AD Instruction Sheets, in the Technical Rules for Boilers, Pressure Vessels and Piping, in the DVS Instruction Sheet, in the Regulations of the Germanischer Lloyd for ships and in the KTA Rules for Nuclear Reactor Components [347, 351]. These codes specify, depending on material, material composition, plate or shell thickness and service temperature, in which cases post-weld heat treatment is necessary and provide data relating to annealing time and annealing temperature, heating and cooling rate as well as minimum annealing and insulation width. The annealing temperature in the German regulations ranges from 530 to 800 °C, the annealing time being at least 15 to 60 minutes. Heating and cooling rates are prescribed dependent on plate or shell thickness. It is unjustified, however, to make the annealing time (after deducting the warming-through time) dependent on the plate or shell thickness, as is the case in many non-German regulations (e.g. in the ASME Code).

Inspection and supervision authorities often see themselves confronted with the question of when it is possible to do without post-weld heat treatment in borderline cases. An assessment table [354] has been proposed for the requirement of vessel construction according to which the following circumstances favour forgoing post-weld heat treatment: Carbon content or carbon equivalent low; primary parts normalized by annealing; plate or shell thickness small; structure flexible; welding area preheated; defect inspection optimal; material particularly ductile.

4.4.3.2.2 Stress relaxation tests

More precise insights into the effect of stress relieving depending on material, initial stress, annealing temperature and annealing time (identical with holding time) can be gained on the basis of stress relaxation tests on rod-like specimens. The particular type of loading applied to the specimens (tension, torsion or bending) requires a method for transferring the test results to different loading conditions in the welded structure. If rough assessments are not adequate, it is necessary to perform a detailed analysis in accordance with Section 4.4.3.2.6.

In the simplest case, the stress relaxation test is performed isothermally at the annealing temperature envisaged for the structure as a long-time test with uniaxially loaded tensile specimens [359]. Different stress relaxation curves over (annealing) time are obtained for different initial stresses while keeping the total strain fixed (Fig. 235). The stress drop designates the conversion of elastic into viscoplastic strain.

The isothermal long-time test described above is not fully satisfactory for assessing hot stress relieving in so far as the initial stress is not applied until after the annealing temperature is reached, whereas, in the structure, the initial stress already drops during heating. Measuring the stress reduction during heating is, however, difficult insofar as the thermal strain, which occurs simultaneously,

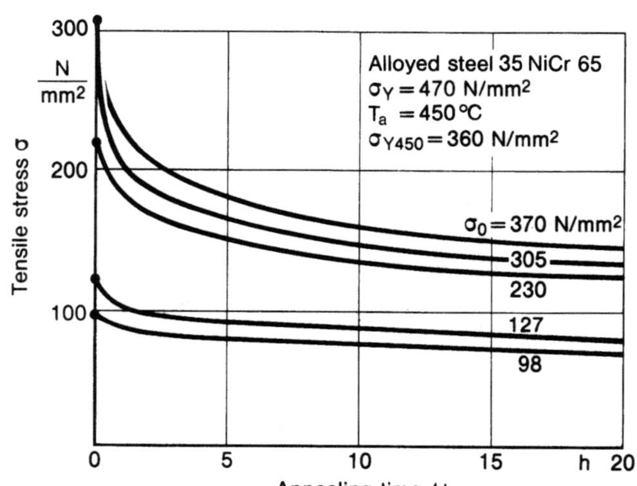

Fig. 235. Stress relaxation curves in isothermal long-time tensile test, NiCr-alloy steel, annealing temperature T_a, different initial stresses σ_0; after Mailänder [359]

must be eliminated. The anisothermal types of test described below overcome this difficulty, each in a characteristic manner, whereas the test results have been found to harmonize with each other [348, 364].

In the anisothermal tensile relaxation test [361, 365–367, 385], the elastic-viscoplastic total strain in the loaded test specimen is maintained at a constant level by means of an extensometer and servo mechanism, proceeding from an unloaded compensator specimen applied in parallel (Fig. 236). The thermal strain displacement of the compensator specimen controls the rotation of the spindle at the crosshead which generates the load. It is necessary to set the temperature at the specimens with an accuracy of ± 1 °C in order to achieve useful results even after severe relaxation. The advantages of this test are the homogeneous distribution of stress in the specimen and the single-valued determination of stress and strain.

In the anisothermal torsion relaxation test [358, 360, 363], the torsional shear stress which is unaffected by the thermal strain serves as the test basis. The shear stress is generated in a torsion rod (normally a solid rod; in special cases

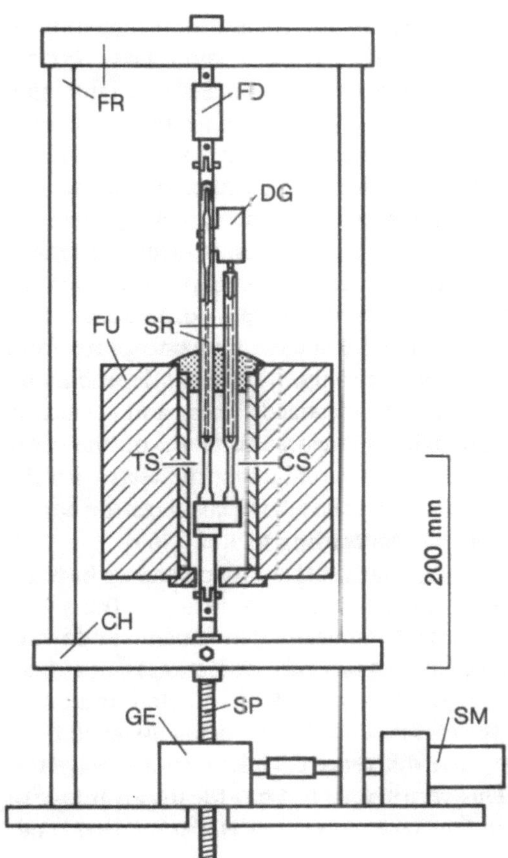

Fig. 236. Anisothermal tensile relaxation test; tensile specimen TS, compensating specimen CS, furnace FU, silicon rods SR, displacement gauge DG (extensometer), force measuring device FD, frame FR, cross head CH, servo-motor SM, gearing GE, spindle SP; after Murry [367]

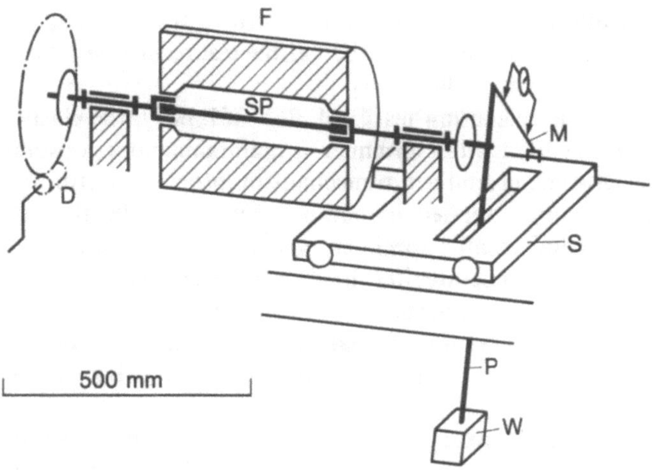

Fig. 237. Anisothermal torsion relaxation test: specimen SP, furnace F, mechanical drive D, pendulum P, weight W, measuring wire M, slide S; after Wellinger [360]

also a hollow rod), the thermal expansion manifests itself as (unrestrained) longitudinal displacement (Fig. 237). The mechanical drive produces a certain twist of the initially cold specimen, the maximum shear stress τ_{max} remaining below the shear yield limit τ_Y, i.e. $\tau_{max} \approx 0.75\ \tau_Y$. This twist is, in terms of forces, in equilibrium with the swing of a weighted pendulum. The swing, once set, is maintained constant during the test (heating, annealing and cooling phase) by a measuring wire, the tension of which is a measure for the unloading moment. The specimen end at the pendulum side is mounted on a slide, which can move longitudinally, to permit unrestrained thermal strain. The expenditure for the torsion test presented is low provided solid rods are used. A drawback is that the torsional shear stress cannot be calculated sufficiently unambiguous from the torsion moment. The values which result are slightly different depending on the assumed radial stress distribution (between elastically linear and plastically constant). Hollow rods with the unambiguous stress calculation, on the other hand, considerably increase the expenditure for conducting the tests with the result that it is then preferable to remain with tensile tests. The relaxation curves of solid and hollow rods differ only slightly according to Ref. [360].

In the (outdated) anisothermal bending relaxation test, two bending bars are mutually braced by a certain amount of deflection generating a transverse bending force and subjected in this state to annealing treatment. Following annealing and cooling, the remaining elastic spring-back deflection is measured. Bending stresses are calculated from the deflections before and after annealing, the distribution of the stresses over the cross-section height being assumed to be linear (as in the purely elastic range). The difference of these stresses designates the stresses relieved by annealing. This very simple test has the disadvantage of the lack of a measuring facility during the test and of the inaccurate determination of the stresses.

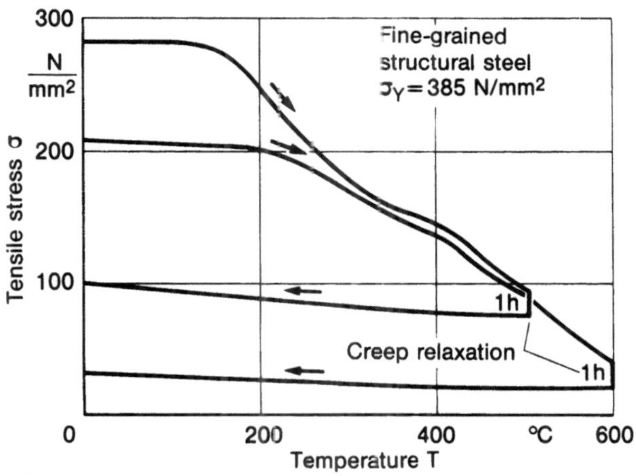

Fig. 238. Anisothermal tensile stress relaxation of fine-grained structural steel; after Ritter and McPherson [361]

Two relaxation curves of a fine-grained structural steel obtained in the anisothermal tensile relaxation test are shown in Fig. 238. Proceeding from the (lower or higher) initial stress, heating is performed to a (lower or higher) annealing temperature; this is maintained for one hour, then recooling is performed. The initial stress is reduced slightly at first with increasing temperature (corresponding to the temperature dependence of the elastic modulus), then more strongly after the high-temperature yield limit is reached (corresponding to the temperature dependence of the yield limit). The stress reduction can be intensified by microstructural transformation processes [367]. The reduction without yield of the initial stress σ_0 to the final stress σ_{0T} according to the elastic modulus E_{20} at 20 °C and E_T at T [°C] follows from the equation:

$$\sigma_{0T} = \sigma_0 \frac{E_T}{E_{20}} . \tag{226}$$

In this connection, a minor deduction for long time relaxation requires to be taken into account in E_T from approximately 200 °C onward if E_T has been determined in a short-time test. Following the stress decrease according to elastic modulus and yield limit the viscoplastic stress decrease takes place, which is relatively small in the considered case of low annealing temperature and annealing time. The rise in stress during recooling is likewise relatively small in Fig. 238 and does not fully correspond to the temperature dependence of the elastic modulus during heating. Too high a recooling stress, on the other hand, is measured in Ref. [366]. The same temperature dependence of the elastic modulus during heating and cooling is required according to traditional concepts.

Relaxation curves determined in the anisothermal torsion relaxation test according to Ref. [348, 362] for different steels are shown in Figs. 239 and 240 (see also Ref. [360, 361]). The increase of stress reduction with annealing temperature and annealing time is shown in Fig. 239. Individual curves from

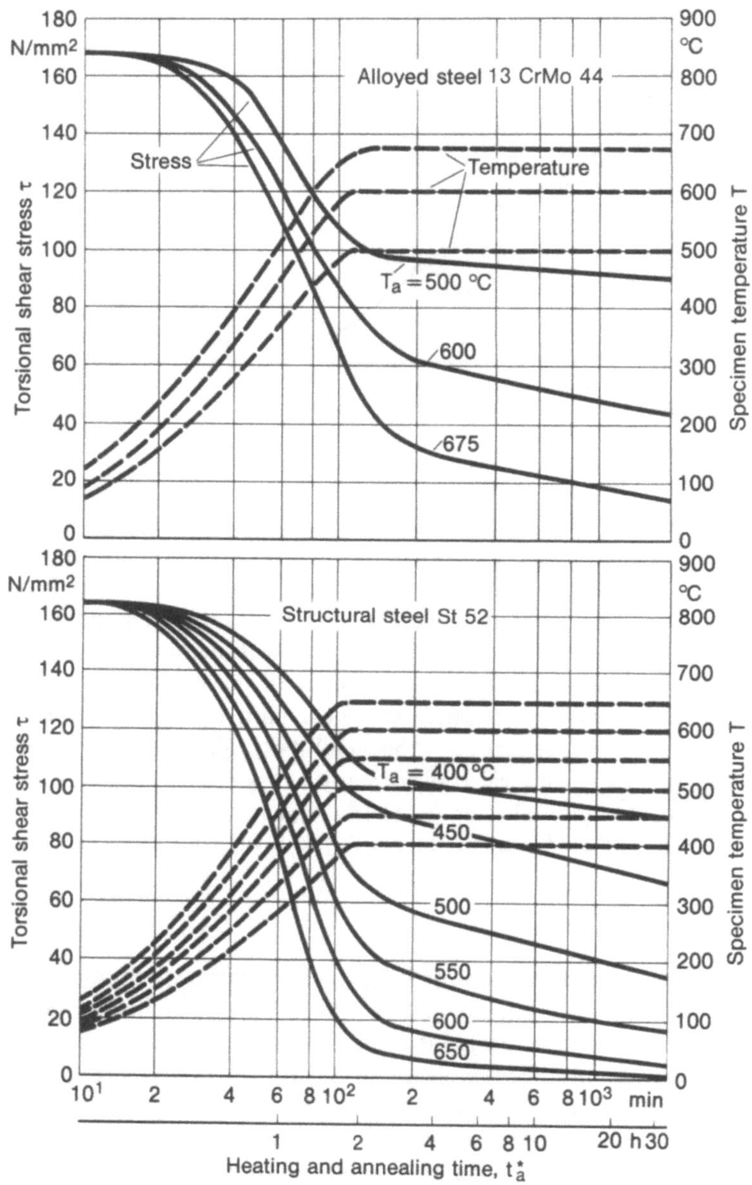

Fig. 239. Stress relaxation in torsion loaded circular rods with fixed twisting angle subjected to a temperature profile over time typical for post-weld heat treatment; after Lange [362]

Ref. [348] for heating and annealing starting from different initial stresses in some cases are presented in Fig. 240. Different steels of the same steel group may display different relaxation behaviour. The conversion of the offset yield limits from $\sigma_{0.2}$ into $\tau_{0.4}$ is of interest for assessing the level of the initial stress relative

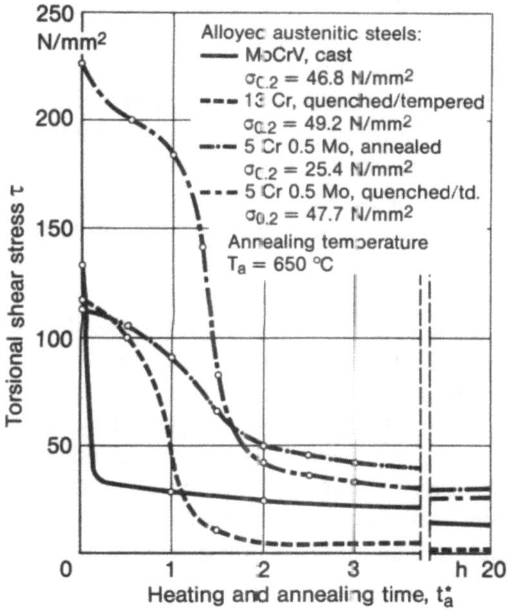

Fig. 240. Stress relaxation in torsion loaded circular rods with fixed partly different twisting angles subjected to heating and annealing; after Tummers [348]

to the yield limit. The following formula is used in this case:

$$\tau_{0.4} = 0.59\,\sigma_{0.2}\,. \tag{227}$$

A further family of relaxation curves obtained in the torsion relaxation test with a structural steel, which was used in the numerical simulation of stress relieving, is shown in Fig. 243a.

Figures 238, 239, 240 and 243 show that, during heating, a reduction in stress occurs principally as a result of the reduction of elastic modulus and yield limit and, during annealing, then mainly as a result of creep. Complete reduction in stress is not possible as annealing time and annealing temperature are limited. After an annealing time of 1 h at an annealing temperature of 600 °C, though, the stress of common steels is already greatly reduced.

4.4.3.2.3 Microstructural change during hot stress relieving

The process described is termed hot stress relieving in distinction to cold stress relieving, which is dealt with in Section 4.4.3.3. Its common designation is "stress relief annealing" (in former times also incorrectly termed "stress-free annealing"). The designation is selected too narrowly insofar as hot stress relieving is linked to microstructural recovery (at $T \approx 0.5T_{\mathrm{m}}$) or even microstructural reconstitution (at $T > 0.5T_{\mathrm{m}}$), and this effect may be more important than that of stress relieving. In the English language, consequently, the term used is the more comprehensive one of "post-weld heat treatment". In the German language, the name "recrystallization annealing" is also used, if there is a one-sided interest in microstructural recovery. In reality, recrystallization annealing is always linked to stress relieving. For the reasons stated, stress relief annealing is considered here from the general aspect of post-weld heat treatment.

The purpose of post-weld heat treatment is to reduce macroscopic residual stresses and to change the microstructure in a favourable direction including reduction of microscopic residual stresses, hardness, strain ageing and hydrogen content of the weld (in the case of ferritic steel). As a consequence of the reduction of (tensile) residual stresses, of hardness, ageing and hydrogen content, the brittle fracture strength, the creep fracture strength and the resistance against (stress) corrosion cracking are enhanced. At the same time, the shape stability of the welded component is assured during machining and when subjected to service loads. As a result of the microstructural change, the ductility of mild and low-alloy steels can be recovered after cold forming with strain ageing; for example, at cracks and defects in the heat-affected zone or at folded edges and other cold-formed areas. The microstructural change of higher-alloy steels applies to the recovery in ductility in the hardened zone or the reduction of hardness peaks. Coarse grain in electroslag welding is eliminated by (normalizing) heat treatment. The desired improvement of the microstructure is often ranked higher in respect of brittle fracture than the reduction in residual stresses obtained [353].

But there are also limits to the positive effects of post-weld heat treatment and, if ill-considered use is made of the process, this results in a deterioration of the welded joint in place of the desired improvement. The reduction in residual stresses may only be slight in the case of triaxial tensile stresses or restrained cooling distortion. Too high an annealing temperature or too long an annealing time, on the other hand, impairs the mechanical material characteristic values, particularly those of low-alloy CMn steels. Tensile strength and yield limit of C and CMn steels are decreased by up to 10%; the transition temperature in the ISO-V-notch impact test is increased by up to 30 °C [369]. In higher-alloy steels, particularly high-temperature steels containing molybdenum and vanadium, annealing may result in precipitation in the heat-affected zone, which leads to a loss in ductility of the grains and shifts the deformation into the grain boundaries with the risk of intercrystalline cracking (relaxation or creep embrittlement, reheat or stress relief cracking).

4.4.3.2.4 Equivalence of annealing temperature and annealing time

In practice, the possible annealing time is limited for reasons of economy, and the annealing temperature, on the other hand, must also not be too high because of the negative effects already stated. As regards the reduction in residual stresses, annealing time and annealing temperature are equivalent quantities within the range of annealing temperatures recommended in Ref. [106, 110, 111]. At a lower temperature, it is necessary to anneal for a significantly longer period than at a higher temperature in order to achieve comparable stress relieving. It is irrelevant in this connection whether annealing is performed in a single operation or in several stages; it is the total annealing time which is the decisive factor [366].

The curves presented in Fig. 241 according to Ref. [355] apply to tensile specimens made of CMn-alloyed steel ($\sigma_{0.1} = 215\ \mathrm{N/mm^2}$, $\sigma_U = 426\ \mathrm{N/mm^2}$), which, after being heated to temperatures between 500 and 650 °C, were subjected to a fixed strain of 0.15% and then given over to relaxation. The initial stress ranged from 146 N/mm² at 500 °C and 43 N/mm² at 650 °C. It was assumed that the stress achieved in the hot state after several hours of relaxation is largely independent of the choice of the fixed strain or initial stress level. The stress achieved after relaxation in the hot state was calculated back to the residual stress in the cold state on the basis of the temperature dependence of the elastic modulus. From the diagram, it is clear that stress relieving, e.g. to 38 N/mm², can be achieved alternatively by 600 °C for 1 h, 575 °C for 4.5 h or by 550 °C for 22 h. The required annealing time thus rises progressively with a dropping annealing temperature and is uneconomically high below 575 °C. It should be noted, however, that heating temperatures around 500 °C in the structure subjected to residual stresses have a marked stress relieving effect alone as a result of the reduced yield limit.

As a further numerical example, the equivalence of 575 °C for 3 h and 625 °C without duration applying to the CMn steel in Ref. [380] is mentioned.

The equivalence of annealing temperature T_a and annealing time t_a in respect of the stress relieving effect is also reflected by the Holloman-Jaffe index

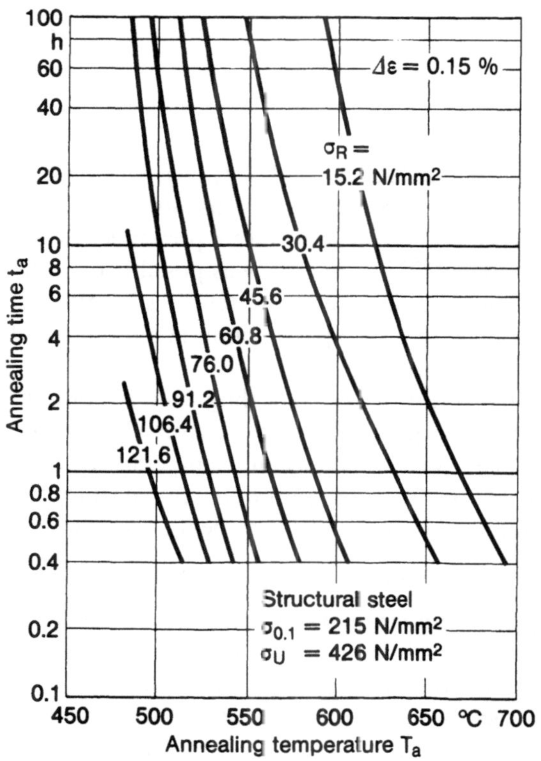

Fig. 241. Equivalence of annealing time and annealing temperature in respect of stress relaxation down to different residual stresses σ_R with a fixed strain $\Delta\varepsilon$ of 0.15%; after Watson [355]

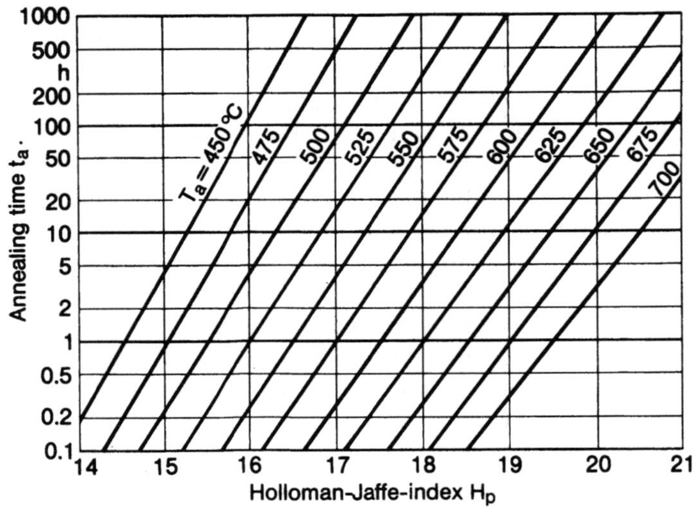

Fig. 242. Equivalence of annealing time and annealing temperature in accordance to the Holloman-Jaffe-index H_p

H_p [369, 371], considering in t_a an allowance both for the heating and cooling phase according to Eriksson (see Ref. [372]), Fig. 242.

$$H_p = T_a(20 + \log t_a)10^3, \tag{228}$$

$$t_a = \frac{T_a}{2.3\, dT/dt(20 - \log dT/dt)}. \tag{229}$$

The Holloman-Jaffe index H_p (with dT/dt [K/h], T_a [K], t_a [h]) was used in Ref. [371] for describing the change in hardness as a consequence of heat treatment for a group of carbon steels. Later, test results were presented for hot stress relieving dependent on this index [372]. Finally, it was possible to verify the correlation with the reduction in yield limit, ultimate tensile strength and notch impact toughness caused by annealing [353]. Reduction in hardness and residual stress have comparable significance in practice.

The equivalence of annealing temperature and annealing time is not confirmed in respect of brittle fracture of large plates with transversely notched longitudinal weld in Ref. [324]. The level of annealing temperature alone determined the reduction in the brittle fracture transition temperature; the reduction in hardness, not the reduction in residual stress, being decisive.

4.4.3.2.5 Creep laws and creep theories relating to hot stress relieving

The term "creep" designates the permanent strain which increases with time for fixed stress or to the permanent strain which increases with time at the expense of elastic strain for fixed (total) strain (stress relaxation). Creep is thus a kind of viscosity of solid bodies. Creep in metals occurs to a greater extent only above the recrystallization temperature. Creep (primarily at high temperature) can be regarded as a process with time-dependent permanent strain in contrast to yielding (primarily at low temperature) as a process with time-independent permanent strain. Whereas, in creep, the viscous strain rate is controlled by the shear stress, in yield, the plastic strain is directly dependent on the shear stress. The momentary creep or yield process can be influenced by the entire strain prehistory (material memory). This is neglected in simplified creep theories, in which creep is made dependent solely on the momentary state variables.

The close connection between creep and yield is accentuated if creep occurs only after the yield limit has been exceeded, creep thus not being possible below the yield limit (untypical for hot stress relieving). The related deformations are designated as viscoplastic. Whereas the plastic deformations occur instantaneously, viscous deformations need time to take place. Plastic deformations predominate at the start of heating for hot stress relieving, whereas only viscous deformations occur during actual annealing [379]. Plastic and viscous strains can be kept separate in the creep theories, with yield laws on the one hand and creep laws on the other, but they can also be united to viscoplastic strains, with creep-yield condition, creep-yield law and creep-yield hardening law. The temperature-dependent form of these laws is required for modelling hot stress relieving processes.

The creep laws and creep theories were developed originally for designing and dimensioning components made from metal (operated at high temperature), concrete and plastic, which are at risk of creep deformation and creep fracture [373–377]. The prime consideration, therefore, was the material behaviour at a given (constant or variable) stress. The starting basis was the creep curves $\varepsilon_c = f(t, \sigma)$ determined dependent on time for differently fixed stresses σ. More suitable for application to hot stress relieving, however, are stress relaxation curves $\sigma = f(t, \varepsilon)$ recorded for differently fixed strains ε [1].

Creep laws for uniaxial stress σ are applied in combination with the rod element model to stress relieving processes [383]. What applies in the simplest case with a constant uniaxial stress is:

$$\varepsilon_c = f(\sigma)\, g(t)\, h(T) \tag{230}$$

or alternatively:

$$\dot{\varepsilon}_c = f(\sigma)\, \dot{g}(t)\, h(T). \tag{231}$$

The stress-, time- and temperature-dependent functions in equations (230) and (231) are defined according to Norton:

$$f(\sigma) = B\sigma^n, \tag{232}$$

according to Bailey:

$$g(t) = Ft^m \qquad (1/3 \leqq m \leqq 1/2), \tag{233}$$

according to McVetty:

$$g(t) = G(1 - e^{-qt}) + Dt \tag{234}$$

and according to Arrhenius:

$$h(T) = Ce^{-\Delta H/RT} \tag{235}$$

introducing the material characteristic values *B*, *F*, *G*, *D*, *C*, *n*, *m*, *q*, the creep activation energy ΔH and the gas constant *R*. The constants *B*, *F* and *C* when inserted in equation (230) are also condensed to a further constant *K*:

$$K = BFC. \tag{236}$$

For the high-tensile steel HT80 it was found $n = 2.4$–3.0 and $m = 0.4$–0.5 at $500 \leq T \leq 600\,°C$ [385]. For the chromium steel ANS5616 used in aeroengines it was found that $K = 1.29 \times 10^{-14}$, $n = 2.4$ and $m = 0.31$ at $500 \leq T \leq 560\,°C$ [389]. The material characteristic values of other materials are reviewed in Ref. [378].

The temperature influence is usually not separated according to equation (230) but is introduced as temperature dependence of the material characteristic values. In respect of variable uniaxial stress σ, the question arises how the transition takes place between the (different) creep curves (which depend on σ), with t being fixed (if time-hardening) with ε_c being fixed (if strain-hardening) or according to another hypothesis.

Creep laws for multiaxial stress are required when applying the finite element method to hot stress relieving processes. The creep law most often used is the (yield-independent) creep law after Norton:

$$\dot{\varepsilon}_{c\,ij} = \frac{3}{2} B\sigma_{eq}^{n-1}\sigma_{d\,ij}. \tag{237}$$

In this power function law, the creep rate $\dot{\varepsilon}_c = d\varepsilon_c/dt$ is set dependent on the deviatoric stress σ_d [154, 194, 379, 380, 386] (von Mises equivalent stress σ_{eq}, material characteristic values *B* and *n*). A corresponding exponential function law (used for structural steel at $T \leq 550\,°C$ [117], material characteristic values *A*, *a*) reads:

$$\dot{\varepsilon}_{c\,ij} = \frac{3}{2\sigma_{eq}} Ae^{a\sigma_{eq}}\sigma_{d\,ij}. \tag{238}$$

The uniaxial version of equation (237) is identical to equation (232), and that of equation (238) reads:

$$\dot{\varepsilon}_c = Ae^{a\sigma}. \tag{239}$$

A complete multiaxial creep-yield law on the basis of the Bingham model is used in Ref. [248].

From the yield and creep laws it follows that stress relieving is controlled by the deviatoric stress portion and therefore reduces directly deviatoric stresses only. The volumetric stresses are reduced only indirectly and to a considerably lesser extent as a consequence of the stress redistribution occurring in parallel to the reduction of the deviatoric stresses. The volumetric stresses occur as purely triaxial tensile stresses only in inhomogeneous stress fields, particularly ahead of crack tips, for instance, but also in the interior of electroslag welds. Relieving the global deviatoric stress state, therefore, also means relieving the local volumetric stress state. This fact is confirmed by the analysis results in Ref. [8] for the triaxial tensile stress state in the interior of an electroslag weld (Fig. 243). The triaxial volumetric stresses in the interior are, after finishing the (long-time) annealing treatment, higher by the factor 2.0 to 2.5 than the predominantly deviatoric uniaxial stresses on the surface of the weld. A direct reduction of the volumetric stresses is admittedly not possible according to the yield and creep laws of the macroscopic theories mentioned, but is, nevertheless, conceivable to some extent in the realm of microstructural reality. For the present, no basic investigations into this question exist.

From the considerations of the paragraph above, it can also be concluded that stress relieving is simulated only in a rough approximative manner when fixing the local total strains (and thus suppressing global deformation), as is occasionally proposed for analysis. The final aim of hot stress relieving, namely the reduction of the local triaxial tensile stresses, is achieved only through

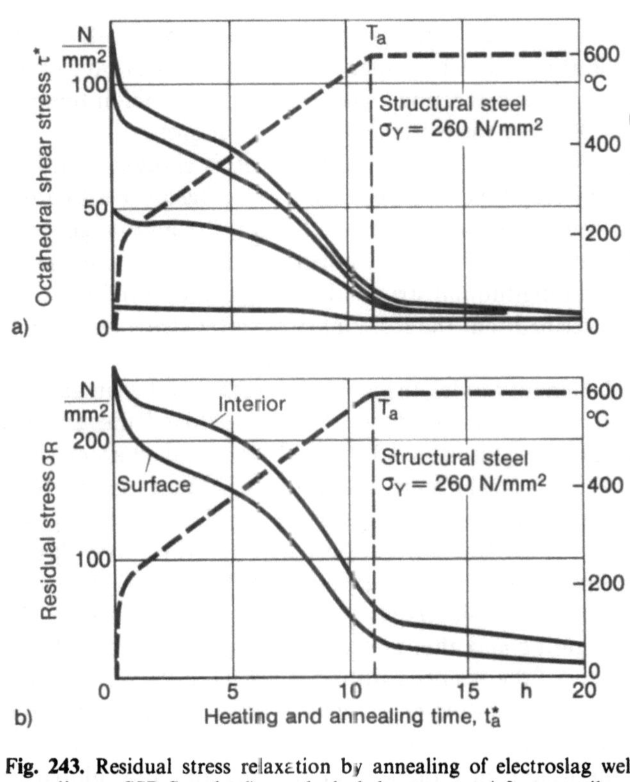

Fig. 243. Residual stress relaxation by annealing of electroslag weld in structural steel (no. 30 according to SSR Standard); octahedral shear stress τ^* from tensile test ($\sigma = 2.12\tau^*$), experimental result (**a**), uniaxial surface stress (transverse to weld) and triaxial interior stress in weld, analytical result (**b**); after Vinokurov [8]

a stress redistribution (connected with local and global deformation). A corresponding statement applies to the constancy observed in Ref. [386] of the ratio of the local stress components which must remain on the yield surface during stress relieving.

4.4.3.2.6 Analysis examples and experimental results relating to hot stress relieving

The analysis of the thermal and viscoplastic processes during hot stress relieving is performed using finite elements and time-incrementally linearized algorithms [190, 373]. The starting basis, as for welding, is the temperature field. Compared to the better known plastic analysis with only formally time-dependent load increments, an essential element of viscoplastic analysis is the computational control via real time steps. Modelling follows the possibilities presented for welding processes applying rod elements, ring elements, plate and shell elements. Calculations of this type can assist in verifying specifications regarding annealing temperature and annealing time in practice.

The rod element model [383] is used to observe the reduction of weld longitudinal residual stresses in austenitic stainless steel 304 and in ferritic high-strength steel HY130. The heating and cooling phase is simulated as purely (elastic-)plastic, the annealing phase as purely viscous. An exponentially time-dependent uniaxial creep law is used. After annealing for several hours at 593 °C, only moderate stress relieving is determined numerically for the high-temperature creep-resistant steel 304 (the measurement indicates more).

Elastic-plastic hot stress relieving of a cylindrical shell is simulated in Ref. [388] by means of layered ring-type shell elements. The basic investigation relates only indirectly to welding residual stresses. In fact, the inhomogeneous residual stress state is considered in a cylindrical shell of finite length, which had been stressed by removing a longitudinal strip and joining the longitudinal shell edges together again. It is verified that the residual stress after stress relieving is 10 to 20% higher than the yield limit at the applied maximum temperature.

With a ring-type solid element model, the residual stresses in the single- and multi-pass circumferential weld of a pipe made of micro-alloyed fine-grained CMn steel were determined by Josefson [379–381] (material characteristic values similar to Fig. 114, $\sigma_Y = 360$ N/mm^2, rising to 640 N/mm^2 in the 50% martensitic part of the heat-affected zone). The processes during welding are simulated as (elastic-)plastic with transformation ($\Delta t_{8/5} = 13$ s), those during annealing as (elastic-)viscoplastic (with creep law after Norton [375]). The weld profile and finite element model for single-pass welding are shown in Fig. 244; the residual stresses after welding and after annealing in Fig. 245. A remarkable fact is the rough consistency of the results with those in Ref. [226, 227] (initial strain model) in the transformation-free cross-section areas. Weld profile and symmetrized finite element model for multi-pass welding are shown in Fig. 246. The residual stresses in the weld centre plane after welding and after global or local annealing are plotted in Fig. 247. The residual stress distribution is fundamentally changed compared to single-pass welding. High tensile residual

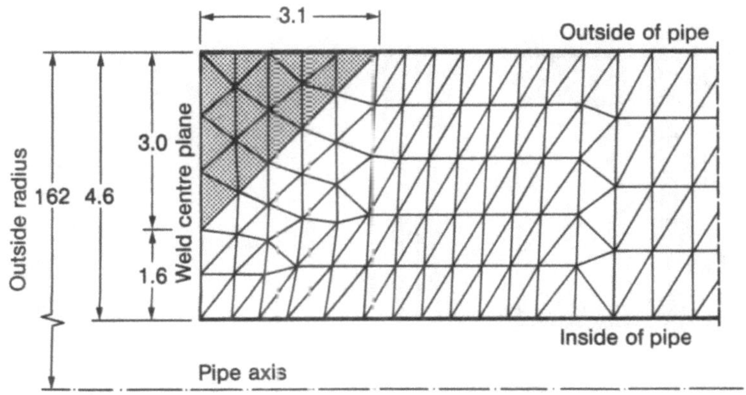

Fig. 244. Finite element cross-section model of single-pass circumferential weld in pipe made of CMn-alloyed steel; after Josefson [379]

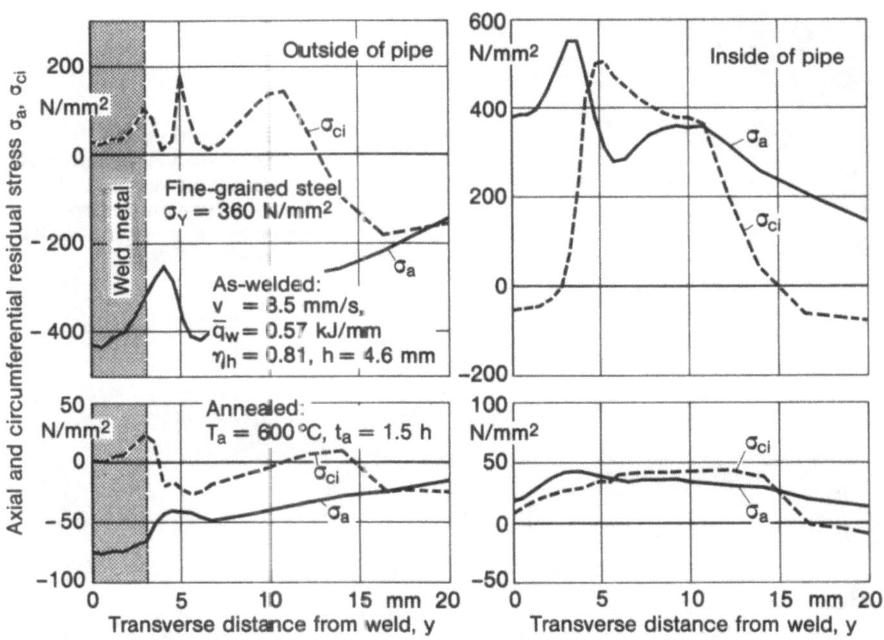

Fig. 245. Axial and circumferential residual stresses, σ_a and σ_{ci}, in outer and inner surface of pipe made of CMn-alloyed steel, as-welded and after annealing, analytical results; after Josefson [380]

stresses occur in the intermediate layers. They are significantly reduced after global annealing (annealing temperature 575 °C; annealing time 2 h). They are less significantly reduced after local annealing on a width of $w_a = 50$ mm on both sides of the weld (maximum annealing temperature on the outer surface of the weld in the weld centre plane 600 °C; annealing time 0.5 h). A remarkable

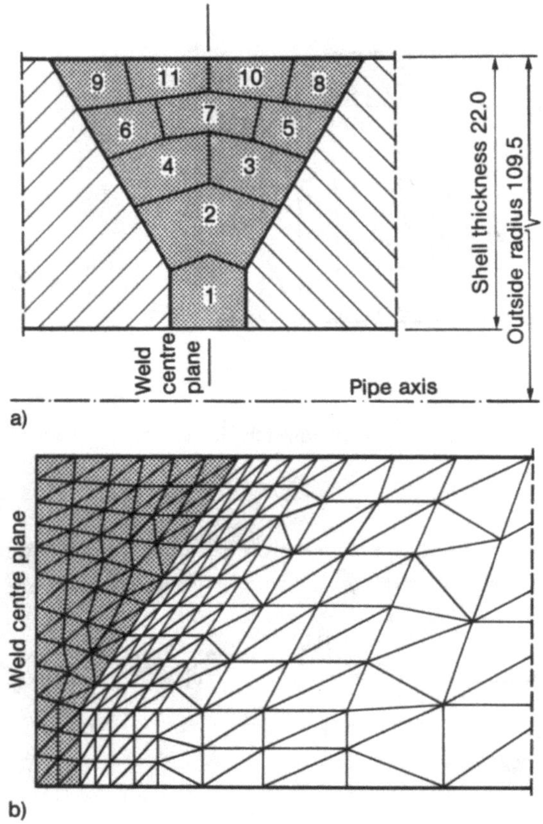

Fig. 246. Multi-pass circumferential weld in pipe made of CMn-alloyed steel (**a**) and finite element cross-section model approximated as symmetric (**b**); after Josefson [379]

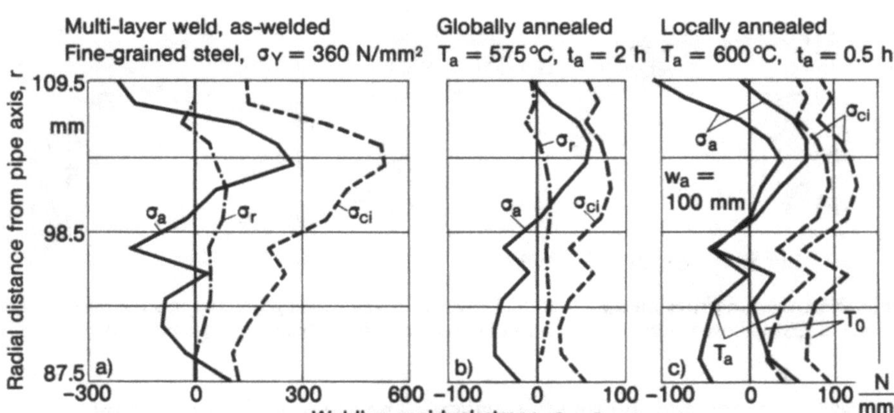

Fig. 247. Welding residual stresses, σ_a, σ_{ci}, σ_r, in weld seam centre plane of pipe made of CMn-alloyed steel, as welded (**a**), globally annealed (**b**) and locally annealed (**c**) (annealing temperature T_a, ambient temperature $T_0 = 0\,°C$, width of annealing zone w_a), analytical results; after Josefson [379]

fact is the relatively marked change in stress during cooling. The investigation was extended to cover steel A508 ($\sigma_Y = 460$ N/mm^2, in the heat-affected zone up to 1060 N/mm^2, steel known as relaxation-embrittling), also calculating the local damage dependent on the equivalent stress (according to the creep fracture theory proposed by Kachanov). In accordance with the analysis result, the damage occurs initially in the martensitic layer of the heat-affected zone, which undergoes a local plastic strain of approximately 0.1%. A high heating rate and annealing temperature prove to be favourable according to comparative analyses.

An early finite element analysis of single- and multi-pass welding and subsequent stress relief annealing of thick-walled CrMo-alloyed steel plates with large thickness has been performed by Ueda [245, 386, 387] (including transformation strain and creep relaxation) (Fig. 248). The investigation by Fidler [384], in which creep relaxation is connected with longtime annealing, is related to austenitic steel.

Measuring results related to the stress relief heat treatment of clad plates and flame-cut edges are referenced and discussed in Section 3.4.4.

The residual stresses after slow heating ($T_{max} = 690$ °C) and slow or quicker cooling of a brick-shaped solid consisting of a low-alloy steel base with single-side austenitic cladding have been investigated using a solid element model [393]. Creep effects are neglected in respect of the missing holding time. The

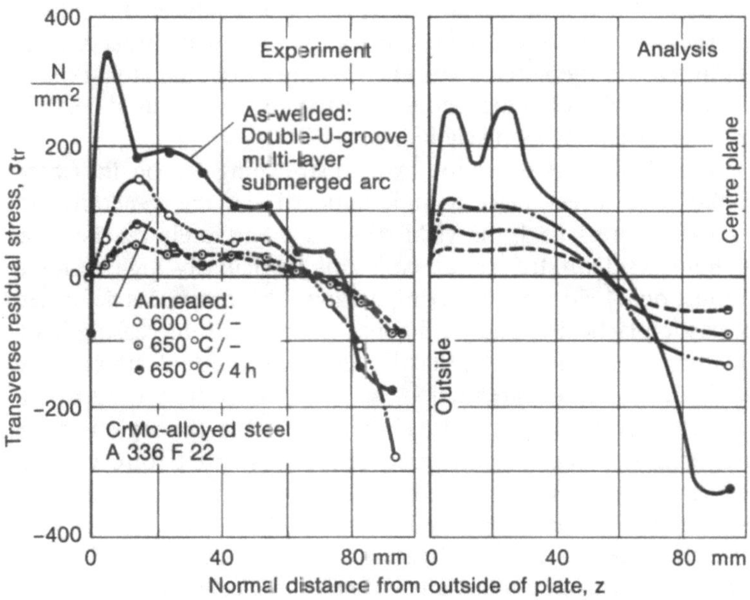

Fig. 248. Transverse residual stresses in multi-pass weld, CrMo-alloyed structural steel A336F22, plate thickness 200 mm comparison of experimental and analytical results; after Ueda, Fukuda and Nakacho [245]

residual stresses and strains after cooling are highly dependent on the cooling rate. The heated model is instantaneously exposed to ambient temperature introducing different heat transfer coefficients. There are tensile stresses in the cladding after slow cooling and compressive stresses after applying higher cooling rates. In the latter case, marked plastic zones occur on the inner and outer surface of the base metal with cladding.

4.4.3.3 Cold stress relieving (cold stretching, flame and vibration stress relieving)

4.4.3.3.1 Rod element model for cold stretching

Whereas the theory of hot stress relieving was explained by proceeding from practice, the reverse order is adopted with cold stress relieving, the reason being the routine nature of hot stress relieving compared to certain problems involved in cold stress relieving. Cold stress relieving includes cold stretching (mechanical cold stress relieving), flame stress relieving (thermal cold stress relieving) and vibration stress relieving. The presentation starts with theoretical ideas regarding cold stretching.

The (welding) residual stresses can be changed by (sufficiently high) loading of the component, especially by overloading (mechanical cold stress relieving, cold stretching). In the (frequent) favourable case they are reduced, in the (less frequent) infavourable case, however, they are also again built up. Consequently, a component containing residual stresses generally displays a reduction of residual stresses, at least of the maximum values, after being subjected to a load. A component originally free of residual stresses, by contrast, displays local residual stresses at least in the notch areas after loading. Residual stresses may also be built up with an inhomogeneously distributed yield limit, e.g. in components with hardened layers or claddings.

The (welding) residual stress reduction by cold stretching can be illustrated as follows. If the load stress field is superimposed on the generally inhomogeneous residual stress field, the yield stress is prematurely reached (i.e. earlier than without residual stresses) at points of particularly high residual stress and equidirectional load stress superimposition. The material is deformed at these points elastic-plastically. The yield zones enlarge with increasing load. Following unloading from the elastic-plastic state, the residual stresses are generally reduced. The cause of the reduction can be explained from the residual stress sources which should be imagined as being located in the region of the residual stress maximum values. As a consequence of the premature yield in these areas, the intensity of the sources is reduced or (to express it differently) the compatibility of the continuum is enhanced.

The more precise quantitative investigation of the process of cold stress relieving proceeds initially from a uniaxially defined rod element model which, in the simplest case, consists of a compound structure of three mutually prestressed rods manufactured of the same non-work-hardening material (after Rühl [396], Soete [397] and Erker [398]). In accordance with the typical distribution of weld-longitudinal residual stresses in non-transforming steels, high tensile stress σ_{te} is set in the inner rod, and half as high compressive stresses σ_c in the two outer rods (Fig. 249d). When applying an external load to the rod compound structure characterized by the tensile stress σ, the yield limit is soon exceeded in the tension rod, whereas the stress in the compression rods can be further increased elastically (up to the yield limit) (Fig. 249a). After unloading before the yield limit has been reached in the tension rod, the residual stress state remains unchanged. After unloading from yielding in all three rods, the residual stresses are completely eliminated. After unloading from yielding in the tension rod only, the tensile residual stress is reduced directly (down to σ_t^*) and the compressive residual stress indirectly (down to σ_c^*, as a consequence of the reduction of the tensile residual stress). The rod compound structure with residual stresses reacts less stiffly than the compound structure without residual stresses after yield has started. After stress reduction, a residual strain remains.

The process of stress relieving described above may also be represented in a diagram in which the rod stresses are plotted over the external load stress σ_L (external load related to total cross-section) (Fig. 250). Here, however, the strain information, is lost.

Further effects, which play a role in practice, can be observed on the prestressed rod compound structure under consideration. If the structure is

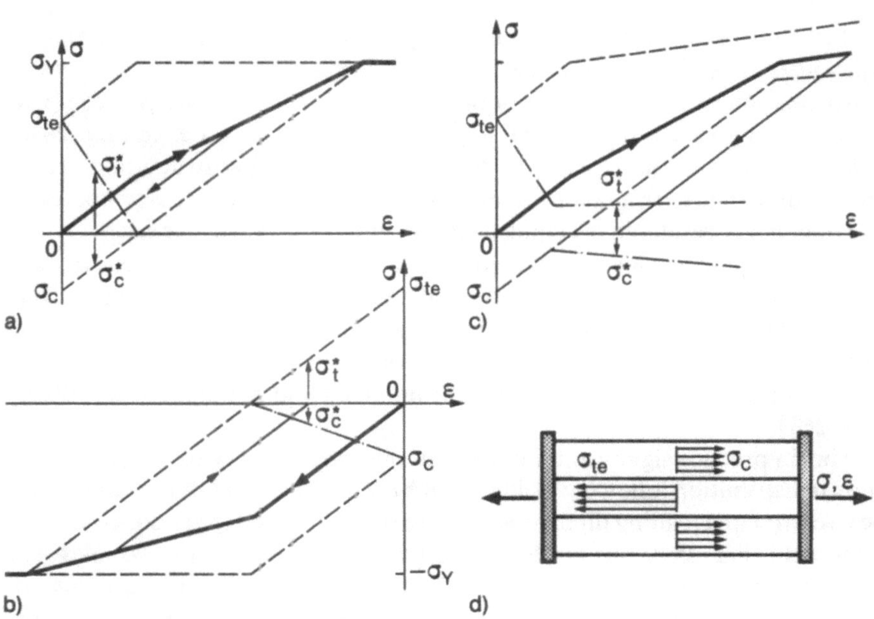

Fig. 249. Three-rod compound structure, prestressed similar to longitudinal stresses in welded joints (residual tensile and compressive stresses, σ_{te} and σ_c) and subsequently loaded (load stress σ, strain ε) (**d**), tensile and compressive loading and unloading resulting in residual stress relief (residual stresses σ_t^* and σ_c^*) (**a**, **b**), loading process with strain hardening, i.e. rising yield limit (**c**); after Vöhringer [401]

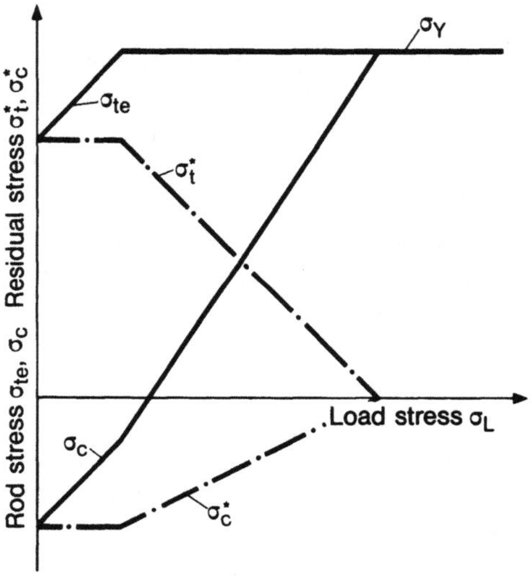

Fig. 250. Three-rod compound structure, prestressed similar to longitudinal stresses in welded joints, loaded and unloaded according to Fig. 249a, rod stresses and residual stresses plotted over load stress; after Soete [397]

subjected to an external compressive load, the compression rods yield first (at a relatively high external load), and the tensile residual stress is reduced indirectly, as a consequence of the elimination of the compressive residual stress (Fig. 249b). Hence, upsetting as well as stretching produces a reduction in residual stresses. If, on the other hand, different yield limits and work hardening capabilities are assigned to the rods in the tensile loaded compound structure (in approximation of the conditions in the weld and heat-affected zone), complete stress relieving as a result of cold stretching is no longer possible (Fig. 249c). Different yield limits may be due to different materials, microstructural states or stress multiaxiality. In practice, therefore, cold stress relieving, like hot stress relieving, never results in a complete elimination of residual stresses.

Plate strip specimens configured according to the three-rod prestressing principle, with transverse or longitudinal weld in the inner strip (the related transverse weld specimen is also termed H specimen according to the arrangement of longitudinal slots and transverse groove), are widespread for investigations into cold and hot stress relieving or into cold and hot crack susceptibility [385, 268].

The (in practice significant) influence of the heterogeneity of the material on stress redistribution following yielding can be represented as follows considering the two principal loading directions, namely transverse and longitudinal loading of the weld (Fig. 251). The mechanical behaviour of a tensile loaded mild steel plate with transverse and longitudinal butt weld is considered in a uniaxially simplified form. The yield limit of the weld σ_{Yw} is higher than the yield limit of the base metal σ_Y. Work-hardening and heat-affected zone are ignored. Initially, the weld-longitudinal residual stress σ_R is set equal to zero. When a load is applied transverse to the weld, the same stress σ occurs in the weld and base metal. The base metal yields finally, whereas the weld metal remains elastic.

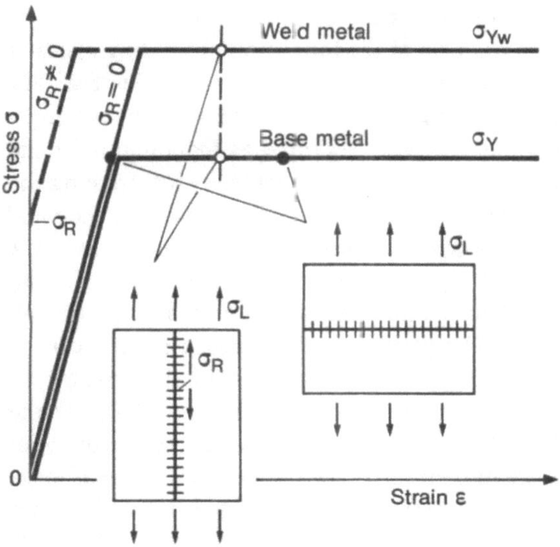

Fig. 251. Stress-strain curves of weld and base metal in tensile loaded mild steel plate with transverse and longitudinal weld, respectively; different yield limits in base metal, σ_Y, and weld metal, σ_{Yw}, influence of weld-longitudinal residual stress; actual stress and strain states after loading marked by black and white dots

After unloading, no residual stresses remain. Brittle fracture is more likely in the base metal. When the load is applied longitudinal to the weld, the same strain ε occurs in the weld and base metal. The base metal yields first although a relatively small plastic strain is sufficient to also cause yield in the weld metal. In this case, the stress in the weld metal is increased compared to the base metal. Following unloading from the partially or fully plastic state, residual stresses remain (in weld-longitudinal direction). Brittle fracture is more likely in the weld metal because of higher stresses in the weld metal. Now, longitudinal welding residual stresses (high tension in the weld metal, lower compression in the base metal) are additionally allowed for. Nothing is changed by the residual stresses when the transverse tensile load is applied. Neither does reduction in the residual stresses occur. When the longitudinal tensile load is applied, the weld metal yields first and then the base metal, followed by a reduction in residual stresses to the values of the originally residual-stress-free plate after loading and unloading. This simple consideration can also be applied to flame cut edges and to claddings.

It is an obvious next step to simulate stress relieving of more complex (uniaxial) weld-longitudinal stress patterns in oblong longitudinally welded components by a compound structure with many rods (in place of only three rods). The methods presented in Section 3.2.2 for calculating longitudinal welding residual stresses can then be applied to the considerably simpler conditions of fixed temperature in cold stress relieving. The loading process is performed numerically in steps (incremental methods); in simpler cases also in a single step. Equilibrium of the rod element forces, compatibility of the rod element strains and observance of the material law (in the simplest case elastic-plastic behaviour without work hardening) are achieved by functional-analysis, graphical and numerical methods.

With the combined graphical and numerical method of Tesar [399], it is possible to reliably determine the reduction of the longitudinal welding residual stress (also with yield stress variable from rod element to rod element, but initially without work-hardening), as is verified by comparison with the results of measurements. The uniaxial model used is also regarded as meaningful for the biaxial welding residual stress state assumed as more realistic by introducing the yield condition after Tresca-Guest-Mohr. That the residual stresses in the yield zone are reduced to the value of "yield stress σ_Y minus load stress σ_L and in the adjoining elastic zone are increased by the position-independent stress increment $\Delta\sigma$ (value dependent on extent of stress relief) is highlighted as a result which can be generalized (Fig. 252). The previously widespread assumption of geometrical similarity between initial and final stress pattern is incorrect.

A (more exact) numerical solution is offered by Schimmöller [400] for the special case of a parabolic residual stress distribution, σ_R, over the width $2w$ of a plate strip (Fig. 253). The pressure maximum at the strip edge is twice as large as the tensile maximum in the strip centre. The superimposed tensile stress σ_L reaches the triple value of the maximum tensile residual stress $\sigma_{R\,max}$. It is thus just below the yield stress σ_Y ($\sigma_L = (12/13)\sigma_Y$). The (greater) tensile and (lesser) compressive residual stress reduction is clearly shown from the results plotted in Fig. 253. Of particular significance is the offset of the zero line penetration point of the residual stress curve in the direction of the edge with increasing stress relief, which is not taken into account in the solution of Ref. [399].

If the rod element models are applied to bending instead of to tensile load, this reveals that a reduction in residual stresses in no way always occurs. For

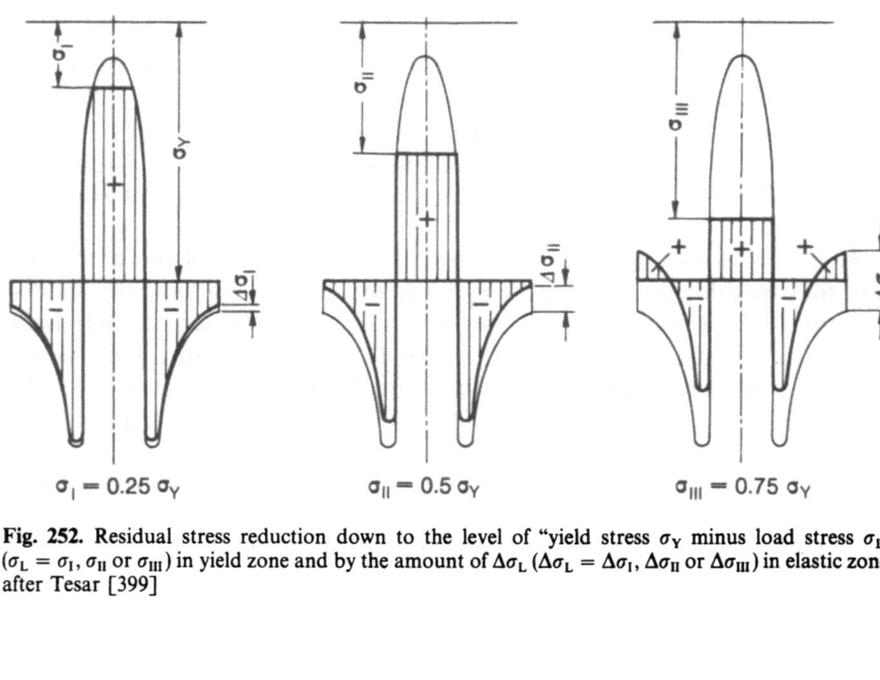

Fig. 252. Residual stress reduction down to the level of "yield stress σ_Y minus load stress σ_L" ($\sigma_L = \sigma_I$, σ_{II} or σ_{III}) in yield zone and by the amount of $\Delta\sigma_L$ ($\Delta\sigma_L = \Delta\sigma_I$, $\Delta\sigma_{II}$ or $\Delta\sigma_{III}$) in elastic zone; after Tesar [399]

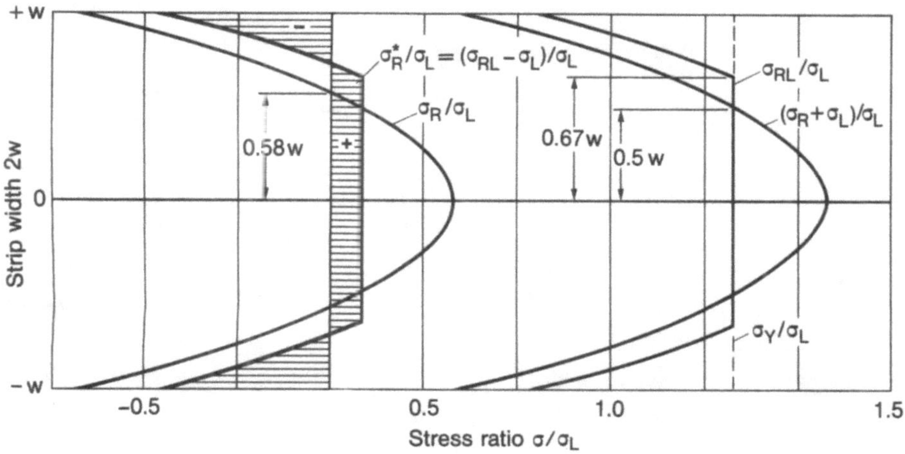

Fig. 253. Parabolic residual stress profile over strip width 2*w*, residual stress σ_R prior to loading, elastic superimposition of residual stress σ_R and load stress σ_L, elastic-plastic stress redistribution σ_{RL}, unloading resulting in residual stress σ_R^*; after Schimmöller [400]

example, bending overload on a plate free of residual stresses causes residual stresses to be produced. However, high residual stresses close to the yield limit may generally be reduced.

4.4.3.3.2 Notch and crack mechanics of cold stretching

Despite their differentiation, the rod element models previously described do not state enough regarding the practical problems which are characterized by the presence of design notches and manufacturing defects (including pores, shrinkholes, cracks). The usual reservation of these theoretical derivations, expressed by the clause "provided the yield stress in the opposite direction is not reached during unloading", is not met at notches or cracks. What happens during cold stress relieving at such more or less sharp notches? The elastic-plastic stress state at the notch or crack root has been the subject of functional-analysis and finite-numerical investigations. In particular, fracture mechanics has been used to investigate the problem of the triaxial tensile stress state below crack tips in thick-walled components. Unfortunately, there are no such investigations available in respect of cold stress relieving; in particular also there exist no finite element solutions, with the consequence that this complex must be considered for the present as being insufficiently cleared. A few fundamental but unquantified notch and crack mechanical considerations are provided below.

The stresses at sharp crack-like notches are distributed as shown in Fig. 254a for the elastic (with maximum stress equal to the yield limit) and elastic-plastic state. As a consequence of triaxial tensile stress underneath the notch root, the stress here may rise to far beyond the uniaxial yield limit (in the case of a crack to a maximum of $2.8\sigma_Y$ with large plate thickness). During unloading from the elastic-plastic state, the purely elastic notch stress state with reverse sign is initially effective. Following tensile loading, compressive residual

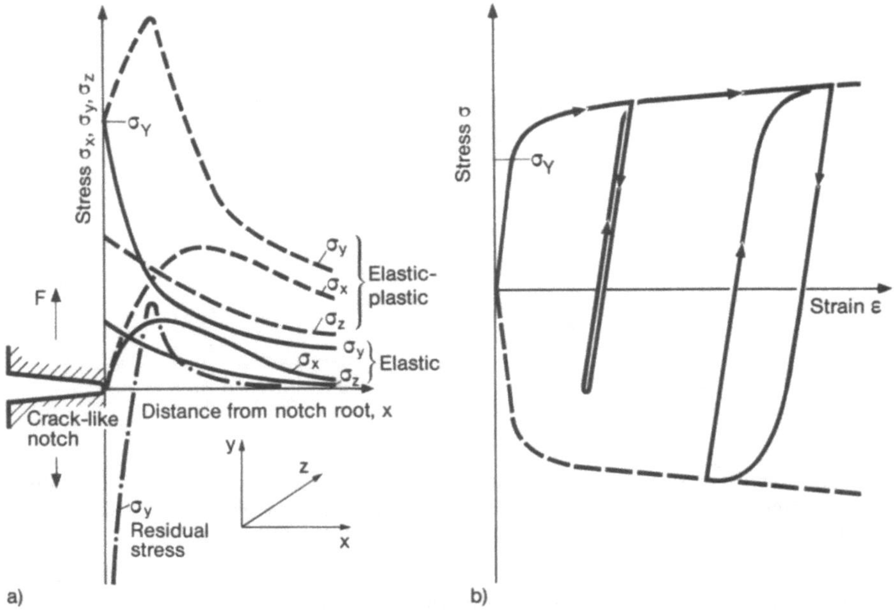

Fig. 254. Stress state at root of crack-like notch subjected to cross tension for elastic conditions ($\sigma \leq \sigma_Y$), for elastic-plastic conditions and after unloading (residual stress σ_y) (**a**), stress-strain hysteresis loops in areas far from and close to notch root (**b**); after Wellinger, Kußmaul and Krägeloh [404]

stresses develop at the notch root which, with appropriate sharpness of the notch, cause elastic-plastic compressive yield. In the stress-strain diagram shown in Fig. 254b, unloading and reloading are presented for the slightly plastically deformed area far from the notch root and for the heavily plastically deformed area close to the notch root. In the former case, purely elastic reloading occurs (strength-favourable behaviour); in the latter, a marked hysteresis loop with premature initiation of yield when subjected to reversed loading (Bauschinger effect) is passed through (strength-unfavourable behaviour). The strong plastic deformation at the notch root also causes (under tension) a certain (strength-favourable) increase in the notch radius.

The conclusion from this consideration is that components with notches or cracks, which were initially free of residual stresses, display residual stresses in the notch root area after being subjected to loading with elastically supported local yield and subsequent unloading. In other words, what has occurred is the opposite of stress relieving. Even if the yield at higher load covers the entire cross-section and extensive stress equalization has taken place, the elastic notch stress is again fully effective during unloading, with the result that even then notch residual stresses remain. On the other hand, a positive element is that sharp notches are rounded by plastic (tensile) deformation.

Components containing global residual stresses show locally increased residual stresses at the notches. When a load is additionally applied, the traditional notch stresses are superimposed on them. Depending on the sign and magnitude of the superimposed stresses, local yield occurs prematurely or delayed. Following unloading, residual stresses remain, which may be increased or diminished. When subjected to reloading, the favourable elastic behaviour without damaging hysteresis loop occurs only in the case of a weak notch effect.

The conditions at the notch root should be assessed particularly from the aspect of the triaxial tensile stress state. The latter occurs below the notch root or crack tip in the (in this area) greatly inhomogeneous load stress or residual stress field, particularly with a large plate thickness. However, the triaxial tensile stress state may also be due to residual stresses within the notch- and crack-free continuum, e.g. in the interior of an electroslag weld. The triaxiality of the stresses produces the local stress increase beyond the yield limit in the interior of the yield zone (up to $2.8\sigma_Y$ in the case of a through-crack in a plate with large thickness). Triaxiality and stress increase together are the reason for the risk of cracking (cold cracking and brittle fracture). The stress relieving measures must therefore be oriented, above all, to eliminating the triaxial tensile stress state. Obviously, though, the triaxial tensile stress state can only be diminished indirectly, namely by reducing the equilibrium stresses of lower degree of multiaxiality in the adjoining areas, because a high degree of multiaxiality impedes direct yield; the situation is in this respect similar in the case of cold stress relieving to that of hot stress relieving. A small plate thickness with an accordingly low degree of stress multiaxiality at the crack tip facilitates stress reduction.

To sum up, it can be stated that uniaxial cold stress relieving is reliably quantified by the rod element model but that only very unreliable statements are justifiable on that basis for the notched and possibly cracked component. In particular, the conditions at crack tips with large plate thickness remain unclear. Plastic deformation which occurs in areas with residual stress concentrations during cold stretching does generally have a stress relieving effect although unfavourable residual stresses may also be generated at notches which were initially free of residual stresses.

4.4.3.3.3 Cold stretching in practice

These theoretical investigations and considerations are not sufficient for assessing cold stretching in a practical situation. They simplify too greatly and quantify too little. In particular, the dependence of the triaxial tensile stress state on the yield and work-hardening properties of the material, on the geometrical parameters and stress state conditions including their limit values, is unknown. The practitioner must therefore proceed from empirical knowledge gained from traditional material and brittle fracture testing when assessing cold stretching, including theoretical considerations, only as an aid in this process. He will take into consideration not only the change in residual stresses as a result of the elastic-plastic deformation, but likewise the change in the mechanical material properties (strength, ductility, fracture toughness) which occur simultaneously. The decision which he is required to take relates not only to the type and level of loading but also to the working temperature during cold stretching. Moreover, defect and crack inspection prior to and following the operation are also a matter for discussion.

The influence of temperature requires to be taken into consideration, especially in the case of metals with a marked temperature embrittlement (and correlated impact embrittlement). Temperature embrittlement occurs only with a cubic body-centred structure of the crystallites, which is connected, moreover, to the phenomena of a marked yield limit and strain ageing. This group of metals includes the common ferritic structural steels.

In order to cold-work a temperature-embrittling metal without the risk of fracture, the temperature must not drop below the transition temperature, which depends not only on the material and its microstructural state but also on component geometry and size (especially plate thickness), type of crack or defect (especially size and sharpness) as well as on the loading speed and residual stress state. The transition temperature of the component is usually unknown because appropriate fracture tests are too expensive. The transition temperature is determined, by contrast, for different specimens and (brittle fracture) test methods. The problem then consists in drawing conclusions from the transition temperature of the specimen on to the (usually higher) transition temperature of the component.

The most extensive statements in this respect can be obtained from the brittle fracture diagram according to Puzak and Pellini [406, 407, 411] (Fig. 255). It should be possible to transfer the rupture nominal stresses determined as a function of temperature by means of certain specimens and test methods (impact load) to the component, assuming the same plate thickness of specimen and component. The curves are graduated according to crack size. The temperature embrittlement appears correspondingly marked, a large (through-)crack size resulting in a relatively high transition temperature and relatively low fracture stresses below the transition temperature (low-stress fractures). The

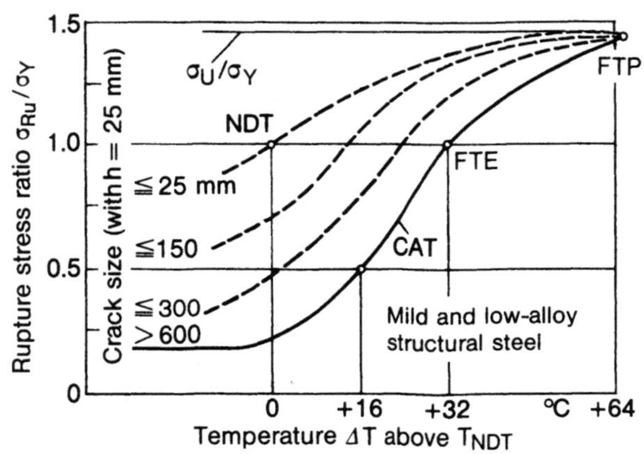

Fig. 255. Brittle fracture diagram according to Puzak and Pellini [406]

reference points required for drawing the diagram are identified by combinations of upper case letters relating to the relevant test method. The most important and reliable reference points are marked as NDT (nil ductility transition) and CAT (crack arrest temperature).

The NDT temperature is determined in the drop-weight test in conformity with ASTM Standard [408, 411]. A steel plate with or without a jointing weld, but in any case with a semicircular saw cut on the tension side, artificially embrittled by a surfacing weld, is deflected by the impact from a dropping weight. The deflection is limited by a mechanical stop so that the yield limit is just reached on the surface of the plate. What is evaluated is the maximum temperature at which the saw cut acting as an initial crack propagates into a complete brittle fracture of the plate (NDT temperature). The test results therefore indicate the temperature below which small cracks propagate in a brittle manner at nominal stresses close to the yield limit.

The CAT curve is determined in the Robertson test [409–411]. In an oblong large-scale specimen, with or without a jointing weld, subjected to transverse tension, an initial crack longitudinal to the specimen and transverse to the tension load is made to propagate. The crack propagates (in the gradient test) in the direction of increasing temperature, proceeding from the undercooled specimen end with the initial crack (− 196 °C) to the other specimen end held at ambient temperature (or even heated). The temperature at the forwardmost tip of the arrested crack (it runs ahead in the interior of the plate and thus forms a "tongue") is the crack arrest temperature. It is usually determined with a large plate thickness and a transverse stress of $0.67\sigma_Y$. The CAT curve of the brittle fracture diagram is obtained from performing tests with different transverse stresses or by estimation, proceeding from the observation of a minimum (temperature-independent) rupture stress below NDT temperature. The CAT curve indicates the temperature at which long (through-)cracks are arrested.

The FTE (fracture transition elastic) and FTP (fracture transition plastic) points of the diagram are determined in special large-scale notched specimen impact tests or large-scale specimen explosion tests as proposed by Pellini [406, 407, 411].

The brittle fracture diagram is normally compiled for the (possibly aged) base material by means of specimens without a (jointing) weld. It can also be determined, however, for the heat-affected zone, which may be embrittled or aged. Curve field and transition temperature are shifted in the direction of higher temperature in the case of embrittlement or ageing.

Proceeding from the CAT curve of the diagram described, it is recommended for the considered group of temperature-embrittling steels to observe a specified temperature difference ΔT, which depends on the load stress level, relative to the NDT temperature T_{NDT}, when applying this stress on the structure. $\Delta T = 32\,°C$ is recommended for stress levels in the range of the yield limit and for only minor plastic strains. $\Delta T = 64\,°C$ is recommended for stress levels above the yield limit and major plastic strains. These values apply if large cracks are likely and if these are to be reliably arrested. If major defects can be reliably excluded by non-

Table 5. NDT temperatures of structural steels, values for design; after Pellini [407]

Base metal	Heat treatment	NDT temperature
Low- and high-strength steels, low-alloy	As-welded	– 20– + 20 °C
	Normalized	– 50– – 10 °C
High-strength steels, medium-alloy	As-welded	– 70– – 40 °C
	Normalized	– 100– – 60 °C

destructive inspection techniques, a smaller ΔT value suffices for stress levels close to the yield limit.

These ΔT values can be applied directly to cold stress relieving. If only a moderate overload is applied in order to verify a sufficiently high ultimate load or in order to reduce residual stress maximum values by local yield, $\Delta T \leq 32\,°C$ (depending on the maximum possible defect size) plus an allowance for the embrittlement or ageing of the heat-affected zone is then sufficient. If more extensive cold straining is applied, e.g. for shape correction, a ΔT value of twice as much plus allowance must be provided. The NDT temperature depends on the steel group and the steel quality class (Table 5). A fine-grained microstructure is favourable. Large components may be less favourable. From the T_{NDT} and ΔT values stated as well as from the recommended additional temperature allowances, it is clear that cold stress relieving should be performed in many cases with preheating. The load, of course, should be applied slowly.

Temperature-embrittling steels are regarded as susceptible to ageing; this means that work-hardening with a reduction in elongation at rupture occurs after cold stretching and storing. Ageing should be regarded as a positive element because further plastic deformations are suppressed as a result of the ageing-related increase in yield stress provided that repeated overload (and likewise a different load) can be excluded. Ageing should be regarded as a negative element if repeated overload or a different load is likely because the crack arrest capability of the steel is reduced due to ageing (the CAT curve is shifted toward a higher temperature).

Cold stress relieving should be assessed significantly differently with metals which are only weakly temperature-embrittling. This group includes high-strength steels. Here, no significantly higher crack arresting capability is attained as a result of increasing the temperature. Cold stress relieving in this material group is only possible with the risk of instantaneous fracture. Reliability for operation is verified if the part has not fractured during cold stress relieving and is not subjected in service to a higher or different load. This reliability verification is intensified in many cases by applying an overload at a particularly low temperature. In the considered group of high-strength (and mostly relatively brittle) materials, the conditions at crack-like defects during cold stress relieving can be assessed on the basis of fracture mechanics, particularly effectively in combination with non-destructive inspection techniques.

Overloading with the cold stress relieving effect may have different objectives in practice [403–405]:

- Overloading as a control method serves as a verification of ultimate load. Defects are indicated as abnormally large local deformation, or, in the most unfavourable case, also as a crack or fracture. The service load remains distant from this by a certain safety margin. The important point is to completely cover the possibly critical service load cases; this means not only the primary loads but also certain secondary loads. The degree of stress relieving which is achieved by overloading is of secondary importance. The pressure test in the case of vessels belongs to this type of overloading.
- Overloading as part of the manufacturing process is aimed at enhancing strength. Local yield produces local hardening and reduction in residual stresses. At service load, no, or only slight, plastic deformations occur as a consequence of overloading. It is important to ensure that an adequate residual deformation capability is retained.
- Severe overloading as part of the manufacturing process serves for shape adaptation (straightening) or shape generation (shaping). Hardening and stress reduction are intensified. High ductility of the material is a prerequisite of the method. This is why the component is usually preheated. Only largely notch-free components are suitable for such treatment. Vessel out-of-roundness can be eliminated or even multi-layer vessel elements adapted in this way.

Practical experience with cold stretching is reported in Ref. [402–404]. The method is only permitted to a restricted extent, at least in Germany [347, 405].

Cold upsetting instead of cold stretching is proposed for the circumferential weld of austenitic steel pipes [427]. The pipe is compressed by clamps from the outside. The hydraulically operated clamps are positioned alongside the weld, i.e. not on the weld itself.

4.4.3.3.4 Flame and induction stress relieving

High weld-longitudinal residual stresses can be reduced more advantageously by thermal than by mechanical cold stress relieving. The current thermal cold stress relieving method is known as flame stress relieving [415].

Flame stress relieving was developed in the US and used initially in that country in marine and large vessel construction around 1950. The reduction in residual stresses which can be achieved with this method was more precisely investigated in Germany [416–419], and was increasingly used, particularly in shipbuilding and large tank construction [347, 419–422].

The plate (or shell) with butt weld which is to be stress relieved (typical plate thickness 20 to 40 mm) is heated to between 150 and 200 °C on both sides of the weld in a 100 to 150 mm wide strip using gas flames from pipe torches which are continuously moved (torch spacing 120 to 270 mm) (Fig. 256). The temperature of the weld zone is increased as a result of this to between 50 and 100 °C. The temperatures on the rear side of the plate are 20 to 30 °C lower and slightly shifted opposite to the movement of the torches. At a distance of 150 to 200 mm

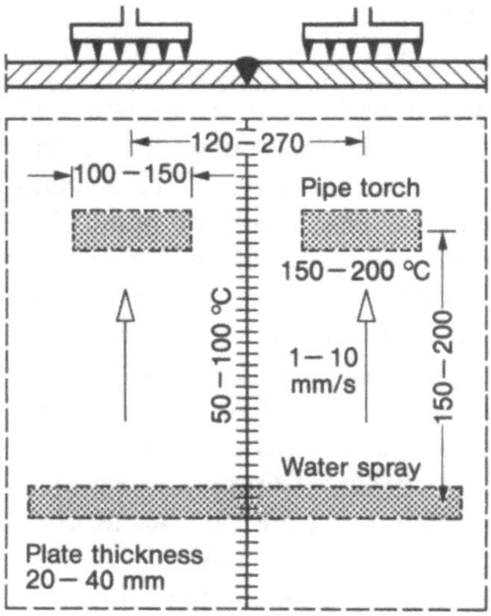

Fig. 256. Flame stress relieving of butt weld: arrangement of pipe torches and water spray; moving direction, spacings and temperatures

after the pipe torches there follows a transverse water spray which cools the plate back to the initial temperature. Pipe torches and water spray are moved at a speed of 1 to 10 mm/s.

As a result of combined flame heating and spray cooling, delimited heat zones are produced on both sides of the weld. The thermal expansion of the heat zones impresses longitudinal tension and transverse compression on the weld area (assuming adequate transverse restraint of the plate). During this process, the yield limit is exceeded, the weld is plastically elongated and the residual stress reduced after cooling. The stress reduction is rather good in the plate surface which undergoes treatment (Fig. 257), but considerably poorer in the plate surface of the opposite side.

In practice the process parameters require to be adapted to the conditions of the respective individual case in order to achieve good results. These measures should be checked by residual stress measurements.

Flame stress relieving displays a decisive advantage compared to mechanical cold stress relieving, due to which the main problem addressed in cold stretching is largely eliminated. Flame stress relieving takes place under transverse pressure provided adequate transverse restraint exists, whereas cold stretching, e.g. in the case of widening of a vessel by internal pressure loading, occurs under transverse tension. Flame stress relieving occurs at a moderately increased temperature, whereas cold stretching is generally performed at the service temperature which is lower. Crack initiation and crack propagation during flame stress relieving can be avoided even with low ductility and high susceptibility to ageing because plastic deformation occurs under transverse compression with significantly smaller tensile stress than under transverse tension. The minimum tensile stress necessary for crack initiation and crack propagation is then not exceeded. Brittle fracture resistance is therefore reliably enhanced even

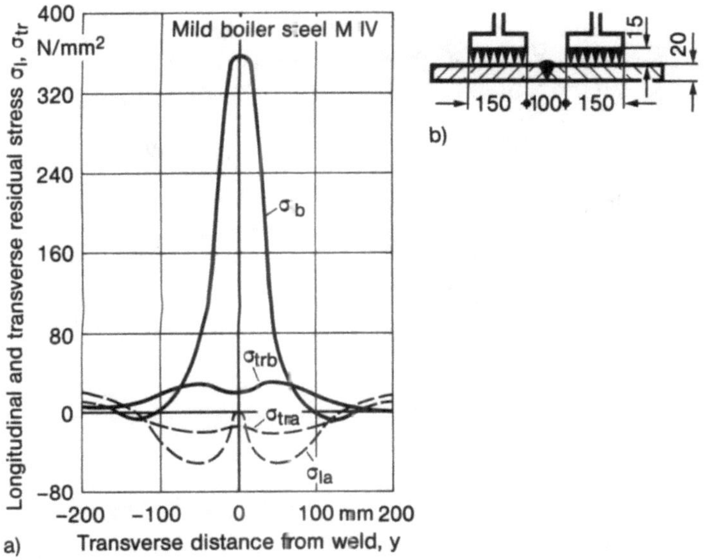

Fig. 257. Longitudinal residual stress reduction in butt weld resulting from flame stress relieving: longitudinal and transverse residual stresses, σ_l and σ_{tr}, before and after treatment (indices *b* and *a*) (**a**), dimensions in cross-section (**b**); experimental results; after Wellinger, Eichhorn and Löffler [418]

in the case of steels with low ductility or high susceptibility to ageing by reducing the residual stresses without any additional material damage.

Flame stress relieving has the disadvantage, compared to hot stress relieving, that no recrystallization takes place.

Induction stress relieving is the term given to a process in which the circumferential weld area of a pipe is inductively short-time heated on the outer surface and simultaneously cooled with water on the inner surface. Provided the process parameters are appropriately selected, it is possible with this method to reduce the longitudinal and transverse tensile stresses at the weld root which enhance the risk of crack initiation or to transform them into compressive stresses. This is verified by Rybicki and McGuire [423] in continuation of Ref. [228, 229] by means of finite element analysis for austenitic steel 304 (Fig. 258). As the high temperatures are reached only for a short time, it is justifiable to ignore the creep strains in the analysis model. As stress relieving occurs in the critical area (of the weld root) in the cold state (in this case with or without plastic strain), induction stress relieving is a special form of thermal cold stress relieving.

4.4.3.3.5 Vibration stress relieving

Vibration stress relieving is used in practice occasionally for reducing residual stresses for stabilizing dimensions and shape during subsequent machining. The component which has cooled down after welding (sometimes it is still hot) is vibrated on an oscillation table or by means of an attached vibrator close to higher natural frequencies of the part (10 to 100 cps) for 5 to 20 min at

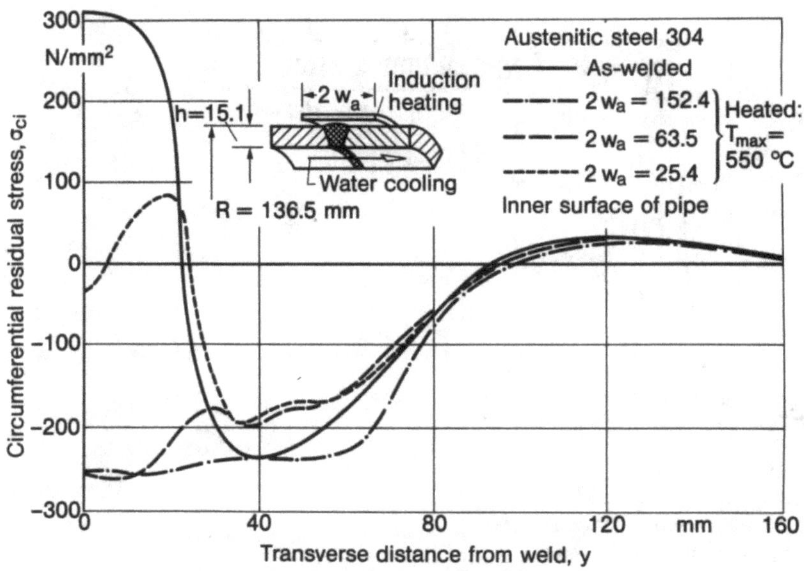

Fig. 258. Circumferential residual stress reduction on inner surface of pipe with circumferential weld resulting from induction stress relieving: pipe outer surface induction heated (different heating widths w_a), pipe inner surface water cooled; analytical results; after Rybicki and McGuire [423]

a relatively small amplitude, whereby the damping of the component, measured by the power consumption of the vibrator, is reduced. The method is controversially assessed because its principles are insufficiently clarified and contradictory results have been published. The literature and the results have been summarized by Gnirß [428] (see also Ref. [429–434]).

A reduction in macroscopic residual stresses (residual stresses of the first order) is a prerequisite for the often verified shape stabilization. Reduction is only possible if the yield limit is exceeded, at least locally. Exceeding the yield limit is enhanced by the following influences:

- high residual stresses, superimposed by equidirected load stresses;
- local increase of load stresses and residual stresses as a result of notches, cracks and defects;
- temporary load stress increase in certain areas as a result of resonance excitation;
- yield from the 0.1% offset yield stress on, which can be significantly lowered both by microscopic residual stresses (residual stresses of the second and third order) and by vibrations (at least for metals which undergo cyclic work softening).

Stresses close to the cyclic 0.1% offset yield stress must obviously at least be achieved in order to effectively reduce the macroscopic (and microscopic) residual stresses. Relaxation processes of a different nature, which are speculated in the literature (e.g. special dislocation mechanisms in Ref. [433]) are, to the extent that they actually occur, already contained in the locally reduced 0.1% offset yield stress. Residual stress reduction is greatest in the first vibration cycles and then rapidly flattens off. A high total number of vibration cycles should be avoided because of the fatigue damage which then occurs.

Global vibration of complex welded components, without knowledge and checking of the residual stress fields before and after stress relieving, is not a particularly promising procedure. An additional impediment to success in the case of welded components are hardened areas of the heat-affected zone in which the 0.1% offset yield stress is greatly raised. Nevertheless, practical application of the method in series production may be possible in individual cases by virtue of the empirically assured findings.

Provided vibration stress relieving is successful, it offers numerous advantages compared to hot stress relieving: low time and energy expenditure, lack of thermal distortion, freedom from mill scale, unchanged mechanical material characteristic values.

4.4.3.4 Hammering, rolling, spot compression and spot heating

Provided the metal is sufficiently ductile, both when cold and when hot, stretching the (butt) weld by *hammering* (progressing in direction of welding) in the cold or still hot state is an effective measure for removing longitudinal and bending shrinkage. At the same time, by reducing the tensile residual stresses the risk of cracking is diminished [412]. Hammering is performed layer after layer with an electric or pneumatic chisel hammer adapted to the weld shape or also by hand with a square set hammer. In the case of flush-welded patches (assembly and repairs) or for fill welding (repair of defects), hammering can be decisive for the feasibility of the measures. Sheet metal dents subjected to radial boundary compression can be removed by hammering the boundary zone and stretching it tangentially. Hammering of less ductile metal, though, enhances the risk of cracking. Crack initiation is promoted by inhomogeneous and locally high impact stress. The cold work ductility of the metal may, therefore, already be exhausted before the first service load is applied. As a counter-measure, preheating or interpass temperature during hammering is recommended. Hammering of welds is common practice in shipbuilding [347].

The warpage and residual stresses on thin sheet metal with butt weld are reduced more gently by cold *rolling* than by cold hammering. The pressure load under the relatively narrow steel rollers (approximately 10 mm wide) acts more evenly and less as an impact. The first rolling operation is performed on the weld; further rolling operations are performed close to and alongside the weld in case of a wide heat-affected zone. The stretching effect produced by rolling while reducing the thickness mainly counteracts the longitudinal and bending shrinkage as well as the longitudinal residual stresses. The optimal compressive force F_r of the roller depends on the roller diameter d_r; and the plate thickness h, as well as on the yield limit σ_Y, and the elastic modulus E of the metal. The influence of the roller width w_r on F_r is negligible in practice. An approximation formula according to Ref. [8] reads:

$$F_r = \sqrt{\frac{10 d_r h \sigma_Y^3}{E}}. \tag{240}$$

Sufficiently small thin sheet components may also be shape-corrected in a press between flat or shaping plates.

Shot peening can produce a favourable compressive residual stress state in the weld toe notch which results in a strong increase in the fatigue strength of the welded joint [413, 414]. A similar effect is achieved by local rolling of the notch root.

Compressing a plate between two dies with a circular cross-section (*spot compression*) produces after unloading the axisymmetrical (in case of the infinite plate) residual stress state shown in Fig. 259, characterized by continuous radial compressive stresses and alternating tangential stresses. The method is suitable for safeguarding weld ends if there is a risk of fatigue failure [414] (Fig. 260), but is scarcely tested in practice. Only longitudinal welds, not transverse welds, can be improved in this way. In the case of spot-welding, this method is known as post-weld compression. A very effective process improvement consists in making the dies to act between hold-downs which prevents a thickening next to the dies and thus permits a higher transverse pressure in the compression spot. The fatigue strength for infinite life of spot-welded joints can thus be increased to the four-fold value.

Short-time flame heating of a circular area (*spot heating*) in the plate to above recrystallization temperature produces compressive yield and, after cooling,

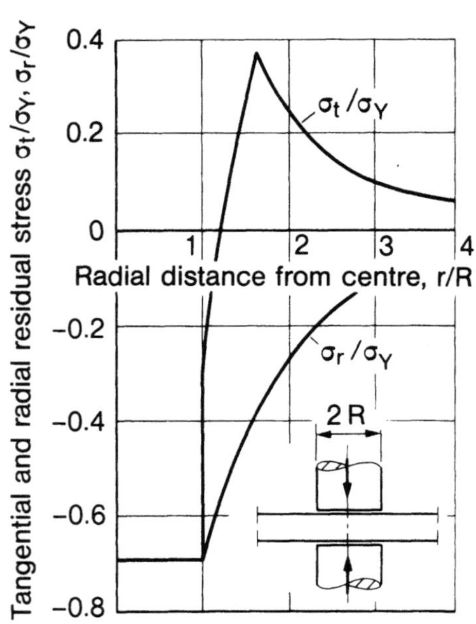

Fig. 259. Residual stresses after spot compression (yield limit σ_Y, radius of compression spot, R); after Gurney [414]

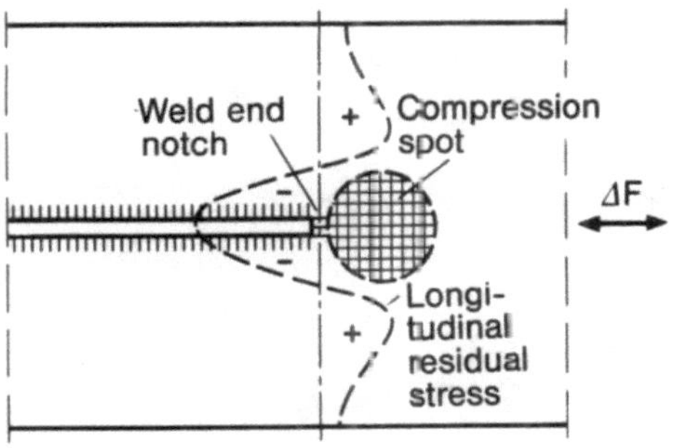

Fig. 260. Compression spot at weld end notch, longitudinal residual stress after unloading; after Gurney [414]

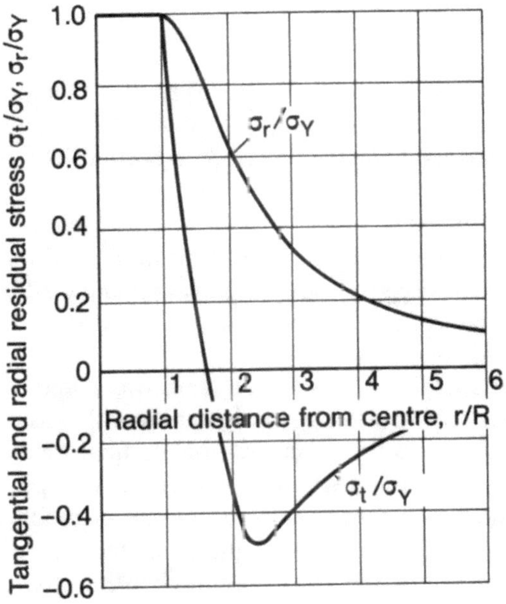

Fig. 261. Residual stresses after spot heating and cooling (yield limit σ_Y, radius of heat spot, R); after Gurney [414]

axisymmetrical (in case of the infinite plate) radial tensile residual stresses as a consequence of contraction of the cooling area and tangential compressive stresses as a consequence of supporting the radial tensle stresses (Fig. 261). The tangential compressive residual stresses can be used to increase the fatigue strength of weld ends [414]. The most effective distance between heat spot and weld end in the case of steel can be set with the red annealing colour ring (280 °C). Good experience has been gained with positioning this ring at 10 to 20 mm distance from the weld end. The heat spot, in contrast to the compression spot, should be positioned at the weld end at right angle to the direction of tension in order to permit the tangential compressive stress to be strength-effective (Fig. 262). The radial tensile stress which occurs simultaneously does not act favourably. Adjustment tests for peak temperature, rate of heating,

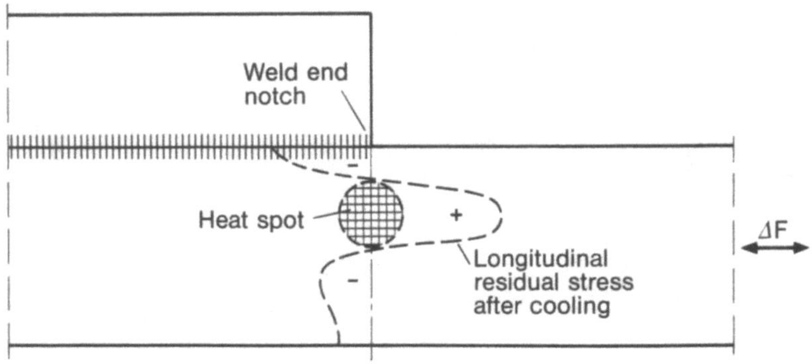

Fig. 262. Heat spot at end of longitudinal weld of gusset plate specimen, longitudinal residual stress after cooling; after Gurney [414]

heating area and distance to weld end are essential when using this method which has been tested little in practice. Use of spot heating to eliminate bending distortion on thin sheet metal (flame straightening) is described in Section 4.4.3.5

4.4.3.5 Hot, cold and flame straightening

Hot straightening after the metal is heated to forging temperature is used for removing abnormally large shape deviations, e.g. for straightening work after cases of catastrophic damage. In this connection, large-area local heating is generally sufficient to eliminate bends or wrinkles.

Cold straightening is performed applying a tensile load (cold stretching) or bending load (see Section 4.4.3.3.3 for cold stretching). Straightening with tensile load is only occasionally applied, e.g. for section bars with longitudinal welds. Post-weld bending-straightening, by contrast, is widespread practice. It is performed using fixed-installation straightening presses or by means of mobile winches or pressing devices. The structural member is upset on the side of bending-compression stress and stretched on the side of bending-tension stress. Because of the spring-back, the member must be bent beyond the specified shape. Cold-straightening requires adequately good cold-work ductility of the metal in the welded joint. The residual stresses following bending-straightening may be increased or reduced depending on the initial state. Disturbing distortion occurs during subsequent machining. Cold-straightened components also tend to later gradual warpage. In both cases residual stresses are released.

Cold straightening also includes hammering and rolling (see Section 4.4.3.4).

In *flame straightening* [415, 424, 425], spot-, strip- or wedge-shaped narrowly delimited areas (heat areas) of the structural member affected by distortion are heated with the flame to a red heat, they are thereby locally hot-compressed and correspondingly shortened during cooling (Fig. 263). If the heat areas are arranged so that they counteract the welding distortion, the latter can be fully or partially compensated by the shrinkage of the heat areas. Flame straightening is thermomechanically a process similar to weld heating and weld cooling. Occa-

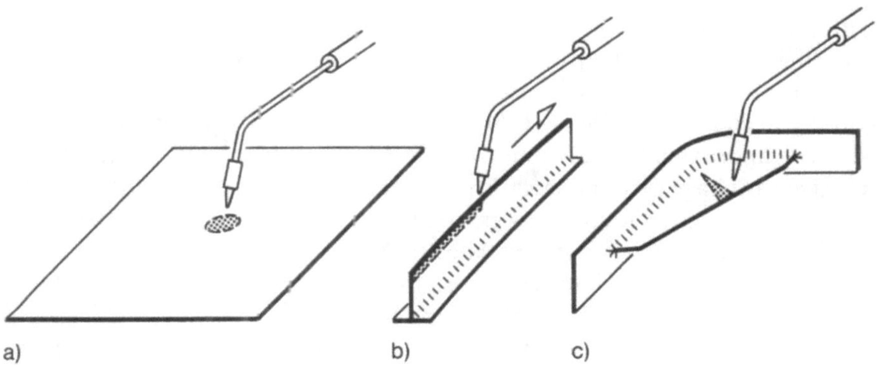

Fig. 263. Flame straightening by means of spot shaped (**a**), strip shaped (**b**) and wedge shaped (**c**) heat areas; after Malisius [5]

sionally, surfacing welds are also used with no other function than to compensate for shrinkage.

The principal fields of application of flame straightening are dents in thin sheet metal and bending shrinkage of girders and other section bars. Dents in thin sheet metal can be eliminated by relatively small heat spots or relatively large circular or ellipsoidal heat rings, occasionally supported by hammering plane in the hot state. Girders or bars with bending shrinkage can be shape-corrected by exterior heat strips in the direction of the bar or by heat wedges perpendicular thereto. Flame straightening has the advantage of only a slight or completely missing additional cold straining in the distortion-causing weld. The welding residual stress state at the weld is left unchanged. The method can also be employed on large components and structures in which mechanical shape correction measures are not possible.

A precondition for successful flame straightening is the rapid and locally delimited heating of the heat areas to a temperature at which the yield limit is greatly reduced (600 to 800 °C for low-alloy steel) so that the heated metal is hot-compressed in the plate plane between the cold surrounding areas resulting in an increase in plate thickness. By contrast, the heat areas can cool down slowly, achieving a temperature balance to the surrounding areas.

Heating with the (acetylene) flame satisfies the demand for rapid and concentrated heating only to an incomplete extent. Electrical heat sources would be more optimal in this respect. However, the flame has the advantage of flexible heat control, visual temperature monitoring (bright red heat) and restricted microstructural change (which does not rule out other types of heat sources in the future). Flame heating can be performed with appropriately perforated shielding plates to enhance the heat concentration.

Only a minor number of theoretically substantiated approximative approaches to flame straightening exist. For instance, the question is posed in restoring dents in thin sheet metal whether heat spots or heat strips are more effective assuming identical heat input. According to Ref. [8] (ibid., p. 282), the question is answered on the basis of the shrinkage areas which are calculated proceeding from the residual stress field at a heat spot (approximative solution) and proceeding from the transverse and longitudinal shrinkage of a heat strip according to equations (168) or (175) and (141). The heat strip according to this analysis is approximately 2.5 times as effective as a heat spot, whereby the transverse shrinkage predominates. The explanation for the greater effectiveness of the heat strip is to be found in its greater transverse shrinkage as well as in the lesser degree of multiaxiality of the related residual stress field. It must remain to be investigated to what extent the ring-shaped heat strip around a sheet metal dent retains its advantage compared to the heat spot.

The question which is posed initially for bending-distorted girders is whether girder-parallel or girder-perpendicular heat strips are more effective (Fig. 264a and 264b). The question is answered according to Ref. [8] (ibid., p. 285) proceeding from equations (176) and (178) for the bending angle φ with F_s according to equation (141) and Δ_{tr} according to equation (168) as well as with the heat strip eccentricity $e = l^*$ in the case of the longitudinal heat strip or $e = l^*/2$ in that of the transverse heat strip. Assuming the same heat input, the transverse heat strip is approximately 1.5 times as effective as the longitudinal heat strip and additionally undergoes lesser cold straining. This behaviour, confirmed in practice, is explained by the relatively large and less restrained transverse shrinkage of the transverse heat strip. The effectiveness of the transverse heat strip is enhanced by a wedge-shaped strip because the wedge strip centre is displaced to the outside compared to $e = l^*/2$. In practice, heat wedges according to Fig. 264c and 264d are also encountered. The plate area which requires to be heated, in other words the optimal wedge angle, is determined by trial and error.

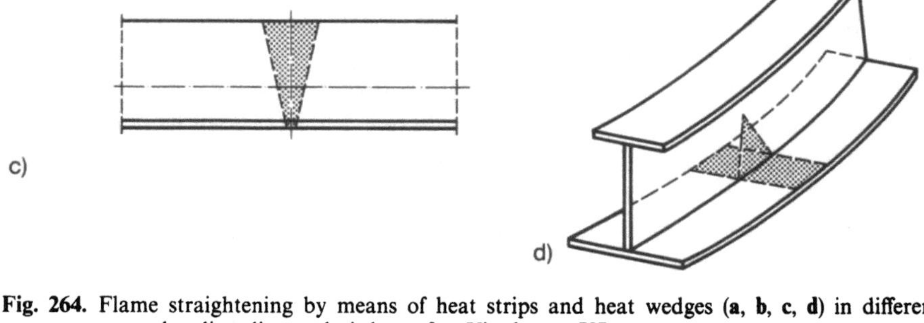

Fig. 264. Flame straightening by means of heat strips and heat wedges (**a**, **b**, **c**, **d**) in different arrangement on bending-distorted girders; after Vinokurov [8]

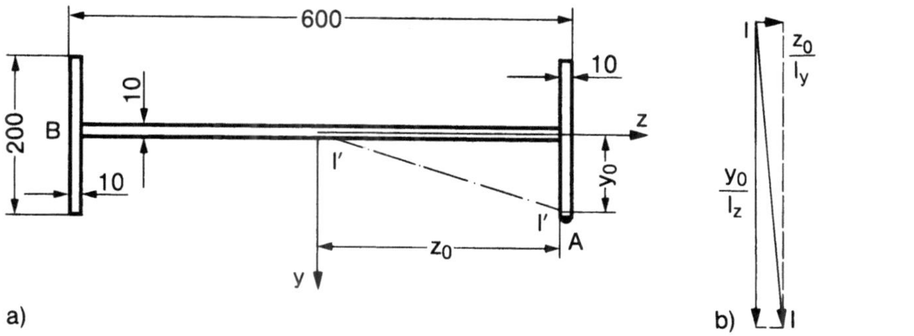

Fig. 265. Flame straightening of bending-distorted I section girder (**a**) by means of heat strip at flange edge A (girder turned from vertical to horizontal in order to save space), resulting bending plane I–I (**b**); after Vinokurov [8]

Bending shrinkage caused by longitudinal heat strips can be approximately determined according to equation (177). If the longitudinal heat strip on the double-symmetrical I section girder is positioned asymmetrically (point A at the flange edge), bending shrinkage vectors occur proportional to y_0/I_z and z_0/I_y in the two principal directions y and z and are superimposed as such (Fig. 265). The resulting bending plane I–I does not approximately coincide, as often maintained, with the connecting plane I′–I′ between heat strip and girder axis. In order to eliminate a certain bending distortion (bending plane I–I specified), what follows therefrom, in a back-calculation, is the distance y_0 on the flange. If the flange width is exceeded in this connection and the heat strip, thermally intensified, is placed on the flange edge instead, the bending distortion in the web direction is then over-compensated. The over-compensation can be reversed by a further heat source in the left-hand flange-web intersection point B. A precondition for the correctness of the approximative calculation according to the beam bending theory, equation (177), is that the girder cross-section remains undeformed, i.e., flanges and web are not displaced or rotated relative to one another. In girders with a general asymmetrical cross-section contour, it is first of all necessary to determine the direction of the principal axes of the cross-section to which the further calculations are then related.

Flame straightening is primarily used on mild and low-alloy structural steels. It can also be performed on stainless alloyed steels provided this does not impair their corrosion resistance. When using the method on aluminium alloys one has to overcome the tendency of this metal to absorb gas at elevated temperatures as well as its particularly intensive heat diffusion. A different type of heat source could partly solve the problem.

5 Survey of strength effects of welding

5.1 Methodical and systematical points of view

The non-detachable joining or coating of components or materials, termed welding, is designed to produce a joint of adequate strength. Strength for the engineer relates to the resistance of the component to function-disturbing failures such as insufficient stiffness, local or global yield or creep, instability phenomena (buckling and postbuckling), cracking (hot and cold cracks), fracture processes (ductile, brittle, creep and fatigue fracture), wear and corrosion.

The usual methodical approach adopted in strength analyses is characterized by separating the structural and material influences [435]. This is the reason for the subdivision in English-speaking countries into "strength of structures" and "strength of materials". The strength investigation on the structural component is characterized by stress and deformation analyses on the basis of continuum and structural theories into which elasticity, yield, creep and hardening characteristic values of the material are introduced as auxiliary quantities which are determined by material testing. Strength limits are expressed by critical (fracture or instability) stresses unless the ultimate load follows directly (i.e. without resort to stresses) from a condition of instability. The strength investigation on the material is characterized by determining material characteristic values in (usually) standardized test procedures and explaining these on the basis of metal physics, e.g. dislocation theory. Welding affects the component strength through residual stresses (mainly of the first order) and distortion and the material strength through microstructural changes (including inhomogeneities and hydrogen diffusion). The separation into structural and material influencing parameters, however, is not, or only partly, accomplished in the term "weldability" and its related testing.

For reasons of systematic treatment of the subject, strength analysis of welded joints and structures is not presented here, but is found in Ref. [436–440], subdivided into fatigue strength and static strength, i.e. into strength when subjected to repeated or to single loading. It is only possible to provide a brief overview here. It is not only because of the scope of the problem that strength matters are dealt separately. Residual stresses (of the first order) act locally on strength in the same manner as load stresses so that is reasonable to treat the related strength matters jointly [439]. A differing opinion can be justified in respect of hot and cold cracks, which may be caused solely by residual stresses, deformations and microstructural changes during welding. In a condensed overview, the effect of welding on strength can be summarized from the classification aspect of crack and fracture phenomena.

5.2 Hot and cold cracks

Hot and cold cracks are relatively short cracks, occurring under certain circumstances in large numbers. They may arise in the fusion and heat-affected zone during or immediately after welding, but also on the occasion of post-weld heat treatment [438, 441].

Hot cracks occur in the fusion zone as longitudinal and transverse cracks, and in the heat-affected zone, proceeding from the fusion boundary area, as transverse cracks. They are initiated as intercrystalline segregations in the range of the solidus temperature if the mixture of solid and liquid phase (liquated grain and subgrain boundaries) is no longer able to cope with the strains or strain rates caused by shrinkage.

Cold cracks occur in the fusion zone and in the heat-affected zone, in the latter primarily as longitudinal cracks (weld root, weld toe and underbead crack) in the range between melting and ambient temperature. They mainly initiate at grain boundaries of the fusion zone boundary area. Cold cracks are greatly promoted by hydrogen diffusion in the microstructure (hydrogen-induced cracking) as well as by hardening of the heat-affected zone. They may still occur even hours after the material has completely cooled down (delayed cracking). Welding residual stresses are essential for initiation of cold cracks. A critical tensile stress is observed which depends on the material and its microstructural state, above which cold cracks may occur. The initiated crack is first stably enlarged before occasionally resulting in unstable instantaneous fracture. Cold crack formation in steels is promoted if the transformation of the austenite into martensite or bainite is shifted to low temperatures ($\leq$ 300 °C), which may be the case with high cooling rates.

The many relevant test methods ("welding tests") cannot usually be clearly assigned to only one of the two types of cracks: rigid restraint cracking test (RRC test), tensile restraint cracking test (TRC test), slot specimen weld test (rectangular or cruciform plate specimen), bead on "finger specimen" test (plate strip package), ring groove weld test, circular bead on plate weld test, implant test, double fillet weld test, edge fillet on cover plate weld test (controlled thermal severity test, CTS test), overlapping plate cruciform specimen, double round bar weld test, bending bar weld test ("Varestraint"), local fusion test; see Ref. [438]. The specimen stress is applied partly by the welding process (self restraint) and partly by external load or deformation. The transferability of the test results to the actual welded component is restricted.

In addition to these "welding tests", there are also welding simulation tests on small uniaxially loaded rod specimens to which the temperature and stress cycle of the considered point of the fusion or heat-affected zone is applied. The hardenability test and the carbon equivalent derived therefrom supplement the hot and cold crack tests.

Hot and cold cracks are avoided by material and manufacturing measures (composition of base and filler metal, preheating and post-weld heat treatment, multi-pass welding).

5.3 Ductile fracture

The term "ductile fracture" relates to the local constriction with subsequent rupture which is observed in the tensile test on smooth specimens manufactured of ductile material. The fracture is preceded by major deformations.

With a high ductility of the material in the welded structure, the influence of welding residual stresses on (ductile fracture) strength is negligibly small, especially as the high ductility is in many cases assured by post-weld heat treatment, which at the same time reduces residual stresses. The residual stresses admittedly appear in a "softer" load-deformation curve during first loading (the structure "settles"). Immediately prior to ductile fracture, though, the residual stresses are removed by plastic deformation and thus have no effect on the rupture load. This gives the justification for ignoring the residual stresses in the stress assessment rules. However, the problem of strength of welded structures is thus only shifted to the verification of ductility, in the case of temperature-embrittling metals in the form of the observance of a minimum service temperature above certain transition temperatures of brittle fracture tests. In normal practice, the ductility of the metal in the welded structure is limited and a service temperature limit cannot be reliably observed, with the result that the residual stresses do in fact have to be incorporated into the strength analysis which then refers to brittle fracture.

5.4 Brittle fracture

The term "brittle fracture" relates to the deformationless or low-deformation rupture which can be observed on specimens and components which are sharply notched or subjected to welding residual stresses. Brittle fractures of this type may also occur under certain conditions at nominal load stresses far below the yield limit ("low-stress fractures"). Brittle fractures may also occur before the first service load is applied, caused solely by residual stresses. The brittle fracture generally originates from incipient cracks or defects which have occurred during manufacture or in service (hot or cold cracks, incipient fatigue cracks, crack-like weld defects).

The inadequate ductility of the material in the (welded) structure, termed brittleness, is not a material characteristic value in the usual sense, but depends on stress state and material parameters. Depending on the stress state, the material is capable of deforming (ductile material) or incapable of deforming (brittle material). The causes of brittleness are therefore distinguished as being related to stress state and material condition.

A necessary stress condition for brittle fracture is the triaxial tensile stress state of adequate intensity and extent. Such stress states may occur at cracks and sharp notches, particularly also, however, as welding residual stress following cooling under more or less rigid restraint from all sides. Large plate thickness permits a larger extent of the stress state and thus favours brittle fracture. Lesser degrees of multiaxiality, particularly compressive stress components, favour plastic deformation. The brittle fracture is thus avoided or turns into a ductile fracture.

In the case of metals with cubic body-centred lattice and marked yield limit (this group of metals includes mild steel), the brittleness depends markedly on temperature and stress rate. With decreasing temperature, a sharp drop in ductility is observed (temperature embrittlement). With increasing stress rate, brittleness is also enhanced (impact embrittlement).

Microstructural material conditions favouring brittle fracture are, on the other hand, coarse grain, hardening, diffusible hydrogen and ageing. Such material states may particularly occur in the heat-affected zone of welded joints with the result that an increased risk of brittle fracture exists in combination with the welding residual stresses acting in this zone. The weld-longitudinal stresses (this being the circumferential stresses of the circumferential weld) are considered to be particularly critical. They may overstress the hardened zone, which in the worst case is predamaged by transverse cracking, as a consequence of residual stresses and load stress redistribution.

The brittle fracture risk of welded components and structures is assessed by means of brittle fracture tests on simplified specimens. The traditional methods with welded or even unwelded (therefore questionable) specimens (the most common being notched impact test, bead bend test, drop weight test, Robertson test, wide plate test) permit rough assessments regarding practical problems, proceeding from the respective transition temperatures of temperature-embrittling materials. Engineering fracture mechanics additionally enable quantitative statements to be made regarding the brittle fracture behaviour of the component or of the structure, proceeding from the test results from relatively small specimens with a crack. In addition to load stresses, residual stresses are fully included in the fracture mechanical analysis as fracture-initiating and propagating quantities. Of course, it is only the crack-opening tensile stress (and not the crack-closing compressive stress) which should be regarded as critical.

The brittle fracture risk can be kept low by a wide range of design, material and manufacturing measures. Reducing welding residual stresses and the lowering of the transition temperature of the component thus achieved is an important partial measure. Welding distortion, by contrast, has no strong influence on the risk of brittle fracture.

5.5 Lamellar tearing type fracture

Lamellar tearing type fracture ("terrace fracture") refers to low-ductility tearing of metal plates in the plate plane along rolled non-metallic inclusions (sulphides, silicates, oxides). The fracture occurs under the action of load and residual tensile stresses perpendicular to the plate plane (tension) and is promoted by diffusible hydrogen in the microstructure as well as by ageing. Tensile stresses perpendicular to the plate plane occur particularly at T and cruciform joints. Lamellar tearing can be counteracted by design measures (reducing the tensile stress perpendicular to the plate plane), higher material purity (i.e. free of rolled inclusions) and avoidance of diffusible hydrogen and ageing.

5.6 Creep fracture

At high service or test temperatures (in the range of the recrystallization temperature), creep is observed at a fixed load and relaxation is observed at a fixed deformation. Primary and secondary creep proceeds in a stable manner at a decreasing, or at least constant deformation rate. In the subsequent range of tertiary creep, the deformation rate increases while the load remains unchanged, which finally results in creep fracture. Stress- and material-related embrittlement may also occur with creep fracture. The material-related embrittlement phenomena at high temperature include, in particular, precipitation processes in the microstructure which may cause temper embrittlement, creep embrittlement or relaxation embrittlement.

Incipient cracks, notches, cross-sectional transitions, weld-longitudinal loading and residual stresses (i.e. the same conditions as with brittle fracture) are considered to be particularly critical.

5.7 Fatigue fracture

The term "fatigue" relates to crack initiation in plastically deformed areas of the component subjected to cyclic loading, to subsequent stable crack propagation under further load cycles and to final instable fracture. The fatigue strength is determined in the Wöhler test applying constant load amplitudes and in the service fatigue strength test applying varying load amplitudes. The fatigue strength of components is dominated by the stress concentrations at notches and cross-sectional transitions, their cyclic amplitude or range being the decisive factor. The statically acting mean stress or prestress plays only a secondary role. The latter may be caused both by an external load or by residual stresses and likewise be increased at the notches and cross-sectional transitions. Whereas the mean stress resulting from external load is generally retained independent of the number of load cycles, the residual stresses may be changed by a single overloading of the component, by the cyclic load itself, by creep and relaxation and by crack formation. In general, tensile residual stresses have an unfavourable, and compressive residual stresses a favourable, effect on fatigue strength.

As regards the influence of welding residual stresses on fatigue strength, a distinction may be made between three groups of welded components [436, 437]:

- components in the as-welded state with high tensile residual stresses (e.g. weld-longitudinal residual stresses);
- components in the stress relieved state (e.g. as a result of post-weld heat treatment);
- components in the surface-treated state with high compressive residual stresses (e.g. as a result of shot peening).

Components with high tensile residual stresses superimposed by equidirected load stresses display a low, approximately constant amplitude strength (this being the cyclic stress amplitude sustainable without fatigue fracture) over the entire load mean stress range. The reduction in residual stresses offers a significant increase in amplitude strength only in the case of compressive load mean stresses. Compressive residual stresses generated as a result of surface treatment increase, on the other hand, the amplitude strength over the entire load mean stress range.

These statements apply to high-cycle fatigue strength and to fatigue strength for infinite life (endurance limit). In the medium- and low-cycle fatigue range, residual stresses are reduced as a result of local yield caused by superimposed load stresses, so that the strength behaviour of the residual-stress-free component is approximated.

From the aspect of microstructural change caused by welding, the fatigue strength of the material may be reduced or increased locally. A reduction occurs as a result of microdefects at the boundary of the fusion zone or by the heat-effected removal of hardenings of different types. Moderate hardening in the heat-affected zone (≈ 350 HV) increases fatigue strength locally. On the other hand, the simultaneously generated residual stresses must be taken into account. Fracture mechanics analysis of crack propagation behaviour in residual stress fields is a topic to be dealt with in this connection.

As the practice-oriented comparative investigations between as-welded and stress relief annealed components are usually conducted with a tensile mean load in the fatigue strength for finite life range, the strength differences determined are relatively small.

The fatigue strength of welded components may be reduced to a greater extent as a result of welding distortion than by welding residual stresses. This is the case if the amount of the bending and angular shrinkage or of the shrinkage dents reaches the plate thickness. The eccentricity of the distorted plate causes, when subjected to a given cyclic tensile or compressive stress, a superimposed cyclic bending stress, which results in premature damage and crack initiation.

Groove edge misalignment, which may be caused by distortion during welding, acts in the same direction.

The fatigue strength of welded components can primarily be increased by design measures which aim at the reduction of notch stress concentrations. Manufacturing measures also play a role, but are of secondary importance (freedom from weld defects, notch root treatment).

5.8 Geometrical instability

Geometrical instability refers to the elastic or elastic-plastic buckling of rods, bars, plates and shells, which may occur already at nominal load stresses below the yield limit: bending buckling of rods, torsion-bending buckling of bars, buckling and post-buckling of plates and shells. In the case of thin sheet metal, the welding residual stresses alone may initiate the instability. With thicker plates and also rods and bars, the welding residual stresses influence the level of the critical load.

The stability limit of rods, bars, plates and shells may be reduced by the compressive residual stresses in the areas alongside longitudinal welds. In the cross-sectional parts with the compressive residual stresses, the compressive yield limit is exceeded prematurely, as a result of which the cross-sectional parts remaining elastic are subjected overproportionally to increased stresses. I section girders or box girders welded longitudinally buckle prematurely because the effective moment of inertia, which determines the buckling load, has to be reduced in accordance with the stress reduction in the plastically deformed parts of the cross-section. Despite the reduced buckling load, however, the ultimate load should remain unchanged in the case of an adequately ductile material because the residual stresses are removed in the post-buckling process. In the case of rods and plates with longitudinal welds subjected to longitudinal compressive load and also in that of cyclindrical shells with circumferential weld subjected to axial compressive load, the reduction in the buckling load as a result of welding residual stresses is up to 10% [440]. On the other hand, the stability limit may be increased as a result of tensile residual stresses. For instance, shear buckling of edge-stiffened sheet metal panels is delayed as a result of heat spots.

The stability limit may be substantially more reduced by welding distortion than by welding residual stresses. Rods, bars, plates and shells, out of shape as a result of welding distortion, buckle prematurely. The bending distortion of rods and bars and the constriction of cylindrical shells with circumferential weld have a particularly strong reduction effect.

The stability limit of welded components can be increased, primarily, by means of design measures. In addition, manufacturing accuracy plays an important role. In special cases, the reduction of residual stresses may also be effective.

5.9 Corrosion and wear

Corrosion is the undesired chemical or, speaking more exactly, electrolytic reaction of the surface of the component with the ambient medium, which results in material removal or cracking on the surface. High tensile stresses in the surface may initiate stress corrosion cracking which is intensified by hydrogen diffusion. If the risk of stress corrosion cracking exists, therefore, tensile welding residual stresses in the surface of the component are damaging. They should be removed in this case or be replaced by compressive residual stresses.

Wear is the undesired mechanical erosion of the surface of the component as a result of small particles breaking loose or also as a result of permanent deformation. Wear is promoted by tensile stresses in the surface. Tensile welding residual stresses in the surface should be avoided where the risk of wear exists.

5.10 Strength reduction during welding

In general, welding is performed while the structure is free from external loads. In some cases though, e.g. assembly and repair, the complete external load or at least part of it is to be sustained without failure during welding. A three-step strength assessment is then necessary [442, 443].

First, the reduction in cross-section in respect of its load-carrying capacity caused by the weld pool and the high temperature area surrounding it should be considered. For longitudinal welds, the reduction is largest shortly after the passage of the heat source through the cross section. It is determined on the basis of the largest transverse extension of the cut-off yield stress isotherm. For transverse welds, the largest longitudinal extension of this isotherm is decisive. The increase in superimposed tensile and bending stresses caused by the reduction in cross-section is calculated. The maximum stresses are assessed in respect of the conventional permissible stresses.

Second, the state of highest thermal distortion immediately after welding, and third, the state of distortion (and residual stress) after complete cooling-down, should be calculated on the basis of the non-reduced cross-section applying equations (174) to (179) and assessed primarily in respect of buckling strength.

For testing the strength reduction during welding, i.e. the weldability of the loaded structural component, large scale loading equipment [444] may be necessary.

Bibliography

Monographs related to temperature fields, residual stresses and distortion caused by welding:

(see also [291, 357, 424, 425, 436, 437, 445–447])

1 Rykalin, N.N.: Berechnung der Wärmevorgänge beim Schweißen. Berlin: VEB Verlag Technik 1957 (Original: Raschety teplovykh protsessov pri svarke. Moscow: Mashgiz 1951)

2 Okerblom, N.O.: Schweißspannungen in Metallkonstruktionen. Halle: VEB Marhold 1959 (Original: Svarochnye napryazheniya i metallokonstruktsii. Moscow: Mashgiz 1955)

3 Okerblom, N.O.: The calculation of deformations of welded metal structures. London: H. M. Stationery Office 1958 (Original: Raschet deformatsii metallokonstruktsii pri svarke. Moscow: Mashgiz 1955)

4 Gunnert, R.: Residual welding stresses. Stockholm: Almqvist & Wiksell 1955

5 Malisius, R.: Schrumpfungen, Spannungen und Risse beim Schweißen. Düsseldorf: DVS-Verlag 1960

6 Hänsch, A.; Krebs, J.: Eigenspannungen und Schrumpfungen in Schweißkonstruktionen. Berlin: VEB Verlag Technik 1961

7 Hänsch, A.: Schweißeigenspannungen und Formänderungen an stabartigen Bauteilen, Berechnung und Bewertung. Berlin: VEB Verlag Technik und Düsseldorf: DVS-Verlag 1984

8 Vinokurov, V.A.: Welding stresses and distortion. Wetherby: British Library 1977 (Original: Svarochnye deformatsii i napryazheniya: metody ikh ustraneniya. Moscow: Mashinostroeniu 1968)

9 Neumann, A.; Röbenack, K.D.: Verformungen und Spannungen beim Schweißen. Düsseldorf: DVS-Verlag 1979

10 Masubuchi, K.: Analysis of welded structures. New York: Pergamon Press 1980

Conference proceedings related to temperature fields, residual stresses and distortion caused by welding:

11 Kihara, H.; Watanabe, M.; Masubuchi, K.; Satoh, K.: Researches on welding stress and shrinkage distortion in Japan. 60th Anniversary Series, Vol. 4. Tokyo: Soc. of Nav. Archit. of Jap. 1959

12 –: Residual stresses in welded constructions and their effects. Abington, Cambr.: The Welding Institute 1977

13 –: Residual stresses and their effects. Abington, Cambr.: Welding Institute 1981

14 Kula, E.; Weiss, V. (Eds.): Residual stress and stress relaxation. 28 Sagamore Army Materials Res. Conf. Proc. New York: Plenum Press 1982

15 Macherauch, E.; Hauk, V. (Eds.): Eigenspannungen, Entstehung-Messung-Bewertung. Oberursel: Deutsche Gesellschaft für Metallkunde 1983

16 Macherauch, E.; Hauk, V. (Eds.): Residual stresses in science and technology. Oberursel: Deutsche Gesellschaft für Metallkunde 1987

17 Sevcuk, O.N.; Pekurovskij, V.J. (Eds.): Mathematische Methoden beim Schweißen (Russian). Kiew: Nauk. dumka, 1981

18 Beck, G.; Denis, S.; Simon, A. (Eds.): International conference on residual stress ICRS2. London: Elsevier Applied Science 1989
19 David, S.A.; Vitek, J.M. (Eds.): Recent trends in welding science and technology. Materials Park, Oh.: ASM International 1990
20 –: Berechnung, Gestaltung und Fertigung von Schweißkonstruktionen im Zeitalter der Expertensysteme, EXPERT '91. Düsseldorf, DVS-Verlag 1991
21 Alder, J.F. (Eds.): The effects of fabrication related stresses. Abington, Cambridge: The Welding Institute 1987.

Temperature field and electrical field, fundamentals:

(see also [1, 8, 190])

22 Carslaw, H.S.; Jaeger, J.C.: Conduction of heat in solids. Oxford: University Press 1973
23 Richter, F.: Die wichtigsten physikalischen Eigenschaften von 52 Eisenwerkstoffen. Stahleisen-Sonderberichte, no. 8, Düsseldorf: Verlag Stahleisen 1973
24 Bowden, R.P.; Williamson, J.B.P.: Electrical conduction in solids, Part 1, Influence of the passage of current on the contact between solids. Proc. Roy. Soc. (Lond.), Ser. A, 246 (1958) July, pp. 1–12
25 Greenwood, J.A.; Williamson, J.B.P.: Electrical conduction in solids, Part 2, Theory of temperature-dependent conductors. Proc. Roy. Soc. (Lond.), Ser. A, 246 (1958) July, pp. 13–31
26 Wilson, E.L.; Bathe, K.J.; Peterson, F.E.: Finite element analysis of linear and nonlinear heat transfer. J. Nucl. Engng. a. Design 29 (1974) pp. 110–124
27 Bathe, K.J.; Khoshgoftaar, M.R.: Finite element formulation and solution of nonlinear heat transfer. J. Nucl. Engng. a. Design 51 (1979) pp. 389–401

Temperature field during seam welding:

(see also [1, 8, 22, 149, 150, 160, 193, 194, 197, 238, 248, 249, 250])

28 Goldak, J.; McDill, M.; Oddy, A.; House, R.; Chi, X.; Bibby, M.: Computational heat transfer for weld mechanics. In: Advances in welding science and technology, pp. 15–20. Metals Park, Ohio: ASMI 1987
29 Rosenthal, D.: Mathematical theory of heat distribution during welding and cutting. Weld. J., Res. Suppl. 20 (1941) no. 5, pp. 220s–234s
30 Rosenthal, D.: The theory of moving sources of heat and its application to metal treatments. Trans. ASME 68 (1946) no. 11, pp. 849–866
31 Christensen, N.; Davies, L.; Gjermundsen, K.: Distribution of temperatures in arc welding. Brit. Weld. J. 12 (1965) pp. 54–75
32 Friedman, E.; Glickstein, S.S.: An investigation of the thermal response of stationary gas tungsten arc welds. Weld. J., Res. Suppl. 55 (1976) pp. 408s–420s
33 Siegfried, W.; Walt, A.: Neue Möglichkeiten für die mathematische Erfassung von Schweißprozessen mit besonderer Berücksichtigung des Lichtbogenschweißens. Schweißtechn. (Zürich) 67 (1977) no. 8, pp. 177–187
34 Sharir, Y.; Grill, A.; Pelleg, J.: Computation of temperatures in thin tantalum sheet welding. Metall. Trans. 11B (1980) pp. 257–265
35 Kou, S.: Simulation of heat flow during the welding of thin plates. Metall. Trans. 12A (1981) pp. 2025–2030
36 Hibitt, H.D.: Application of the finite element method to welding problems. Proc. U.S.–Jap. Semin. on Interdisciplinary Finite Element Analysis, Cornell Univ., Ithaca, Aug. 1978, Publ. 1981
37 Nickel, R.E.; Hibbitt, H.D.: Thermal and mechanical analysis of welded structures. Nucl. Engng. a. Design 32 (1975) pp. 110–120
38 Goldak, J.; Chakravarti, A.; Bibby, M.: A new finite element model for welding heat sources. Metall. Trans. AIME 15B (1984) no. 6, pp. 299–305
39 Nguyen, K.G.; Lesinski, K.J.: Verfahren zur Berechnung von Temperaturfeldern für Schweißnähte beliebiger Form. Schweiß. u. Schneid. 42 (1990) no. 10, pp. 512–514

40 Jhaveri, P.; Moffatt, W.G.; Adams, C.M.: The effect of plate thickness and radiation on heat flow in welding and cutting. Weld. J., Res. Suppl. 41 (1962) no. 1, pp. 12s–16s
41 Goldak, J.; Bibby, J.; Downey, D.; Gu, M.: Heat and fluid flow in welds. In: Advanced joining technologies (Ed.: T.H. North), pp. 69–82. London: Chapman and Hall 1990
42 Grosh, R.J.; Trabant, E.A.: Arc-welding temperatures. Weld. J., Res. Suppl. 35 (1956), no. 8, pp. 396s–400s
43 Kim, Y.-S.; Eagar, T.W.: Temperature distribution and energy balance in the electrode during GMAW. In [19], pp. 13–18
44 McDill, J.M.; Goldak, J.A.; Oddy, A.S.; Bibby, M.J.: Isoparametric quadrilaterals and hexahedrons for mesh-grading algorithms. Comm. Appl. Num. Meth. 3 (1987) pp. 155–163
45 Goldak, J.; Bibby, M.; Moore, J.; House, R.; Pakl, B.: Computer modelling of heat flow in welds. Metall. Trans. B. 17B (1986), pp. 587–600
46 Tekriwal, P.; et al.: Finite element modelling of heat transfer for tungsten arc welding. Metal Constr. 19 (1987), no. 10, pp. 599r–606r
47 Wells, A.A.: Heat flow in welding. Weld. J., Res. Suppl. 31 (1952) no. 5, pp. 263s–267s

Temperature field and electrical field during spot welding:

(see also [160, 212–217, 220])

48 Greenwood, J.A.: Temperatures in spot welding. Brit. Weld. J. 8 (1961) no. 6, pp. 316–322
49 Bentley, K.P.; Greenwood, J.A.; Knowlson, P.M.: Baker, R.G.: Temperature distributions in spot welds. Brit. Weld. J. 10 (1963), pp. 613–619
50 Ruge, J.; Hildebrandt, P.: Einfluß von Temperaturverteilung und Werkstoffeigenschaften auf das Haften der Elektroden beim Widerstandspunktschweißen von Aluminium und Aluminiumlegierungen. Schweiß. u. Schneid. 16 (1964) no. 4, pp. 115–124. Reviewed in [74], ibid. pp. 89–94
51 Rice, W.; Funk, E.J.: An analytical investigation of the temperature distributions during resistance welding. Weld. J., Res. Suppl. 46 (1967) no. 4, pp. 175s–186s
52 Pelli, Z.; Zoller, H.; Galli, G.: Berechnung der Temperaturverteilung beim Anschmelzen einer Platte. Schweiß. u. Schneid. 26 (1974) no. 8, pp. 299–303
53 Kaiser, J.G.; Dunn, G.J.; Eager, T.W.: The effect of electrical resistance on nugget formation during spot welding. Weld. J., Res. Suppl. 61 (1982) no. 6, pp. 167s–174s
54 Nied, H.A.: The finite element modeling of the resistance spot welding process. Weld. J., Res. Suppl. 63 (1984) no. 4, pp. 123s–132s
55 Schwab, R.: Ein Rechenprogramm zur numerischen Temperaturfeldberechnung beim Widerstandsschweißen unter besonderer Berücksichtigung von Transformatorkennlinie, Stromflußverteilung und thermoelektrischen Effekten. Schweiß. u. Schneid. 38 (1986) no. 1, pp. 22–25
56 Schwab, R.: Numerische Berechnung von Temperaturen beim Widerstandsschweißen am Beispiel des Kollektorschweißens. Schweiß. u. Schneid. 38 (1986) no. 8, pp. 365–369
57 Houchens, A.F.; Page, R.E.; Yang, W.H.: Numerical modeling of resistance spot welding. In: Numerical modeling of manufacturing processes, pp. 117–129. New York: ASME 1977
58 Reddy, G.P.; Sharma, S.: Simulation of the spot welding process. In [19], pp. 59–68
59 Öhlschläger, E.: Untersuchungen zum Laserpunktschweißvorgang und zur Tragfähigkeit lasergeschweißter Metallverbindungen. Diss. TU Berlin 1986
60 Gould, J.E.: An examination of nugget development during spot welding using both experimental and analytical techniques. Weld. J., Res. Suppl. 66 (1987) no. 1, pp. 1s–10s
61 Cho, H.S.; Cho, Y.J.: A study of the thermal behavior in resistance spot welds. Weld. J., Res. Suppl. 68 (1989) no. 6, pp. 236s–244s
62 Roberts, W.L.: Resistance variations during spot welding. Weld. J., Res. Suppl. 30 (1951) no. 11, pp. 1004–1019
63 Dickson, D.W.; Franklin, J.E.; Stanya, A.: Characterization of spot welding behavior of dynamic electrical parameter monitoring. Weld. J., Res. Suppl. 59 (1980) no. 6, pp. 170s–176s
64 Savage, W.F.; Nippes, E.F.; Wassell, F.A.: Dynamic contact resistance of series spot welds. Weld. J., Res. Suppl. 57 (1978) no. 2, pp. 43s–50s
65 Archer, G.R.: Calculations for temperature response in spot welds. Weld. J., Res. Suppl. 39 (1960) no. 12, pp. 327s–330s
66 Dickinson, D.; Tsai, C. L.; Jammal, O.: Modelling of resistance spot nugget growth – applications for the automotive industry. SAE Paper 900738. Warrendale, Pa.: SAE 1990 (identical with [67])

67 Tsai, C.L.; Dickinson, D.; Jammal, O.A.: Study of nugget formation in resistance spot welding using finite element method. In [19], pp. 43–52 (identical with [66])

Temperature field during friction welding:

(see also [448–450])

68 Potente, H.; Tappe, P.: Theoretische und experimentelle Analyse des Rotationsreibschweißens von Kunststoffen. Schweiß. u. Schneid. 41 (1989) no. 8, pp. 392–396
69 Potente, H.; Kaiser, M.: Ein mathematisch-physikalisches Modell zur Beschreibung des stationären Schweißprozesses beim Reibschweißen von Kunststoffen. Schweiß. u. Schneid. 42 (1990) no. 7, pp. 342–347

Temperature field during flame heating:

(see also [1])

70 Stutzenberger, W.: Konvektiver Wärmeübergang von Brenngas-Sauerstoff-Flammen an gekühlte und sich erwärmende Metalloberflächen. Schweißtechn. Forschungsber. no. 14. Düsseldorf: DVS-Verlag 1987
71 Stutzenberger, W.; Kremer, H.: Untersuchung des instationären Erwärmens metallischer Werkstoffe durch Einzel- und Mehrfachflammen der Autogentechnik. Schweiß. u. Schneid. 41 (1989) no. 6, pp. 289–293

Welding heat sources:

(see also [1])

72 Lancaster, J.F. (Ed.): The physics of welding. Oxford: Pergamon Press 1986
73 Schellhase, M.: Der Schweißlichtbogen, ein technologisches Werkzeug. Düsseldorf: DVS-Verlag 1985
74 Ruge, J.: Handbuch der Schweißtechnik, Vol. 2: Verfahren und Fertigung. Berlin: Springer 1980
75 Killing, R.: Handbuch der Schweißverfahren, Part 1, Lichtbogenschweißverfahren. Düsseldorf: DVS-Verlag 1984
76 Pfeifer, L.: Fachkunde des Widerstandsschweißens. Essen: Girardet 1969
77 Anders, W.: Einfluß der Schweißverfahren und der Schweißbedingungen auf die Wärmeeinbringung beim Schweißen. Schweißtechn. (Berl.) 18 (1968) no. 7, pp. 326–328
78 Dilthey, U.; Killing, R.: Beitrag zur Berechung des Wärmeeinbringens beim Metall-Schutzgasschweißen mit Impulslichtbogen. Schweiß. u. Schneid. 39 (1987) no. 10, pp. 495–497; Commentaries: Schweiß. u. Schneid. 40 (1988) no. 3, pp. 146–147; 40 (1988) no. 9, pp. 457–458; 41 (1989) no. 5, pp. 245–246
79 Beckert, M.; Neumann, A.: Grundlagen der Schweißtechnik-Energiequellen. Berlin: VEB Verlag Technik 1988
80 Davis, A.C.: The science and practice of welding, Vol. 1 and 2. Cambridge: Cambridge University Press 1989

Weld pool modelling:

(see also [72, 129])

81 Szekely, J.: Transport phenomena in welds with emphasis on free surface phenomena. In [19], pp. 3–12
82 Zacharia, T.; David, S.A.; Vitek, J.M., DebRoy, T.: Modeling the effect of surface active elements on weld pool fluid flow, heat transfer and geometry. In [19], pp. 25–30
83 Matsunawa, A.; Yokoya, S.: Fluid flow and its effect on penetration shape in stationary arc welds. In [19], pp. 31–36
84 Kou, S.: Grain structure development in the fusion zone. In [19], pp. 137–146
85 Alexiades, V.: Shrinkage effects on solidification. In [19], pp. 171–176

86 Sudnik, W.: Untersuchung und Projektierung der Schmelzschweißtechnologien mit Hilfe von physikalisch-mathematischen Modellen und Computer. Schweiß. u. Schneid 43 (1991) no. 10, pp. 588–590
87 Matsunawa, A.: Role of surface tension in fusion welding. IIW Doc. 212-618-85; Trans. IWRI 11 (1982) no. 2, pp. 145–154 (Part 1), 12 (1983) no. 1, pp. 123–130 (Part 2), 13 (1984) no. 1, pp. 147–156 (Part 3)
88 Heiple, C.R.; Roper, J.R.; Stagner, R.T.; Alsen, J.J.: Surface active element effects on the shape of GTA, laser and electron beam welds. Weld. J., Res. Suppl. 62 (1983) no. 3, pp. 72s–77s
89 Athey, D.R.: A mathematical model for fluid flow phenomena in weld pools. J. Fluid Mech. 98 (1980) pp. 787–801
90 Oreper, G.M.; Szekely, J.: Heat and fluid flow phenomena in weld pools. J. Fluid. Mech. 147 (1984) pp. 53–79
91 McLay, R.; Carey, G.F. Coupled heat transfer and viscous flow and magnetic effects in weld pool analysis. Int. J. Num. Meth. Engng. 9 (1989) pp. 713–730
92 Zacharia, T.; Eraslan, A.H. Aidun, D.K.: Modelling of non-autogenous welding. Weld. J. 67 (1988) pp. 18s–27s
93 Steffens, H.-D.; Thier, H.; Killing, R.; Sievers, E.-R.; Li, Z.: Auswirkung der Eisenbegleitelemente auf die Schmelzzonengeometrie beim vollmechanischen Wolfram-Inertgasschweißen von austenitischen Stählen. Schweiß. u. Schneid. 42 (1990) no. 7, pp. 338–342
94 Mills, G.S.: Fundamental mechanisms of penetration in GTA welding. Weld. J., Res. Suppl. 58 (1979) no. 1, pp. 21s–24s
95 Heiple, C.R.; Roper, J.R.: Mechanisms for minor element effects on GTA fusion zone geometry. Weld. J., Res. Suppl. 63 (1982) no. 4, pp. 97s–102s
96 Heiple, C.R.; Burgardt, P.: Interaction between impurities and welding variables in determining GTA weld shape. Weld. J., Res. Suppl. 65 (1986) no. 6, pp. 150s–155s
97 Hammerschmid, P.: Bedeutung des Marangoni-Effekts für metallurgische Vorgänge. Stahl u. Eisen 107 (1987) no. 1, pp. 61–66
98 Ohji, T.; Nishiguchi, K.: Mathematical modelling of a molten pool in arc welding of thin plate. Technology Reports of the Osaka Univ. 33 (1983) no. 1688, pp. 35–43
99 Pardo, E.; Weckman, D.C.; The interaction between process variables and bead shape in GMA welding: a finite element analysis. In [19], pp. 391–395
100 Na, S.-J.; Ruge, J.: Temperaturfelder beim Schweißen von Aluminiumdruckguß. Schweiß. u. Schneid. 42 (1990) no. 7, pp. 327–332
101 Eager, T.W.: An iconoclast's view of physics of welding – rethinking old ideas. In [19], pp. 341–346
102 Blumschein, E.: Kurzlichtbogenschweißen auf dem Computer. Schweiß. u. Schneid. 42 (1990) no. 11, pp. 553–556
103 Blumschein, E.: Software für die Schweißnahtform. Schweiß. u. Schneid. 43 (1991) no. 3, pp. 154–157
104 Pandey, S.; Parmax, R.S.: Mathematical models for predicting bead geometry and shape relationship for MIG welding of aluminium alloy 5083. In [19], pp. 37–41
105 Hashimoto, T.; Matsuda, F.: Piercing mechanism of the weld pool in electron beam welding. Trans. Nat. Res. Inst. Met. 7 (1965), pp. 117–185

Microstructural change, heat affected zone:

(see also [45, 46])

106 Rose, A.: Die Bedeutung der Zeit-Temperatur-Umwandlungsschaubilder und ihre Anwendung in der Schweißtechnik. Schweiß. u. Schneid. 8 (1956) no. 11, pp. 442–449
107 Rose, A.: Schweißbarkeit und Umwandlungsverhalten der Stähle. Forschungsberichte Nordrhein-Westfalen no. 1534. Köln Opladen: Westdeutscher Verlag 1965
108 Rose, A.: Schweißbarkeit der hochfesten Baustähle, Einfluß der Schweißbedingungen auf das Werkstoffverhalten. Stahl u. Eisen 86 (1966) no. 11, pp. 663–672
109 Rose, A.; Hougardy, H.: Atlas zur Wärmebehandlung der Stähle, Vol. 2. Düsseldorf: Verlag Stahleisen 1972
110 Hofmann, W.; Müller, R.: Umwandlung und Schweißbarkeit von niedriglegierten Stählen. Schweiß. u. Schneid. 8 (1965) no. 7, pp. 237–240

111 Hofmann, W.; Burat, F.: Beitrag zur Schweißbarkeit unlegierter und niedriglegierter Bau- und Vergütungsstähle. Schweiß. u. Schneid. 14 (1962) pp. 289–299
112 Berkhout, C.; Leut, P.v.: Anwendung von Spitzentemperatur-Abkühlzeit (STAZ)-Schaubildern beim Schweißen hochfester Stähle. Schweiß. u. Schneid. 20 (1968) no. 6, pp. 256–260
113 Ruge, J.; Gnirß, G.: Untersuchung der mechanischen Eigenschaften in der Wärmeeinflußzone von Schweißverbindungen an synthetischen Proben aus hochfestem Baustahl. Schweiß. u. Schneid. 23 (1971) no. 7, pp. 255–258
114 Gnirß, G.; Ruge, J.: Simulation von Schweißtemperaturzyklen und ihre Anwendung zur Beurteilung der Schweißeignung eines Feinkornstahles. Schweiß. u. Schneid. 27 (1975) no. 6, pp. 221–224
115 Hrivnak, I.; Stembera, W.; Mraz. L.: Umwandlungsschaubilder für höherfeste Baustähle und Schweißgut. Bratislava: VUZ 1980
116 Seyffarth, P.: Atlas Schweiß-ZTU-Schaubilder. Berlin: VEB Verlag Technik and Düsseldorf: DVS-Verlag 1982
117 Siegfried, W.: Neue Gesichtspunkte für die Berechnung des Gefüges in der Wärmeeinflußzone von Schweißverbindungen. Schweiß. u. Schneid. 35 (1983) no. 12, pp. 595–601
118 Tanaka, K.; Iwasaki, R.; Nagaki, S.: On T-T-T- and C-C-T-diagrams of steels, a phenomenological approach to transformation kinetics. Ing.-Arch. 54 (1984) pp. 81–90
119 Ruge, J.: Handbuch der Schweißtechnik. Vol. 1: Werkstoffe. Berlin: Springer 1980
120 Watt, D.; Coon, L.; Bibby, M.; Goldak, J.; Henwood, C.: Modelling microstructural development in weld heat-affected zones. Acta Metall. 36 (1988) no. 11, pp. 3029–3035
121 Henwood, C.; Bibby, M.; Goldak, J.; Watt, D.: Coupled transient heat transfer, microstructure weld computations. Acta Metall. 36 (1988) no. 11, pp. 3037–3046
122 Frank, G.: Schweiß-ZTU-Schaubilder und Eigenschaftsdiagramme von Baustählen mit Hilfe von Computern. Düsseldorf: DVS-Verlag 1990
123 Frank, G.: Berechnen und Darstellen von Schweiß-Zeit-Temperatur-Umwandlungsschaubildern für Kohlenstoff-Mangan-Stähle mit Computern. Schweiß. u. Schneid. 41 (1989) no. 11. pp. 586–589
124 Buchmayr, B.: WEZ-Kalkulator, ein Softwarepaket zum Berechnen schweißtechnischer Kennwerte. Schweiß. u. Schneid. 41 (1989) no. 3, pp. 69–75
125 Buchmayr, B.; Cerjak, H.: Mathematical description of HAZ behaviour of low-alloyed structural steels. In: Weld quality – the role of computers, pp. 43–50. Oxford: Pergamon Press 1988
126 Seyffarth, P.: Schweißtechnisches rechnergestütztes Beratungssystem WELDWARE. Paper supplementing Ref. [20]
127 Buchmayr, B.: WEZ-Kalkulator – ein modernes Hilfsmittel für den Schweißingenieur. In [20], pp. 56–60
128 Easterling, K.E.: Modelling the weldability of structural steels. In: Welding metallurgy of structural steels, pp. 45–62. AIME 1987
129 Kou, S.: Welding metallurgy. New York, John Wiley 1987
130 Savage, W.F.: Solidification, segregation and weld imperfections. Weld. in the World 18 (1980) pp. 89–114
131 Pätzold, U.; Hou, C.-X.; Ruge, J.: Bestimmung von Gefügeinhomogenitäten in der Wärmeeinflußzone von Schweißverbindungen. Schweiß. Schneid. 41 (1989) no. 3, pp. 134–137
132 Leblond, J.B.; Devaux, J.; Devaux, J.-C.: Modellierung metallurgischer Umwandlungsprozesse. CAD-CAM Report no. 6 (1990), pp. 78–89
133 Easterling, K.: Introduction to physical metallurgy of welding. London: Butterworth 1985
134 Wegrzyn, J.: Physics and Metallurgy of welding (Polish). Gliwice: Publisher TH 1990
135 Easterling, K.E.: Microstructure and properties of the heat affected zone. In [19], pp. 177–188
136 Reed, R.C.; Bhadeshia, H.K.: Complete reaustenitisation in multirun steel weld deposits. In [19], pp. 205–209
137 Umgeher, A.; Cerjak, H.: Investigations on the HAZ-microstructures and properties of the high strength aluminium alloy 7075. In [19], pp. 279–283
138 Buchmayr, B.; Kirkaldy, J.S.: Modeling of the temperature field, transformation behavior, hardness and mechanical response of low alloy steels during cooling from the austenite region. J. Heat Treat. 8 (1990) no. 2, pp. 127–136
139 Mitter, M.: Umwandlungsplastizität und ihre Berücksichtigung bei der Berechnung von Eigenspannungen. Berlin: Gebr. Borntraeger 1987
140 Denis, S.; Gautier, E.; Simon, A.: Modelling the mechanical behaviour of steels during phase transformation – a review. In [18], pp. 393–398

141 Leblond, J.B.; Devaux, J.: Interactions between transformation plasticity and strain hardening phenomena in steels. In [13], pp. 410–415
142 Sjöström, S.; Denis, S.; Gautier, E.; Simon, A.: FEM calculation of the micromechanics of a diffusional transformation. In [18], pp. 416–422
143 Klueh, R.L.; Canonico D.A.: Microstructure and tensile properties of 2 1/2 Cr-1Mo steel weldments with varying carbon contents. Weld. J., Res. Suppl. 55 (1976) no. 8, pp. 253s–260s
144 Evans, G.M.: The effect of heat input on the microstructure and properties of C-Mn all-weld-metal deposits. Weld. J., Res. Suppl. 61 (1982) no. 4, pp. 125s–130s
145 Chung, S.H.; Takahasi, H.; Suzuki, M.: Microstructural gradient in HAZ and its influence upon the HAZ fracture toughness. Weld. in the World 16 (1978) no. 11/12, pp. 248–261
146 Nakao, Y.; Oshige, H.; Noi, S.: Distribution of microstructure in HAZ of multi-pass welded high strength steel. Faculty of Engineering, Osaka Univ., 1984
147 Challenger, K.D.; Brucker, R.B.; Elger, W.M.; Sorek, M.J.: Microstructure–thermal history correlations for HY-130 thick section weldments. Weld. J., Res. Suppl. 63 (1984) no. 8, pp. 254s–273s
148 Abson, D.J.; Kosik, O.; Uvira, J.L.; Jonas, J.J.: Correlation between room-temperature hardness and flow stress in hot worked metals. In: The science of hardness testing and its research application (Eds.: J.H. Westbook and H. Conrad), pp. 91–110. ASM, Metals Park 1978
149 Ashby, M.F.; Easterling, K.E.: A first report on diagrams for grain growth in welds. Acta Metall. 30 (1982) no. 12, pp. 1969–1978
150 Ashby, M.F.; Easterling, K.E.: The transformation hardening of steel surfaces by laser beams – I. Hypo-eutectoid steels. Acta Metall. 3 (1984) no. 11, pp. 1935–1948
151 Ion, J.C.; Easterling, K.E.; Ashby, M.F.: A second report on diagrams of microstructure and hardness for heat affected zones in welds. Acta Metall. 32 (1984) no. 11, pp. 1949–1962
152 Leblond, J.B.; Devaux, J; Devaux, J.C.: Mathematical modelling of transformation plasticity in steels. I – Case of ideal plastic phases: Int. J. Plasticity 5 (1989) pp. 551–572
153 Leblond, J.B.; Devaux, J; Devaux, J.C.: Mathematical modelling of transformation plasticity in steels II – Coupling with strain hardening phenomena. Int. J. Plasticity 5 (1989) pp. 573–591
154 Rammerstorfer, F.G.; Fischer, D.F.; Witter. W.; Bathe, K.J.; Snyder, M.D.: On thermo-elastic-plastic analysis of heat treatment processes including creep and phase changes. Comput. a. Struct. 13 (1981) pp. 771–779
155 Greenwood, G.W.; Johnson. R.H.: The deformation of metals under small stresses during phase transformation. Proc. Roy. Soc. (Lond.), Ser. A, 283 (1965) pp. 403–422

Heat input, cooling rate, cooling time, dwell time:

(see also [1])

156 Eichhorn, F.; Niederhoff, K.: Streckenenergie als Kenngröße des Wärmeeinbringens beim mechanisierten Lichtbogenschweißens. Schweiß. u. Schneid. 24 (1972) no. 10, pp. 399–403
157 Röbenack, K.D.; Hüther, G.; Röhling, S.: Bestimmung des effektiven Wärmenutzungsfaktors beim Schmelzschweißprozeß. Schweißtechn. (Berl.) 27 (1977) no. 12, pp. 562–563
158 Anders, W.: Einfluß der Schweißverfahren und der Schweißbedingungen auf die Wärmeeinbringung beim Schweißen. Schweißtechn. (Berl.) 18 (1968) no. 7, pp. 326–328
159 Siegried, W.: Die Erfassung der Einflußgrößen des Schweißprozesses durch die Bestimmung dimensionsloser physikalischer Parameter. Schweiß. u. Schneid. 17 (1965) no. 11, pp. 595–604
160 Myers, P.S.; Uyehara, O.A.; Borman, G.L.: Fundamentals of heat flow in welding. Weld. Res. Counc. Bull., no. 123, July 1967
161 Uwer, D.; Degenkolbe, J.: Temperaturzyklen beim Lichtbogenschweißen, Berechnung von Abkühlzeiten. Schweiß. u. Schneid. 24 (1972) no. 12, pp. 485–489
162 Uwer, D.; Degenkolbe J.: Temperaturzyklen beim Lichtbogenschweißen, Einfluß des Wärmebehandlungszustandes und der chemischen Zusammensetzung von Stählen auf die Abkühlzeit. Schweiß. u. Schneid. 27 (1975) no. 8, pp. 303–306
163 Stahl-Eisen-Werkstoffblatt SEW 088 Beiblatt: Schweißgeeignete Feinkornbaustähle, Richtlinien für die Verarbeitung, besonders für das Schmelzschweißen, Ermittlung der Abkühlzeit $t_{8/5}$ zur Kennzeichnung von Schweißtemperaturzyklen. Düsseldorf: Verlag Stahleisen 1987
164 Uwer, D.; Degenkolbe, J.: Kennzeichnung von Schweißtemperaturzyklen hinsichtlich ihrer Auswirkung auf die mechanischen Eigenschaften von Schweißverbindungen. Schweißtechn. (Zürich) 71 (1981) no. 2, pp. 45–53

165 Degenkolbe, J.; Uwer, D.; Wegmann, H.: Characterisation of weld thermal cycles with regard to their effect on the mechanical properties of welded joints by the cooling time $t_{8/5}$ and its determination. IIW-Doc. IX-1336-84
166 Inagaki, M.; Nakamura, H.; Okada, H.: Eine Untersuchung über den Abkühlungsprozeß beim Lichtbogenschweißen, verdecktes Lichtbogenschweißen und UP-Schweißen. Rep. Nat. Res. Inst. f. Met. (Tok.) 7 (1964) no. 4, pp. 296–308
167 Inagaki, M.; Nakamura, H.; Okada, A.: Untersuchungen über Abkühlungsprozesse beim Schweißen mit umhüllten Elektroden und beim UP-Schweißen. J. Jap. Weld. Soc. (Tokyo) 34 (1965) no. 10, pp. 1064–1075
168 Inagaki, M.; Okada, A.: Über den Abkühlungsprozeß beim Lichtbogenschweißen mit örtlicher Erwärmung durch eine Flamme. Rep. Nat. Res. Inst. f. Met. (Tokyo) 9 (1966) no. 4, pp. 245–253
169 Adkins, G.; et al.: Effective heat input study, phase 1, final report. Weld. Res. Counc., Proj. Rep. 42 (1987) no. 1/2, pp. 19–22
170 Radaj, D.: Kritische Anmerkungen zur Schweißrichtlinie für Feinkornbaustühle. Schweiß. u. Schneid. 40 (1988) no. 4, pp. 192–193

Hydrogen diffusion:

(see also [132])

171 Pochodnja, I.K.; Demtschenko, V.F. a. L.I.: Matematitscheskoe modelirovanie povedenija gasov v soarnich soedinenijach. Kiev: Naukova Dumka 1979
172 Yurioka, N.; Ohshita, S.; Nakamura, H.; Asano, K.: An analysis of effects of microstructure, strain and stress on the hydrogen accumulation in the weld heat affected zone. IIW-Doc. IX-1161-80
173 Karkhin, V.A.; Michailov, V.G.: Wasserstoffverteilung beim Schweißen von einlagigen Nähten (Russian). Avtom. Svarka (1985) no. 6, pp. 39–42
174 Karkhin, V.A.; Michailov, V.G.: Wasserstoffverteilung beim Schweißen von mehrlagigen Nähten (Russian). Trudi LPI (1983) no. 395, pp. 7–13
175 Michailov, V.G.: Modellierung der Verteilung des Wasserstoffs und Stickstoffs in Schweißverbindungen aufgrund der Thermodiffusion (Bulgarian). Technitscheska misal (1989) no. 1, pp. 109–114
176 Dickehut, G.; Ruge, J.: Wasserstoffverteilung in der Schweißnaht – Theorie zur Berechnung. Schweiß. u. Schneid. 40 (1988) no. 6, pp. 289–292
177 Ruge, J.; Dickehut, G.: Wasserstoffverteilung in der Schweißnaht, Verteilung in Unterpulvereindraht- und Unterpulvertandemschweißungen. Schweiß. u. Schneid. 40 (1988) no. 8, pp. 393–396
178 Michailov, V.G.; Ruge, J.; Thomas, K.: Berechnung der Wasserstoffverteilung beim Schweißen. Schweiß. Schneid. 43 (1991) no. 11, pp. 655–658
179 Michailov, V.G.; Karkhin, V.A.: Using the finite element method in predicting the constructional steels susceptibility to cold cracking during welding. Second Intern. Conf. "Computer technology in welding", Cambridge, 8–9 June 1988. Organised by The Welding Institute, Abington 1988 (paper submitted but not published)

Elastic and inelastic thermomechanics:

180 Melan, E.; Parkus, H.: Wärmespannungen infolge stationärer Temperaturfelder. Wien: Springer 1953
181 Parkus, H.: Instationäre Wärmespannungen. Wien: Springer 1959
182 Boley, B.A.; Weiner, J.H.: Theory of thermal stresses. New York: J. Wiley 1960
183 Nowacki, W.: Thermoelasticity. Oxford: Pergamon Press 1962
184 Ziegler, H.: An introduction to thermomechanics. Amsterdam: North-Holland 1977

Elastic-plastic continua:

185 Hill, R.: The mathematical theory of plasticity. Oxford: Clarendon Press 1950
186 Hoffman, O.; Sachs, G.: Introduction to the theory of plasticity for engineers. London: McGraw Hill 1953

187 Johnson, W.; Mellor, P.W.: Plasticity for mechanical engineers. Princetown: Van Norstrand 1962
188 Mendelsohn, A.: Plasticity, theory and application. New York: J. Wiley 1968

Elastic-plastic finite elements:

189 Argyris, J.H.; Scharpf, D.W.; Spooner, J.B.: Die elastoplastische Berechnung von allgemeinen Tragwerken und Kontinua. Ing.-Arch. 37 (1969) no. 5, pp. 326–352
190 Zienkiewicz, O.C.; Cheung, Y.K.: The finite element method in structural and continuum mechanics. London: McGraw-Hill 1967
191 Owen, D.R.J.; Hinton, E.: Finite elements in plasticity, theory and practice. Swansea, Wales: Pineridge Press 1980
192 Bathe, K.J.: Finite Element procedures in engineering analysis. Englewood Cliffs: Prentice Hall 1982

Basic equations in welding residual stress analysis:

(see also [8, 181])

193 Argyris, J.H.; Szimmat, J.; William, K.J.: Computational aspects of welding stress analysis. Comput. Meth. in Appl. Mech. a Engng. 33 (1982) pp. 635–666
194 Karlsson, L.: Thermal stresses in welding. In: Thermal stresses, Vol. 1 (Ed.: R.B. Hetnarski), pp. 299–389. Amsterdam: North-Holland 1986
195 Radaj, D.: Matrizenverschiebungsmethode für temperaturabhängig elastisch-plastische Tragwerke und Kontinua. Acta Mechanica 14 (1972) pp. 71–78
196 Radaj, D.: Vollständige Spannungs-Dehnungs-Temperaturänderungsbeziehung für die Schweißeigenspannungsberechnung mit finiten Elementen. Schweiß. u. Schneid. 27 (1975) no. 10, pp. 394–396
197 Farong, Y.; Huadong, S.: Transient temperature fields and residual stress fields of metallic materials under welding. In [18], pp. 504–509

Thermomechanical material characteristic values:

(see also [10, 23, 72, 138, 140, 193, 215, 234, 238, 249])

198 Toulokian, Y.S.: Thermophysical properties of high temperature solid materials. New York: Macmillan 1967
199 –: Handbuch der Kennwerte von metallischen Werkstoffen, Vols. 1 and 2, Düsseldorf: DVS-Verlag 1990

Status reports on finite element analysis of welding residual stress:

(see also [194])

200 Radaj, D.: Berechnung der Schweißeigenspannungen und Schweißformänderungen mit elastisch-plastischen finiten Elementen. Schweiß. u. Schneid. 27 (1975) no. 7, pp. 245–249
201 Radaj, D.: Welding stress analysis with elastoplastic finite elements. Trans. 3rd Int. Conf. Struct. Mech. React. Technol., London 1975, Vol. 5, Part M, pp. 1–17
202 Masubuchi, K.: Report on the state-of-the-art of numerical analysis of stresses, strains and other effects produced by welding. IIW-Doc. X–738–74
203 Radaj, D.: Finit-Element-Berechnung von Temperaturfeld, Eigenspannungen und Verzug beim Schweißen. Schweiß. u. Schneid. 40 (1988) no. 6, pp. 269–276
204 Radaj, D.: Finite element analysis of welding residual stress. In [18], pp. 510–516
205 Ueda, Y.; Murakawa, H.: Applications of computer and numerical analysis techniques in welding research. Trans. JWRI 13 (1984) no. 2, pp. 337–346
206 Goldak, J.: Modeling thermal stresses and distortions in welds. In [19], pp. 71–82

Rod element model:
(see also [2, 8])

207 Tall, L.: Residual stresses in welded plates - a theoretical study. Weld. J., Res. Suppl. 43 (1964) no. 1, pp. 10s–23s
208 Tall, L.; Feder, D.: Längsschweißung in Platten und ihr Einfluß auf die Grenzlast von geschweißten Stahlstützen. Schweiß u. Schneid. 17 (1965) no. 3, pp. 99–107
209 Radaj, D.: Berechnung der Schweißeigenspannungen in Stäben mit Längsnähten. Schriftenreihe Schweiß u. Schneid. 2 (1971) no. 7, pp. 1–8
210 Radaj, D.: Schweißeigenspannungen und Verwerfungen: In [438], vol. 1, pp. 146–173
211 Karthik, V.G.; Radhakrishnan, V.M.: Analysis of heat affected zone and residual stresses in welds. Scand. J. Metallurgy 16 (1987) no. 6, pp. 262–266

Ring element model of heat and weld spot:
(see also [48–57, 59–67])

212 Gurney, T.R.: Residual stresses in a large circular disc caused by local heating and cooling at its centre. J. Strain Anal. 6 (1971) no. 2, pp. 89–98
213 Watanabe, M.; Satoh, K.: Thermal stress and residual stress of circular plate heated at its center. J. Soc. Nav. Archit. Jap. 86 (1954) pp. 185–197
214 Pelli, T.: Berechnung der beim Schweißen entstehenden inneren Spannungen und ihre Abhängigkeit von der Wärmeabgabe und der Form des Spannungs-Dehnungs-Diagrammes. Schweiß. u. Schneid 27 (1975) no. 5, pp. 215–217
215 Lindh, D.V.; Tocher, J.L.: Heat generation and residual stress development in resistance spot welding. Weld. J., Res. Suppl. 46 (1967) no. 8, pp. 351s–360s
216 Nguyen, C.T.: Eigenspannungen in Widerstandspunktschweißverbindungen. Schweiß. u. Schneid. 32 (1980) no. 1, pp. 10–12
217 Schröder, R.; Macherauch, E.: Berechnung der Wärme- und Eigenspannungen bei Widerstandspunktschweißverbindungen unter Zugrundelegung unterschiedlicher mechanisch-thermischer Werkstoffdaten. Schweiß. u. Schneid. 35 (1983) no. 6, pp. 270–276
218 Gorissen, E.: Schweißtemperatur und Schweißeigenspannungen bei Großrohren. Arch. f. d. Eisenhüttenw. 37 (1966) no. 1, pp. 49–55
219 Wellinger, K.; Krägeloh, E.: Letters to the editor concerning Ref. [218]. Arch. f. d. Eisenhüttenw. 39 (1968) no. 4, pp. 311–318
220 Popkovskii, V.A.; Berezienko, V.P.: Effect of resistance spot welding conditions on the distribution of residual stresses. Weld. Int. 2 (1988) no. 12, pp. 1058–1061

Ring element model of ring weld and circular rod butt weld:
(see also [8, 37])

221 Hibitt, H.D.; Marcal, P.V.: Numerical thermomechanical model for the welding and subsequent loading of a fabricated structure. J. Comput. a. Struct. 3 (1973) pp. 1145–1174
222 Marcal, P.V.: Weld problems. In: Structural Mechanics Computer Programs, pp. 191–206. Charlottesville: University Press 1974
223 Yu, H.J.: Berechnung von Abkühlungs-, Umwandlungs-, Schweiß-sowie Verformungseigenspannungen mit Hilfe der Methode der Finiten Elemente. Diss. Univ. Karlsruhe, 1977
224 Cyr, N.A.; Teter, R.D.; Stocks, B.B.: Finite element thermoelastoplastic analysis. J. Struct. Div., Proc. Am. Soc. Civ. Eng. 98 (1972) no. ST7, pp. 1585–1603
225 Nied, H.A.: Interface displacement characteristics of upset welding. In [19], pp 453–459

Ring element model of circumferential weld in cylindrical or spherical shell:
(see also [379, 380, 423, 426])

226 Fujita, Y.; Nomoto, T.; Hasegawa, H.: Welding deformations and residual stresses due to circumferential welds at the joint between cylindrical drum and hemispherical head plate. IIW-Doc. X-985-81

227 Fujita, Y.; Nomoto, T.; Hasegawa, H.: Deformations and residual stresses in butt welded pipes and shells. Nav. Archit. a. Ocean Engng. (Soc. of Nav. Archit. of Jap.) 18 (1980) pp. 164–174 and IIW-Doc. X–963–80
228 Rybicki, E.F.; Schmueser, D.W.; Stonesifer, R.W.; Groom, J.J.; Mishler, H.W.: A finite element model for residual stresses and deflections in girth-butt welded pipes. J. Press. Vessel Technol. (ASME) 100 (1978) no. 10, pp. 256–262
229 Rybicki, E.F.; Stonesifer, R.B.: Computation of residual stresses due to multipass welds in piping systems. J. Press. Vessel Technol. (ASME) 101 (1979) no. 5, pp. 149–154
230 Hepworth, J.K.: Finite element calculation of residual stresses in welds. In: Proc. Int. Conf. on Numerical Methods for Non-linear Problems, pp. 51–60. Swansea, Wales: Pineridge Press 1980
231 Lobitz, D.W.; McClure, J.D.; Nickell, R.E.: Residual stresses and distortions in multi-pass welding. In: Numerical modelling of manufacturing processes, ASME booklet PVP-PB-025. New York: ASME 1977

Membrane plate element model in plate plane:

(see also [194, 287])

232 Karlsson, L.; Akesson, B.A.: Plane stress field induced by a concentrated heat source moving perpendicular toward free edge of semi-infinite plate. J. Appl. Mech. (ASME) 96 (1974) pp. 825–827
233 Akesson, B.; Karlsson, L.: Prevention of hot cracking of butt welds in steel panels by controlled additional heating of the panels. Weld. Res. Int. 6 (1976) no. 5, pp. 35–52
234 Karlsson, L.: Plane stress fields induced by moving heat sources in butt-welding. J. Appl. Mech. (ASME) 99 (1977) pp. 231–236
235 Andersson, B.: Thermal stresses in a submerged-arc welded joint considering phase transformations. J. Engng. Mater. a. Technol. (ASME) 100 (1978) no. 10, pp. 356–362
236 Andersson, B.; Karlsson, L.: Thermal stresses in large butt-welded plates. J. Therm. Stress. 4 (1981) no. 3/4, pp. 491–500
237 Jonsson, M.; Karlsson, L.; Lindgren, L.E.: Deformations and stresses in butt-welding of large plates with special reference to the mechanical material properties. J. Engng. Mater. Technol. (ASME) 107 (1985) no. 4, pp. 265–270
238 Jonsson, M.; Karlsson, L.; Lindgren, L.E.: Deformations and stresses in butt-welding of large plates. In: Numerical methods in heat transfer, Vol. 3 (Ed.: Lewis, R.W.), pp. 35–57. London: J. Wiley 1985
239 Fujita, Y.; Nomoto, T.: Studies on thermal stresses in welding with special references to weld cracking. In: Prepr. 1st Int. Symp. on Precaution of Cracking in Welded Structures, pp. IC6.1–IC6.12. Tokyo: Jap. Weld. Soc. 1971
240 Fujita, Y.; Nomoto, T.; Hasegawa, H.: Thermal stress analysis based on initial strain method. Nav. Archit. a. Ocean Engng. (Soc. of Nav. Archit. of Jap.) 17 (1979) pp. 174–183
241 Lindgren, L.E.; Karlsson, L.: Deformations and stresses in welding of shell structures. Int. J. Num. Meth. Engng. 25 (1988) pp. 635–655

Membrane plate element model transverse to plate plane:

(see also [193])

242 Ueda, Y.; Yamakawa, T.: Analysis of thermal elastic plastic behaviour of metals during welding by finite element method (Japanese). J. Jap. Weld. Soc. 42 (1973) no. 6, pp. 61–71
243 Ueda, Y.; Takahashi, E.; Fukuda, K.; Nakacho, K.: Transient and residual stresses in multi-pass welds. IIW-Doc. X–698–73
244 Satoh, K.; Terai, K.; Yamada, S.; Matsui, S.; Ohkuma, Y.; Kinoshita, T.: Theoretical study on transient restraint stress in multi-pass welding. Trans. Jap. Weld. Soc. 6 (1975) no. 1, pp. 42–52
245 Ueda, Y.; Fukuda, K.; Nakacho, K.: Basic procedures in analysis and measurement of welding residual stresses by the finite element method. In [12], pp. 27–37
246 Ueda, Y.; Nakacho, K.: Simplifying methods for analysis of transient and residual stresses and deformations due to multipass welding. Trans. Jap. Weld. Res. Inst. 11 (1982) pp. 95–103

247 Guth, W.; Szimmat, J.: Numerical and experimental determination of residual stresses in multi-layer welds. In [16], pp. 1025–1031
248 Argyris, J.H.; Szimmat, J.; Willam, K.J.: Finite element analysis of arc welding processes. In: Numerical methods in heat transfer, Vol. 3 (Ed. R.W. Lewis), pp. 1–34. London: J. Wiley 1985
249 Friedman, E.: Thermomechanical analysis of the welding process using the finite element method. J. Press. Vessel Technol. (ASME) 97 (1975) no. 8, pp. 206–213
250 Papazoglou, V.J.; Masubuchi, K.: Numerical analysis of thermal stresses during welding including phase transformation effects, J. Press. Vessel Technol. (ASME) 104 (1982) no. 8, pp. 198–203
251 Szimmat, J.: Eigenspannungsberechnung beim Mehrlagenschweißen mit der Methode der finiten Elemente. Inst. f. Computer-Anwendungen, Univ. Stuttgart, Bericht no. 16, 1987
252 Ueda, Y.; Yamakawa, T.: Thermal stress analysis of metals with temperature dependent mechanical properties. In: Mechanical behavior of materials, Vol. III, pp. 10–20. Soc. Mat. Science, Jap. 1972
253 Carmet, A.; Debiez, S.; Devaux, J.; Pont, D.; Leblond, J.B.: Experimental and numerical study of residual stresses and strains in an electron-beam-welded joint. In [18], pp. 491–496
254 Kussmaul, K.; Guth, W.: Evaluation of residual stresses in multilayer welds. In [18], pp. 612–618

Solid element model:

(see also [241])

255 Ueda, Y.; Kim, Y.C.; Garatani, K.; Yamakita, T.: Mechanical characteristics of repair welds in thick plate – distribution of three dimensional welding residual stresses and plastic strains and their production mechanisms. Quart. J. Jap. Weld. Soc. 4 (1986) no. 3, pp. 533–539
256 Tekriwal, P.; Mazumder, J.: Thermomechanical analysis of residual strains and stresses in a GMA weld. In [19], pp. 91–95
257 Oddy, A.S.; Goldak, J.A.; McDill, J.M.J.: Transformation effects in the 3d finite element analysis of welds. In [19], pp. 97–101
258 Josefson, L.; et al.: Transient and residual stresses in a single-pass butt welded pipe. In [18], pp. 497–503
259 Karlsson, R.I.; Josefson, B.L.: Three dimensional finite element analysis of temperatures and stresses in a single pass butt-welded pipe. J. Pressure Vessel Technol. (ASME) 112 (1990) no. 1, pp. 76–84
260 Oddy, A.S.; McDill, J.M.J.; Goldak, J.A.: Consistent strain fields in 3d finite element analysis of welds. J. Pressure Vessel Technol. (ASME) 112 (1990) no. 8, pp. 309–311

Shrinkage force model and restraint intensity:

(see also [2, 7, 8, 10, 280, 297])

261 Buttenschön, K.: Beulen von dünnwandigen Kastenträgern aufgrund von Schweißeigenspannungen. Schweiß. u. Schneid. 24 (1972) no. 6, pp. 217–221
262 Wells, A.A.: The mechanics of brittle fracture. Weld. Res. 7 (1953) no. 4, pp. 34r–56r
263 Vaidyanathan, S.; Todaro, A.F.; Finnie, I.: Residual stresses due to circumferential welds. J. Engng. Mater. a. Technol. (ASME) 95 (1973) pp. 233–237
264 Guan, Q.; Liu, J.D.: Residual stress and distortion in cylindrical shells caused by a single pass circumferential butt-weld. IIW-Doc. X–929–79
265 Hoffmeister, H.; Nölle, P.; Schimmel, P.: Entwicklung und Aussagen eines instrumentierten Einspannschweißversuchs. Arch. Eisenhüttenw. 49 (1978) no. 3, pp. 151–154
266 Watanabe, M.; Satoh, K.; Matsui, S.: Effect of restraint on root cracking of steel welds. Weld. Res. Abroard 10 (1964) no. 8, pp. 63–75
267 Satoh, K.: Restraint stress strain versus cold cracking in RRC test of high strength steel. In [12], pp. 283–293
268 Mar, E.; Graville, B.A.: Effects of welding procedure on contraction and reaction stress in multipass welds. Weld. Res. Int. 9 (1979) no. 3, pp. 1–11
269 Ueda, Y.; Fukuda, K.; Kim, Y.C.: Restraint stress and strain due to slit weld in rectangular plate. Trans. JWRI 7 (1978) no. 1, pp. 11–16

Residual stress source model:

(see also [11, 240])

270 Reissner, H.: Eigenspannungen und Eigenspannungsquellen. ZAMM 11 (1931) no. 1, pp. 1–8
271 Rieder, G.: Spannungen und Dehnungen im gestörten elastischen Medium. Z. Naturforsch. 11a (1956) pp. 171–173
272 Schimmöller, H.: Bestimmung von Eigenspannungen in ebenen plattierten Werkstoffen, Part 1, Rechnerische Grundlagen. Materialprüfung 14 (1972) no. 4, pp. 115–122
273 Schönbach, W.: Zur Ermittlung der Eigenspannungen in stab- und scheibenartigen Bauteilen infolge von Quernähten und punktförmigen Eigenspannungsquellen. Stahlbau 12 (1962) pp. 365–378

Residual stresses, experimental results:

(see also [4, 7, 8, 10, 15, 16, 228, 229, 235, 253, 262, 331, 394])

274 Koch, H.: Beiträge zu den Fragen der Formabweichungen und Eigenspannungen beim Schmelzschweißen. In: Schweißen und Schneiden, Forschritte in den Grundlagen und in der Anwendung, pp. 130–135. Düsseldorf: DVS-Verlag 1966
275 Klöppel, K.: Sicherheit und Güteanforderungen bei geschweißten Konstruktionen. In: Stahlbau, Vol. 1, pp. 61–93. Köln: Stahlbau-Verlag 1961
276 Rappe, H.A.: Betrachtungen zu Schweißeigenspannungen. Schweiß. u. Schneid. 26 (1974) no. 2, pp. 45–50
277 Wohlfahrt, H.: Schweißeigenspannungen, Entstehung – Berechnung – Bewertung. In [15], pp. 85–116
278 Onoue, H.: On the reaction stress of welded butt joint. IIW-Doc. X–370–66
279 Poje, R.: Schweißeigenspannungen in Elektronenstrahl- und Unterpulver-Schweißverbindungen von Stählen bei verschiedenen Blechdicken. Diss. RWTH Aachen, 1984
280 Omar, M.: Zur Wirkung der Schrumpfbehinderung auf den Schweißeigenspannungszustand und das Sprödbruchverhalten von unterpulvergeschweißten Blechen aus StE460N. Forschungsber. no. 109, BAM Berlin, 1985
281 Schimmöller, H.: Bestimmung von Eigenspannungen in ebenen plattierten Werkstoffen, Part 2, Eigenspannungen in warmgewalzten und sprenggeschweißten austenit-plattierten Stahlblechen. Materialprüfung 14 (1972) no. 11, pp. 380–387 (Part 1: [272])
282 Hillman, G.; Hofmann, W.; Schimkat, H.: Measurement of residual stresses in steel plates clad with stainless steel. In: Corrosion of reactor materials. Wien: Int. Atom. Energy Agency 1962
283 Tatsukawa, I.; Oda, I.: Residual stress measurements on explosive stainless clad steel. Trans. Jap. Weld. Soc. 2 (1971) no. 2, pp. 26–34
284 Rao, K.N.; Ruge, J.; Schimmöller, H.: Bestimmung der durch Brennschneiden von Stahlblechen verursachten Eigenspannungen. Forsch. Ing. Wes. 36 (1970) no. 6, pp. 192–200
285 Ruge, J.; Schimmöller, H.: Berechnung von Eigenspannungen in brenngeschnittenen Stahlblechen aus StE36. In: DVS-Ber. no. 22, pp. 67–71. Düsseldorf: DVS-Verlag 1971
286 Poalas, K.; Graf, U.; Kosteas, D.: Messungen von Schweißeigenspannungen an Aluminiumbauteilen im ermüdungskritischen Bereich. Schweiß. u. Schneid. 40 (1988) no. 8, pp. 389–393
287 Mahin, K.W.; et al.: Residual strain distributions in gas tungsten arc welds: a comparison between experimental results and weld model predictions. In [19], pp. 83–89
288 Troive, L.; et al.: Finite element simulations of the bending of a flat plate to U-shaped beam cross-section and the welding to rectangular hollow cross-section and neutron diffraction determination of residual stresses. In [19], pp. 107–111
289 Dahl, W.; Krebs, H.: Determination and assessment of residual stresses in a multilayer submerged arc welded joint. In [18], pp. 683–689
290 Jonsson, M.; Josefson, B.L.: Experimentally determined transient and residual stresses in a butt-welded pipe. J. Strain Anal. Eng. Design 23 (1988), pp. 25–31

Welding distortion:

(see also [3, 5, 7, 8, 10, 234, 266, 267, 280])

291 Malisius, R.: Die Schrumpfung geschweißter Stumpfnähte. Braunschweig: Vieweg 1936
292 Spraragen, W.; Clausen, G.E.: Shrinkage distortion in welding. Weld. J., Res. Suppl. 16 (1937) no. 7, pp. 29s–39s

293 Spraragen, W.; Ettinger, W.G.: Shrinkage distortion in welding. Weld. J., Res. Suppl. 29 (1950) no. 6, pp. 292s–294s, no. 7, pp. 323s–335s
294 Gilde, W.: Beitrag zur Berechnung der Querschrumpfung. Schweißtechn. (Berl.) 7 (1957) no. 1, pp. 10–14
295 Richter, E.; Georgi, G.: Nahtquerschnitt und Schrumpfng. ZIS-Mitt. (1970) no. 2, pp. 148–160
296 Leggatt, R.H.; White, J.D.: Predicting shrinkage and distortion in welded plate. In [12], pp. 119–132
297 Satoh, K.; Matsui, S.: Reaction stress and weld cracking under hindered contraction. IIW-Doc. IX-574-68
298 Graville, B.A.; Beynon, G.: Free contraction in butt, T, and corner joints. Weld. Res. Int. 3 (1973) no. 2, pp. 68–83
299 –: Weld distortion handbook. Columbus, Ohio: Battelle Mem. Inst. 1969 (not publ.)
300 –: Control of distortion in welded fabrication. Abington. Cambridge: The Welding Institute 1968
301 Uelze, A.: Schweißverformungen beim Bau von Reisezugwagen. Schweiß. Schneid. 43 (1991) no. 3, pp. 133–137

Measuring methods for residual stress and distortion, similarity relations:

(see also [4, 8, 10, 18, 194, 272, 281, 284, 287, 288])

302 Tietz, H.D.: Grundlagen der Eigenspannungen. Leipzig: VEB Deutscher Verlag für Grundstoffindustrie 1983
303 Hauk, V.; Macherauch, E. (Eds.): Eigenspannungen und Lastspannungen, moderne Ermittlung, Ergebnisse, Bewertung. Härterei Techn. Mitt., Beih. 1982
304 Peiter, A.: Eigenspannungen I. Art, Ermittlung und Bewertung. Düsseldorf: Triltsch 1966
305 Fink, K.; Rohrbach, C. (Eds.): Handbuch der Spannungs- und Dehnungsmessung. Düsseldorf: VDI-Verlag 1958
306 Thürlimann, B.: Der Einfluß von Eigenspannungen auf das Knicken von Stahlstützen. Schweiz. Arch. (1957) pp. 388–404
307 Stäblein, F.: Spannungsmessungen an einseitig abgelöschten Knüppeln. Krupp. Monatsh. 12 (1931) pp. 93–99
308 Treuting, R.G.; Read, W.T.: A mechanical determination of biaxial residual stress in sheet materials, J. Appl. Phys. 22 (1951) no. 2, pp. 130–134
309 Kunz, H.G.: Ermittlung von Formänderungen und Eigenspannungen durch mechanische Feindehnungsmessungen. Industriebl. 61 (1961) pp. 652–659
310 Mathar, J.: Determination of initial stresses by measuring the deformation around drilled holes. Trans. ASME 56 (1934) no. 4, pp. 249–259
311 Rendler, N.J.; Vigness, I.: Hole-drilling strain-gage method of measuring residual stresses. Proc. SESA 23 (1966) no. 2, pp. 577–586
312 ASTM-Standard E 837-85: Standard test method for determining residual stresses by the hole driling strain-gage method
313 –: Measurement of residual stresses by the hole-drilling strain-gage method. Raleigh, N.C.: Measurement Groups 1986
314 Kelsey, R.A.: Measuring non-uniform residual stresses by the hole-drilling method. Proc. SESA 14 (1956) no. 1, pp. 181–194
315 Häusler, H.; König, G.; Kockelmann, H.: On the accuracy of determining the variation with depth of residual stresses by means of the hole-drilling method. In [16], pp. 257–264
316 Ueda, Y.; Fukuda, K.; Tanigawa, M.: New measuring method of three dimensional residual stresses based on the theory of inherent strain. Trans. Jap. Weld. Res. Inst. 8 (1979) no. 2, pp. 89–96
317 Ueda, Y.; Kim, Y.C.: New measuring method of three-dimensional residual stresses using effective inherent strains as parameters. In [16], pp. 199–206
318 Heyn, E.; Bauer, O.: Über Spannungen in kaltgereckten Metallen. Z. Metallogr. 1 (1911) pp. 16–50
319 Mesnager, M.: Méthode de détermination des tensions existant dans un cylindre circulaire. Compt. Rond., Acad. d. Sci. 169 (1919) pp. 1391–1393
320 Sachs, G.: Nachweis innerer Spannungen in Stangen und Rohren. Z. Metallkd. 19 (1927) no. 9, pp. 352–359

321 Bollenrath, F.: Das Verhalten von Schweißspannungen in Behältern bei innerem Überdruck. Stahl u. Eisen 57 (1937) no. 15, pp. 389–398
322 Peiter, A.: Spannungsmeßpraxis, Ermittlung von Last- und Eigenspannungen. Braunschweig: Vieweg 1986
323 Macherauch, E.: Grundlagen und Probleme der röntgenografischen Ermittlung elastischer Spannungen. Materialprüf. 5 (1963) pp. 14–26
324 Macherauch, E.; Müller, P.: Das $\sin^2\psi$-Verfahren der röntgenografischen Spannungsmessung. Z. angew. Phys. 13 (1961) pp. 305–312
325 Glocker, R.: Materialprüfung mit Röntgenstrahlen. Berlin: Springer 1958
326 Neff, H.: Grundlagen und Anwendung der Röntgen-Feinstruktur-Analyse. München: Oldenbourg 1959
327 Noyan, J.C.; Cohen, J.B.: Residual stress, measurement by diffraction and interpretation. Berlin: Springer 1987
328 Peiter, A.; et al.: Flächige Eigenspannungsanalyse nach dem Abtrennverfahren in wärmebehandelten und geschweißten Blechen aus St37. Schweiß. u. Schneid. 40 (1988) no. 11 p. 551–555. Letter to the editor by D. Radaj in Schweiß. u. Schneid. 41 (1989) no. 5, pp. 244
329 Peiter, A.: Abtrennverfahren zum Ermitteln von Eigenspannungen im Blech. Bänder, Bleche, Rohre 29 (1988) no. 1, pp. 30–33
330 Fischer, F.D.; Hinteregger, E., Rammerstorfer, F.G.: Numerische Simulation einer experimentellen Spannungsanalyse. Materialprüfung 32 (1990) no. 6, pp. 181–185
331 Hosbons, R.R.; Ibrahim, E.F.; Holden, T.M., Root, J.H.: The use of neutron diffraction to determine non-destructively the residual strain and texture in welds. In [19], pp. 103–106
332 Gross, P.M.: Residual stress vs fatigue life of welded austenitic stainless steel pressure vessels. In [19], pp. 815–819
333 Crostack, H.-A.; Reimers, W.: Residual stress profile from grain to grain in a welding zone. In [18], pp. 58–64
334 Schajer, G.S.: Judgement of residual stress field uniformity when using the hole-drilling method. In [18], pp. 71–77
335 Karjalainen, P.L.; Moilanen, M.; Rautioaho: Evaluating the residual stresses in welding from Barkhausen noise measurement. Materialprüfung 22 (1980), pp. 196–200
336 Curtis, G.J.: The use of surface elastic waves in examining surface mechanical properties. In: Ultrasonic testing (Ed. J. Szilard), pp. 297–380. London: John Wiley 1982
337 Crecraft, D.I.: Ultrasonic measurement of stress. In: Ultrasonic testing (Ed. J. Szilard), pp. 437–458. London: John Wiley 1982
338 Arai, Y.; Kobayashi, H.; Suzuki, M.: Measurement of welding residual stresses by acoustoelastic technique – evaluation of principal direction of material anisotropy. Proc. VI Int. Cong. Exp. Mech., Vol. I., pp. 624–629. Portland, Soc. Exper. Mech. 1988
339 Kobayashi, H.; et al.: Non-destructive measurement of welding residual stresses by acoustoelastic technique and prediction of fatigue crack growth. In [18], pp. 387–392
340 Schajer, G.S.: Measurement of non-uniform residual stresses using the hole-drilling method, part I and II. Engng. Mat. Technol. (ASME) 110 (1988), pp. 338–343 and 344–349

Weldability, welding suitability, dimensional tolerances for welded structures:

(see also [74, 119])

341 DIN 1910 Parts 1–4 and 10–12: Schweißen und Schneiden, Begriffe und Verfahren. Berlin: Beuth 1977–1983
342 DIN 8528 Parts 1 and 2: Schweißbarkeit. Berlin: Beuth 1973 and 1975. See also: Euronorm EN 45–1974 and ISO-Recommendation R 581–1967
343 Radaj, D.: Bewertung der Schweißeignung des Werkstoffs aus der Sicht der Schweißeigenspannungen und der Schweißformänderungen. In: DVS-Ber. no. 108, 1st German-Chinese Conf. on „Neue Entwicklungen und Anwendungen in der Schweißtechnik“, Beijing 1987, pp. 27–29. Düsseldorf: DVS-Verlag 1987. Engl.: IIW-Doc. IX–1456–86
344 Cottrell, C.L.M.: Hardness equivalent may lead to a more critical measure of weldability. Met. Constr. 16 (1984) no. 12, pp. 740–744
345 DIN 8570: Freimaßtoleranzen für Schweißkonstruktionen. Berlin: Beuth 1974–1976

Annealing for stress relief, preheating:

(see also [415])

346 DIN 32527: Wärmen beim Schweißen, Löten, Schneiden und bei verwandten Verfahren; Begriffe, Verfahren. Berlin: Beuth 1984
347 DVS-Merkblatt 1002, Part 2: Verfahren zur Verringerung von Schweißeigenspannungen. Düsseldorf: DVS-Verlag 1986
348 Tummers, G.E.: Summary report on stress relaxation data. Brit. Weld. J. 10 (1963) pp. 292–303
349 Schimmöller, H.: Eigenspannungen. In [74], pp. 277–315
350 Müller, H.H.: Stand und Entwicklungstendenzen des Wärmebehandelns von Schweißverbindungen. Schweiß. u. Schneid. 24 (1972) no. 9, pp. 381–383
351 Müller, H.H.: Das internationale Regelwerk über die örtliche Wärmebehandlung von Schweißnähten. Düsseldorf: DVS-Verlag 1986
352 Alf, F.; Müler, H.H.: Örtliches Entspannen von Schweißnähten durch induktive Wärmebehandlung. Merkblatt 238 d. Beratungsst. f. Stahlverwend. Düsseldorf 1969
353 Toyooka, T.; Terai, K.: On the effects of postweld heat treatment. Weld. J., Res. Suppl. 52 (1973) no. 6, pp. 247s–254s
354 Schaar, K.: Uberlegungen zur Befreiung vom Spannungsarmglühen geschweißter Bauteile. Techn. Überwach. 9 (1968) no. 12, pp. 410–414
355 Watson, S.I.: Stress relief of mild-steel welded structures. Brit. Weld. J. 4 (1957) pp. 422–423
356 Burdekin, F.M.: Local stress relief of circumferential butt welds in cylinders. Brit. Weld. J. 10 (1963) pp. 483–490
357 Burdekin, F.M.: Heat treatment of welded structures. Abington, Cambr.: The Welding Institute 1969
358 Vinokurov, V.A.: Tempering of welded structures. Weld. Prod. (1967) no. 3, pp. 7–12
359 Mailänder, R.: Die Verminderung der Eigenspannungen durch Anlassen. Stahl u. Eisen 51 (1931) no. 22, pp. 662–671
360 Wellinger, K.: Temperatur-Spannungslinien bei gleichbleibender Verdrehung und zunehmender Temperatur zur Bestimmung von Resteigenspannungen. Mitt. d.V.G.B. (1951) no. 15, pp. 335–338
361 Ritter, J.C.; McPherson, R.: Anisothermal stress relaxation in a carbon-manganese steel. J. Iron a. Steel Inst. 208 (1970) no. 10, pp. 935–941
362 Lange, G.: Entspannungsversuche an Kesselbaustählen und Stahlguß. Schweiß. u. Schneid. 19 (1967) pp. 454–458
363 Brozzo, P.: Quelques considérations sur les résultats d'essais de relaxation par torsion de quadre aciers typiques résistant à chaud. Soud. Tech. Conn. 16 (1962) pp. 98–102
364 Tummers, G.E.: Comparison of two methods for the determination of stress relaxation in steel. Weld. in the World 4 (1966) pp. 92–103
365 Leymonie, C.: Interprétation des essais de relaxation anisotherme. Mém. Sci. Rev. Mét. 71 (1974) pp. 609–620
366 Leymonie, C.: Utilization of anisothermal stress relaxation tests. Int. Conf. on Engng. Aspects of Creep. Pap. C243/80. Univ. Sheffield 1980, pp. 115–120
367 Murry, G.; Constant, A.: Contribution à l'étude de la relaxation des contraintes dues au soudace. Rev. Mét. 62 (1965) no. 2, pp. 127–137
368 Lauprecht, W.; Emrich, P.; Speth, W.: Austauschbarkeit von Temperatur und Zeit beim Entspannen von Schweißverbindungen aus Stahl. Schweiß. u. Schneid. 22 (1970) no. 8, pp. 321–323
369 Gulvin, T.F.; Scott, D.; Haddrill, D.M.; Glen, J.: The influence of stress relief on the properties of C- and CMn-pressure vessel plate steels. West Scotl. Iron a. Steel Inst. J. 80 (1972/73) no. 621, pp. 149–175
370 Schaar, K.: Bedeutung der Untersuchungen über die Austauschbarkeit von Zeit und Temperatur beim Spannungsarmglühen für die Stahlnormung und die Vorschriften für überwachungsbedürftige Anlagen. Schweiß. u. Schneid. 22 (1970) no. 8, pp. 323–324
371 Holloman, J.H.; Jaffe, L.D.: Time temperature relations in tempering steel. Trans. Am. Inst. Min. Met. Eng. 162 (1945) pp. 223–228
372 Ulff, C.: Inverkan av avspänningsglodgning ordinära tryckkärlsstads mekaniska egenskaper. Jernkont. Annal. 154 (1970) pp. 53–62. B.I.S.I.- Transl. 9098: The effects of stress relieving on the mechanical properties of ordinary pressure vessel steels

373 Mendelson, A.; Hirschberg, M.M.; Manson, S.S.: A general approach to the practical solution of creep problems. J. of Basic Engng. (ASME) 81 (1959) No. 12, pp. 585–598
374 Oding, I.A.: Creep and stress relaxation in metals. London: Oliver & Boyd 1965
375 Penny, R.K.; Marriott, D.L.: Design for creep. London: McGraw-Hill 1971
376 Kraus, H.: Creep analysis. New York: J. Wiley 1980
377 Bloyle, J.T.; Spence, J.: Stress analysis for creep. London: Butterworth 1983
378 –: Aerospace Structural Metals Handbook. New York: Dep. of Def. 1976
379 Josefson, B.L.: Stress redistribution during annealing of a multi-pass butt-welded pipe. J. Press. Vessel Technol. (ASME) 105 (1983) pp. 165–170
380 Josefson, B.L.: Residual stresses and their redistribution during annealing of a girth-butt-welded thinwalled pipe. J. Press. Vessel Technol. (ASME) 104 (1982) no. 8, pp. 245–250
381 Josefson, B.L.: Stress redistribution during local annealing of a multi-pass butt-welded pipe. J. Press. Vessel Technol. (ASME) 108 (1986) no. 5, pp. 125–130
382 Josefson, B.L.: Reheat cracking during stress relief annealing, a numerical approach. In: Mechanical behaviour of materials (Eds.: J. Carlson and N.-G. Ohlsson), pp. 191–197. Oxford: Pergamon Press 1984
383 Agapakis, J.E.; Masubuchi, K.: Analytical modeling of thermal stress relieving in stainless and high strength steel weldments. Weld. J., Res. Suppl. 36 (1984) no. 6, pp. 187s–196s
384 Fidler, R.: The effect of time and temperature on residual stresses in austenitic welds. J. Press. Vessel Technol. (ASME) 104 (1982) pp. 210–214
385 Tanaka, J.: Decrease of residual stress, change in mechanical properties and cracking due to stress relieving heat treatment of HT80 steel. Weld. in the World 10 (1972) no. 1/2, pp. 54–67
386 Ueda, Y.; Takahashi, E.; Fukuda, F.; Sakamoto, K.; Nakacho, K.: Multipass welding stresses in very thick plates and their reduction from stress relief annealing. Trans. Jap. Weld. Res. Inst. 5 (1976) no. 2, pp. 79–89
387 Ueda, H.; Fukuda, K.: Analysis of welding stress reliefing by annealing based on finite element method. Trans. Jap. Weld. Res. Inst. 4 (1975) no. 1, pp. 39–45
388 Cameron, I.G.; Pemberton, C.S.: A theoretical study of thermal stress relief in thin shells of revolution. Int. J. f. Numeric. Meth. in Engng. 11 (1977) pp. 1423–1437
389 Niering, E.: Rechnerische Untersuchung zum örtlichen Spannungsarmglühen von Reparaturschweißungen und Verfahrensrichtlinien. Schweiß. u. Schneid. 39 (1987) no. 9, pp. 451–457
390 Klöppel, K.; Reuschling, D.: Berechnung der Wärmeeigenspannungen an den Rundnähten rotationssymmetrisch vorgewärmter Kreiszylinderschalen. In: Tragfähigkeitsermittlung bei Schweißverbindungen, Vol. I, pp. 98–107. Düsseldorf: DVS-Verlag 1968
391 Frank, G.: Berechnung von Vorwärmtemperaturen beim Schweißen. Schweiß. u. Schneid. 40 (1988) no. 4, pp. 169–171
392 Horn, W. u. H.J.; Marfels, W.: Wärmebehandlung von Stahl. Düsseldorff: DVS-Verlag 1987
393 Donzella, G.: Numerical analysis of residual stress state in clad components. In [18], pp. 625–630
394 Wiewecke, F.; Wohlfahrt, H.: Comparison of the residual stress distributions after stress relief annealing of welded sheets of different high strength structural steels at different annealing temperatures and times. In [18], pp. 670–676
395 Kannatey-Asibu, E.: Slit-beam laser welding. In [19], pp. 443–451

Cold stretching, hammering, shot peening, spot compression:

(see also [347])

396 Rühl, H.: Die Tragfähigkeit metallischer Baukörper, Berlin: Ernst & Sohn 1952
397 Soete, V.: Die Schweißnaht im Übergangsgebiet der elastisch-plastischen Verformung. Schweiß. u. Schneid. 8 (1956) no. 11, pp. 430–435
398 Erker, A.: Einfluß der Eigenspannungen und des Werkstoffzustandes auf die Betriebssicherheit. Schweiß. u. Schneid. 8 (1956) no. 11, pp. 436–442
399 Tesar, S.: Verlagerung von Eigenspannungen durch Außenlast Schweiß. u. Schneid. 22 (1970) no. 4, pp. 145–150
400 Schimmöller, H.: Rechnerische Behandlung der Verlagerung von Eigenspannungen in einem Stabmodell durch Überbelastung. Z. f. Werkstofftechn. 3 (1972) no. 6, pp. 301–306
401 Vöhringer, O.: Abbau von Eigenspannungen. In [15], pp. 49–83

402 Nichols, R.W.: Overstressing as a means of reducing the risk of subsequent brittle failure. Brit. Weldg. J. 15 (1968) no. 1, pp. 21–42 and no. 2, pp. 75–84
403 Rädeker, W.: Anwendung einer gezielten Überbelastung zur Verringerung der Sprödbruchgefahr. Schweiß. u. Schneid. 22 (1970) pp. 178–183
404 Wellinger, K.; Kußmaul, K.; Krägeloh, E.: Überbelastung zur Verringerung der Sprödbruchgefahr, eine kritische Betrachtung. Schweiß. u. Schneid. 23 (1971) pp. 297–301
405 VdTÜV-Merkblatt Rohrleitungen 1060: Richtlinien für die Durchführung des Stress-Tests. Herford: Maximilian-Verlag
406 Pellini, W.S.; Goode, R.J.; Puzak, P.O.; Lange, E.A.; Huber, R.W.: Review of concepts and status of procedures for fracture safe design of complex welded structures involving metals of low to ultra-high strength levels. U.S. Nav. Res. Lab. NRL Rep. 6300, pp. 1–84, Washington 1965
407 Pellini, W.S.: Guidelines for fracture-safe and fatigue-reliable design of steel structures. Abington, Cambr.: The Welding Institute 1983
408 ASTM Standard E 208–66 T: Conducting drop weight test to determine nil-ductility
409 Robertson, T.S.: Bittle fracture of mild steel. Engng. 71172 (1951) Oct. 5, pp. 445–448
410 Robertson, T.S.: Propagation of brittle fracture in steel. J. Iron Steel Corp. 175 (1953) no. 4, pp. 361–374
411 Radaj, D.: Werkstoffgütenachweis. In [438], Vol. 1, pp. 87–144
412 Reumont, G.A.v.: Einfluß des Wärmehämmerns auf die mechanischen Eigenschaften der Schweißverbindung. Schweiß. u. Schneid. 19 (1967) no. 12, pp. 575–577
413 Maddox, S.J.: Improving the fatigue strength of welded joints by peening. Met. Constr. 17 (1985) no. 4, pp. 220–224
414 Gurney, T.R.: Fatigue of welded structures. Cambridge: University Press 1979

Flame stress relieving, flame straightening, spot heating:

(see also [346, 347, 414])

415 DIN 8522: Fertigungsverfahren der Autogentechnik, Übersicht. Berlin: Beuth 1980
416 Wellinger, K.: Möglichkeiten des Abbaus von Schweißeigenspannungen. Schweiß. u. Schneid. 5. (1953) Sonderh., pp. 157s–162s
417 Pfender, M.: Ausgleich und Steuerung von Eigenspannungen durch autogenes Erwärmen. Schweiß. u. Schneid. 5 (1953) Sonderh., pp. 62s–69s
418 Wellinger, K.; Eichhorn, F.; Löffler, F.: Versuche über den Abbau von Schweißeigenspannungen durch überlagerte Wärmespannungen. Schweiß. u. Schneid. 7 (1955) pp. 7–14
419 Kunz, H.G.: Autogenes Entspannen im Schiffbau. Hansa 93 (1956) no. 24/25, pp. 1155–1163
420 Kunz, H. G.: Die Entspannungsbehandlung von Schweißverbindungen. Industriebl. (1963) no. 4, pp. 191–198
421 Kunz, H.G.: Stand und Anwendungsumfang des Flammentspannens. BEFA-Mitt. no. 1. Hurth: Beratungsst. f. Autogentechn. 1976
422 Bernard, P.; Schreiber, G.: Verfahren der Autogentechnik. Düsseldorf: DVS-Verlag 1973
423 Rybicki, E.F.; McGuire, P.A.: The effects of induction heating conditions on controlling residual stresses in welded pipes. J. Engng. Mater. Technol. (ASME) 104 (1982) pp. 267–273
424 Pfeiffer, R.: Richten und Umformen mit der Flamme. Düsseldorf: DVS-Verlag 1983
425 Glizmanenko, D.L. (Ed.): Gazoplannaya obrabotka metallov (Flammbearbeitung der Metalle). M. Proftekhizdat 1962
426 Ueda, Y.; Nakacho, K.; Shimizu, T.: Improvement of residual stresses of circumferential joint of pipe by heat-sink welding. J. Press. Vessel Technol. (ASME) 108 (1986), pp. 14–23.
427 Danko, J.C.: A Review of weld residual stresses in austenitic stainless steel pipes. In [19], pp. 113–118

Vibration stress relieving:

(see also [347])

428 Gnirß, G.: Rütteln und Vibrationsentspannen, Theorie und praktische Anwendung. Techn. Überwach. 28 (1986) no. 11, pp. 439–442

429 Bühler, H.; Pfalzgraf, H G.: Untersuchungen übre die Verminderung von Schweißspannungen durch mechanisches Rütteln. Schweiß. u. Schneid. 16 (1964) no. 5, pp. 178–183
430 Wozney, G.P.; Crawner G.R.: An investigation of vibrational stress relief in steel. Weld. J., Res. Suppl. 33 (1968) no. 9, pp. 441s–449s
431 Rich, S.: Quantitative measurement of vibratory stress relief. Weld. Engng. (1969) no. 3, pp. 44–45
432 Wohlfahrt, H.: Zum Eigenspannungsabbau bei der Schwingbeanspruchung von Stählen. Härterei Techn. Mitt. 28 (1973) no. 11, pp. 288–293
433 Rappen, A.; Verringerung von Schweißeigenspannungen durch Vibration zur Erzielung von Maß- und Formgenauigkeit von Maschinenteilen. DVS-Ber. no. 74, pp. 191–202. Düsseldorf: DVS-Verlag 1982
434 Sedek, P.: Können mechanische Schwingungen das Spannungsarmglühen geschweißter Maschinenelemente ersetzen? Schweiß. u. Schneid. 35 (1983) no. 10, pp. 483–486

Strength effects of welding:

(see also [10, 18, 414])

435 Radaj, D.: Grundzüge einer Festigkeitslehre für Schweißkonstruktionen. Schweiß. u. Schneid. 39 (1987) no. 10, pp. 498–502
436 Radaj, D.: Gestaltung und Berechnung von Schweißkonstruktionen, Ermüdungsfestigkeit. Düsseldorf: DVS-Verlag 1985
437 Radaj, D.: Design and analysis of fatigue resistant welded structures. Cambridge: Abington Publishing 1990
438 Radaj, D.: Festigkeitsnachweise, Vol. 1: Grundverfahren, Vol. 2: Sonderverfahren. Düsseldorf: DVS-Verlag 1974
439 Macherauch, E.; Kloos, K.H.: Bewertung von Eigenspannungen. In [303], pp. 175–194
440 Bornscheuer, F.W.: Einfluß der Schweißeigenspannungen auf die Traglast von Stäben, Platten und Schalen aus Stahl. Schweiß. u. Schneid. 37 (1985) no. 8. pp. 351–356
441 Matsuda, F.: Solidification crack susceptibility of weld metal. In [19], pp. 127–136
442 Hünersen, G.: Nachweisführung für Stahltragglieder beim Schweißen unter Last. Informationen des VEB MLK 27 (1988) no. 1, pp. 22–28
443 Hünersen, G.: Schweißen an belasteten Bauteilen. In [20], pp. 202–204
444 Kleistner, H.; Stahlfeld, G.; Wilken, K.: Möglichkeiten der Prüfung von Großproben und Bautteilen mit der Großprüfanlage. In: DVS-Berichte no. 131. Düsseldorf, DVS-Verlag 1990

Related books published after copy deadline:

445 Buchmayr, B.: Computer in der Werkstoff- und Schweißtechnik, Anwendung von mathematischen Modellen. Düsseldorf, DVS-Verlag 1991
446 Schimmöller, H.: Analytische Behandlung von Eigenspannungszuständen auf der Grundlage der Elastizitätstheorie. Düsseldorf: VDI-Verlag 1990
447 Rappaz, M.; Özgü, M.R.; Mahin, K.M. (Eds.): Modeling of casting, welding and advanced solidification processes – V. Warrendale, Pa.: The Minerals, Metals & Materials Society 1991

Supplement to "Temperature field during friction welding":

448 Imshennik, K.P.; Kragelski, I.V.: Heating in friction welding. Weld. Prod. (1973) no. 10, pp. 76–79
449 Adam, P.: Ablauf der Verbindungsbildung beim Schwungradreibschweißen von hochwarmfesten Legierungen , Temperaturverlauf und Wulstbildung. Schweiß. u. Schneid. 31 (1979) no. 7, pp. 279–283
450 Adam, P.: Phänomenologischer Ablauf der Verbindungsbildung beim Schwungradreibschweißen von hochwarmfesten Legierungen. Schweiß. u. Schneid. 32 (1980) no. 10, pp. 76–79

By the same author:

Radaj, D.: Festigkeitsnachweise. Bd. I u. II. Düsseldorf: DVS-Verlag 1974
Radaj, D.; Schilberth, G.: Kerbspannungen an Ausschnitten und Einschlüssen. Düsseldorf: DVS-Verlag 1977
Radaj, D.: Gestaltung und Berechnung von Schweißkonstruktionen – Ermüdungsfestigkeit. Düsseldorf: DVS-Verlag 1985
Radaj, D.: Wärmewirkungen des Schweißens – Temperaturfeld, Eigenspannungen, Verzug. Berlin: Springer 1988
Radaj, D.: Design and analysis of fatigue resistant welded structures. Cambridge: Abington Publishing 1990

Subject index

MIX
Papier aus verantwortungsvollen Quellen
Paper from responsible sources
FSC® C105338

If you have any concerns about our products,
you can contact us on
ProductSafety@springernature.com

In case Publisher is established outside the EU,
the EU authorized representative is:
Springer Nature Customer Service Center GmbH
Europaplatz 3, 69115 Heidelberg, Germany

Printed by Libri Plureos GmbH
in Hamburg, Germany